Biofilms in the Dairy Industry

The Society of Dairy Technology (SDT) has joined with Wiley-Blackwell to produce a series of technical dairy-related handbooks providing an invaluable resource for all those involved in the dairy industry, from practitioners to technologists, working in both traditional and modern large-scale dairy operations. For information regarding the SDT, please contact Dr Liz Whitley, Larnick Park, Higher Larrick, Trebullett, Launceston, Cornwall, PL15 9QH, UK.
email: execdirector@sdt.org

Other volumes in the Society of Dairy Technology book series:

Probiotic Dairy Products (ISBN 978 1 4051 2124 8)
Fermented Milks (ISBN 978 0 6320 6458 8)
Brined Cheeses (ISBN 978 1 4051 2460 7)
Structure of Dairy Products (ISBN 978 1 4051 2975 6)
Cleaning-in-Place (ISBN 978 1 4051 5503 8)
Milk Processing and Quality Management (ISBN 978 1 4051 4530 5)
Dairy Fats (ISBN 978 1 4051 5090 3)
Dairy Powders and Concentrated Products (ISBN 978 1 4051 5764 3)
Technology of Cheesemaking, Second Edition (ISBN 978 1 4051 8298 0)
Processed Cheese and Analogues (ISBN 978 1 4051 8642 1)
Membrane Processing – Dairy and Beverage Applications (ISBN 978 1 4443 3337 4)
Milk and Dairy Products as Functional Foods (ISBN 978 1 4443 3683 2)

Biofilms in the Dairy Industry

Edited by

Koon Hoong Teh
Massey University, Palmerston North, New Zealand

Steve Flint
Massey University, Palmerston North, New Zealand

John Brooks
Auckland University of Technology, Auckland, New Zealand

Geoff Knight
Food Process Hygiene Solutions, Melbourne, Victoria, Australia

WILEY Blackwell

Registered Office
John Wiley & Sons, Ltd, The Atrium, Southern Gate, Chichester, West Sussex, PO19 8SQ, UK

Editorial Offices
9600 Garsington Road, Oxford, OX4 2DQ, UK
The Atrium, Southern Gate, Chichester, West Sussex, PO19 8SQ, UK
111 River Street, Hoboken, NJ 07030-5774, USA

For details of our global editorial offices, for customer services and for information about how to apply for permission to reuse the copyright material in this book please see our website at www.wiley.com/wiley-blackwell.

Library of Congress Cataloging-in-Publication data applied for.

ISBN: 9781118876213

A catalogue record for this book is available from the British Library.

Wiley also publishes its books in a variety of electronic formats. Some content that appears in print may not be available in electronic books.

Cover images: Top and bottom left photos by Sara Burgess with special acknowledgement to Manawatu Microscopy & Imaging Centre. Bottom right photo by Geoff Knight.

Set in 10/12.5pt Times by SPi Global, Pondicherry, India

1 2015

Contents

About the Editors

Dr Koon Hoong Teh

Dr Koon Hoong Teh graduated from Massey University in 2013 with a PhD in Food Technology. His work focused on the dairy biofilms found on milk tankers and their effect on the quality of dairy products. Prior to his doctorate study, his Masters research project was on biofilm formation by *Campylobacter jejuni* in a mixed bacterial population. He generated six scientific papers, five from his PhD and one from his Masters research project, and presented his works in national and international conference. Currently, he works as a rumen microbiologist. His research interest includes biofilms associated with food quality and safety, and culturing novel and previously uncultured rumen microorganisms. He is also a member of the New Zealand Microbiology Society.

Professor Steve Flint

Steve Flint is Professor of Food Safety and Microbiology and Director of the Food Division of the Institute of Food Nutrition and Human Health at Massey University, Palmerston North, New Zealand. Steve leads a team of postgraduate research students studying a variety of food safety and quality issues with an emphasis on understanding biofilm development and control. Approximately half these projects are associated with the dairy industry. Future research will focus on bacterial interactions in biofilms and mechanisms of biofilm dispersion. Steve has more than 100 scientific publications and more than 100 presentations at national and international scientific conferences. He lectures in food safety and microbiology and does consultancy work for food manufacturers. Steve is a fellow of the New Zealand Institute of Food Science and Technology, president of the New Zealand Microbiological Society and a certified food scientist with the Institute of Food Technology.

Professor John Brooks

John Brooks is a microbiologist, specialising in food microbiology. On graduation, he spent a period of time working at ICI UK, helping to develop the methanol-based Single Cell Protein process. He did a PhD in biochemical engineering at Sydney University, continuing his work on C1 metabolism. He then took up a position at Massey University, teaching food microbiology, and remained there for 30 years, eventually concentrating on biofilm research. John is now Adjunct Professor at Auckland University of Technology. He has consulted extensively for the food and process industries and is a Fellow of NZIFST.

Geoff Knight

Geoff graduated with a B. Appl. Sci. (Hons) in 1991 from La Trobe University (Bendigo) with a major in Microbiology. He initially worked on wastewater microbiology before moving to the food industry, where he developed an interest in biofilms. Geoff worked with The University of Tasmania and the Dairy Process Engineering Centre on research projects for the Australian dairy industry. In 1998, he joined CSIRO Division of Animal, Food and Health Sciences (formerly Food Science Australia), where he continued to study biofilms in food systems. His work at CSIRO has included investigating the impact of biofilm formation and cleaning-in-place procedures on the contamination of milk powders by thermophilic bacteria. More recently, his work has focused on the persistence of bacteria, including pathogens such as *Listeria monocytogenes* and *Cronobacter* species, in biofilms on environmental surfaces in dairy manufacturing plants.

List of Contributors

Editors

John Brooks
School of Applied Sciences, Faculty of Health and Environmental Sciences
Auckland University of Technology
Auckland
New Zealand

Steve Flint
Institute of Food, Nutrition and Human Health
Massey University
Palmerston North
New Zealand

Geoff Knight
Food Process Hygiene Solutions
Melbourne, Victoria
Australia

Koon Hoong Teh
Institute of Food, Nutrition and Human Health
Massey University
Palmerston North
New Zealand

Contributors

Rod Bennett
Institute of Food, Nutrition and Human Health
Massey University
Palmerston North
New Zealand

Phil Bremer
Department of Food Science
University of Otago
Dunedin
New Zealand

Sara Burgess
Institute of Food, Nutrition and Human Health
Massey University
Palmerston North
New Zealand

Michael Dixon
Institute of Food, Nutrition and Human Health
Massey University
Palmerston North
New Zealand

Robin Hankin
School of Computer and Mathematical Sciences
Auckland University of Technology
Auckland
New Zealand

Norshhaidah Mohd Jamaludin
Institute of Fundamental Sciences
Massey University
Palmerston North
New Zealand

Isabel Li
School of Applied Sciences, Faculty of Health and Environmental Sciences
Auckland University of Technology
Auckland
New Zealand

Kieran Mellow
Institute of Fundamental Sciences
Massey University
Palmerston North
New Zealand

Jon Palmer
Institute of Food, Nutrition and Human Health
Massey University
Palmerston North
New Zealand

Shanthi Parkar
Plant & Food Research
Palmerston North
New Zealand

Brent Seale
School of Applied Sciences, Faculty of Health and Environmental Sciences
Auckland University of Technology
Auckland
New Zealand

Ben Somerton
Institute of Food, Nutrition and Human Health
Massey University
Palmerston North
New Zealand

Xuemei Tang
Institute of Food, Nutrition and Human Health
Massey University
Palmerston North
New Zealand

Siti Norbaizura Md Zain
Institute of Food, Nutrition and Human Health
Massey University
Palmerston North
New Zealand

Foreword

Microbial biofilms have held a fascination for me since my first introduction to them as an undergraduate student. Since those early days, many scientific publications on this topic have shown the width and breadth of their complexity. Despite these extensive studies, so much still remains to be discovered. This is particularly true from a practical and manufacturing perspective. Publications in the scientific literature are one thing, but practical experiences in the real manufacturing environment often differ substantially from the sterile laboratory setting. Professor Steve Flint has had a long and successful career in both academia and in the dairy manufacturing industry, and has the ability to successfully translate academic biofilm studies and observations into practical applications for the industry. Professor John Brooks is one of New Zealand's most respected food safety experts, and he and academics including Professors Phil Bremer, Brent Seale, Jon Palmer and a group of collaborative experts have meshed their experiences and expertise to create a comprehensive book on dairy biofilms. Thus, the theme of this book can best be described as an amalgamation of the available fundamental and theoretical science on bacterial biofilms and the practical experiences from a food manufacturing environment, specifically focusing on dairy production. Overviews on the roles that the microbial surface, the attachment surface and the composition of the growth medium (i.e. the dairy product) play in bacterial surface attachment and biofilm formation are presented. These concepts are interwoven with general theories on how bacterial biofilms form, and how control is maintained, especially for foodborne pathogens. Some practical examples of microorganisms in real dairy manufacture, e.g. *Streptococcus* and thermophiles, and in selected processes, e.g. ultrafiltration and dairy wastewater treatment, are discussed in detail. As a result, both academic and nonacademic audiences can learn greatly from these chapters.

Dr Denise Lindsay
Senior Research Scientist, Fonterra, New Zealand

Preface to the Technical Series

For more than 60 years, the Society of Dairy Technology (SDT) has sought to provide education and training in the dairy field, disseminating knowledge and fostering personal development through symposia, conferences, residential courses, publications and its journal, the *International Journal of Dairy Technology* (previously published as the *Journal of the Society of Dairy Technology*).

In recent years, there have been significant advances in our understanding of milk systems, probably the most complex natural food available to man. At the same time, improvements in process technology have been accompanied by massive changes in the scale of many milk processing operations, and the manufacture of a wide range of dairy and related products.

The Society has embarked on a project with Wiley-Blackwell to produce a Technical Series of dairy-related books to provide an invaluable source of information for practising dairy scientists and technologists, covering the range from small enterprises to modern large-scale operation. This thirteenth volume in the series, on *Biofilms*, provides a timely and comprehensive review of a natural threat to the integrity of manufacturing processes as well as the quality and shelf life of dairy products. These problems are not limited to dairy operations but are also found in other food manufacturing operations and much of the principles covered in the chapters can be applied elsewhere. Biofilms can also be used beneficially, for instance in the bioremediation of effluent streams.

Andrew Wilbey
Chairman of the Publications Committee, SDT

Preface

The dairy industry has grown in size, sophistication and quality to satisfy an international demand for food and food ingredients. The major risk to product quality and economic manufacture is microbial contamination, predominantly due to the release of microorganisms and their metabolites from biofilms forming on the surfaces of equipment used in the handling of milk and the manufacture of milk products. The ultimate origin of the microorganisms is the raw milk, but the conditions through the manufacturing process provide specific niches ideal for the propagation of biofilms. The composition of these biofilms varies according to the conditions at any particular point in the manufacturing process. Microbial groups from psychrotrophs to thermophilic spore-forming bacteria form biofilms at specific zones in the manufacturing process. In some situations, the conditions are so selective that only a single species is detected. In other areas, interactions between species that can enhance biofilm development, spore production and the production of metabolites such as enzymes occur, all representing a threat to product quality.

Our understanding of the factors involved in the development of biofilms in the dairy industry has focused on the processes leading to microbial attachment in a dairy environment, conditions supporting biofilm growth and potential damage to product quality, the release of microorganisms from biofilm communities and the effect of cleaning systems on controlling biofilms. This has led to engineering solutions to limit the amount of surface area available for biofilm growth, replicating key pieces of equipment (e.g. evaporators) to enable frequent cleaning without stopping manufacture, improved cleaning systems and changes in plant operation – especially temperature – to limit biofilm growth and prevent activities such as spore production.

This book represents the result of 15 years of research into dairy biofilms involving researchers across several universities and research organisations. The content covers methods used in the detection and analysis of the microflora comprising dairy biofilms, information on the environments within the dairy industry that support biofilm development and a critical analysis of control methods used for biofilm control. Dairy industry managers, researchers and students will find this book useful in providing a fundamental understanding of problems relating to biofilms in the dairy industry and in offering some solutions and suggestions for improvement in managing a dairy manufacturing plant.

Dr Koon Hoong Teh
Prof. Steve Flint
Prof. John Brooks
Dr Geoff Knight

Preface

Acknowledgements

The editors and authors of this book wish to acknowledge Jessica Childs for her contribution in preparing graphics for several figures used in this book. Jessica was able to take our concepts and mould them into images that have added a unique aspect to this publication.

We also thank Matt Levin for setting up a virtual meeting room that enabled us to bring all the authors together for video conferencing during the preparation of this publication.

John Brooks and Geoff Knight deserve special mention for their proofreading of all of the chapters, which has provided some consistency and polish.

Owen McCarthy assisted in the final proofreading of Chapter 11.

This book was a true team effort from all concerned and could not have been achieved without the passion and dedication of everyone involved.

1 Introduction to Biofilms: Definition and Basic Concepts

Phil Bremer[1], Steve Flint[2], John Brooks[3] and Jon Palmer[2]

[1]Department of Food Science, University of Otago, Dunedin, New Zealand

[2]Institute of Food, Nutrition and Human Health, Massey University, Palmerston North, New Zealand

[3]School of Applied Sciences, Faculty of Health and Environmental Sciences, Auckland University of Technology, Auckland, New Zealand

1.1 Definition of biofilms

In 2012, the term 'biofilm' was defined by the International Union of Pure and Applied Chemistry (IUPAC), Polymer Division as an 'Aggregate of micro-organisms in which cells that are frequently embedded within a self-produced matrix of extracellular polymeric substances (EPS) adhere to each other and/or to a surface'. IUPAC included the following notes after the definition:

Note 1: A biofilm is a fixed system that can be adapted internally to environmental conditions by its inhabitants.

Note 2: The self-produced matrix of EPS, which is also referred to as slime, is a polymeric conglomeration generally composed of extracellular biopolymers in various structural forms.

The idea behind the development of this definition was to provide a terminology usable, without any confusion, in the various domains dealing with biorelated polymers, namely, medicine, surgery, pharmacology, agriculture, packaging, biotechnology and polymer waste management (Vert *et al.*, 2012).

Bearing this definition in mind, in this book we use the term 'biofilm' to refer to 'microorganisms attached to and growing, or capable of growing, on a surface'. This definition is broader than the IUPAC definition, as it includes cells or spores that are attached to a surface but have yet to produce a biofilm matrix. We have included attached cells not within a matrix in order to acknowledge that in many instances the act of attaching induces phenotypic changes to a cell. We have included the phrase 'growing or capable of growing' to reinforce the point that many of the unique features associated with biofilms arise as a result of the

Biofilms in the Dairy Industry, First Edition.
Edited by Koon Hoong Teh, Steve Flint, John Brooks and Geoff Knight.

growth and replication of microorganisms on a surface, such as the production of EPS and the development of a complex three-dimensional structure.

In this chapter, we briefly discuss the importance of biofilms to the dairy industry, before introducing their general features, including their development, composition and structure, the advantages they confer to microorganisms living in them and how they may be controlled. This chapter serves as an introduction to the other chapters in the book, and includes cross-references to more detailed information on dairy-specific features in other chapters.

1.2 Importance of biofilms in the dairy industry

On a global basis, the dairy industry produces a wide range of perishable (milk and cream) and semiperishable foods (cheese, butter and yoghurt) and food ingredients (milk powders, whey protein concentrates and caseinates). Microbial contamination of dairy products is of great concern to the dairy industry. Strict adherence to microbiological guidelines is essential to maintain product quality, functionality and safety (see Chapter 4) and to allow companies to remain competitive in the international market.

Those microorganisms associated with bovine raw milk and dairy manufacturing plants that are of particular interest to the dairy industry can be divided into three major categories, namely, spoilage, pathogenic and beneficial microorganisms. Spoilage microorganisms can have an impact on the quality and sensory properties of milk and other dairy products, through the production of metabolic byproducts and/or extracellular enzymes. Pathogenic microorganisms (see Chapter 9) have the potential to cause human illness and to have significant economic repercussions. Beneficial microorganisms generally belong to a diverse group loosely termed 'lactic acid-producing bacteria' (LAB) and are used as starter cultures for the manufacture of cheese, yoghurt and other fermented dairy products. A subgroup of LAB that is becoming more commonly used in fermented dairy products, such as yoghurt, is the probiotic bacteria, which include strains of *Lactobacillus* and *Bifidobacterium* (Jamaly *et al.*, 2011; Quigley *et al.*, 2013).

Biofilms have become a major issue within the dairy industry and are now recognised as sources, or potential sources, of contamination by spoilage or pathogenic microorganisms, which can decrease product safety, stability, quality and value. Many manufacturing processes provide unique niches, within processing equipment, where bacteria are able to grow and survive. Examples are thermoresistant streptococci in pasteurisation equipment (see Chapter 6) and thermophilic spore-forming bacteria in milk powder production equipment (see Chapter 7). Within the last 2–3 decades the importance of biofilms in the processing environment has also been recognised, particularly around drains and other locations that are difficult to reach and where cleaning and sanitation applications may be inadequate to eliminate bacteria present within biofilms.

In dairy manufacturing plants, biofilms can be divided into two categories: process biofilms, which are unique to processing plants and form on surfaces in direct contact with flowing product; and environmental biofilms, which form in the processing environment, such as in niches where cleaning and sanitation is poor and around drains. Process biofilms differ from environmental biofilms in two key ways. First, in a process biofilm, one or a few species may dominate, as the unit operation employed (e.g. pasteurisation equipment) may select for particular groups of bacteria (e.g. thermoduric). Second, process biofilms are frequently characterised by rapid growth rates. An example of this is the increase in numbers from 'not detectable' to 10^6 bacteria per cm^2 within 12 hours of operation that occurs in the regeneration section of a pasteurisation plant (Bouman *et al.*, 1892). In contrast, environmental biofilms can take several days or weeks to develop (Zottola & Sasahara, 1994).

1.3 Biofilm formation

The development of a biofilm on a surface follows a logical series of steps, in which the first step is the initial contact of the free-living microorganism with the surface. The initial interaction of cells with a surface is influenced by a wide range of chemical, physical and biological cues, as outlined in detail in Chapter 2. In general, the initial interactions are influenced by: (i) the surface topography, chemistry (functional groups, surface charge, presence of antibacterial compounds) and free energy (hydrophobicity); (ii) environmental conditions, including temperature, pH, nutrients and the presence of other microorganisms, which can either inhibit or enhance contact; (iii) processing factors such as fluid velocity and shear force; and (iv) the various mechanisms employed by the cell (quorum sensing, nutrient sensing, production of EPS) and the cell surface structures (such as pili, flagella, fimbriae, adhesins) to interact with the surface (Figure 1.1).

Once on or near a surface, a bacterium has to commit to adopting either an attached or a planktonic lifestyle based on a series of signals or cues it receives (Karatan & Watnick, 2009). An obvious cue for settlement is nutrient concentration, with high or low concentrations of nutrients

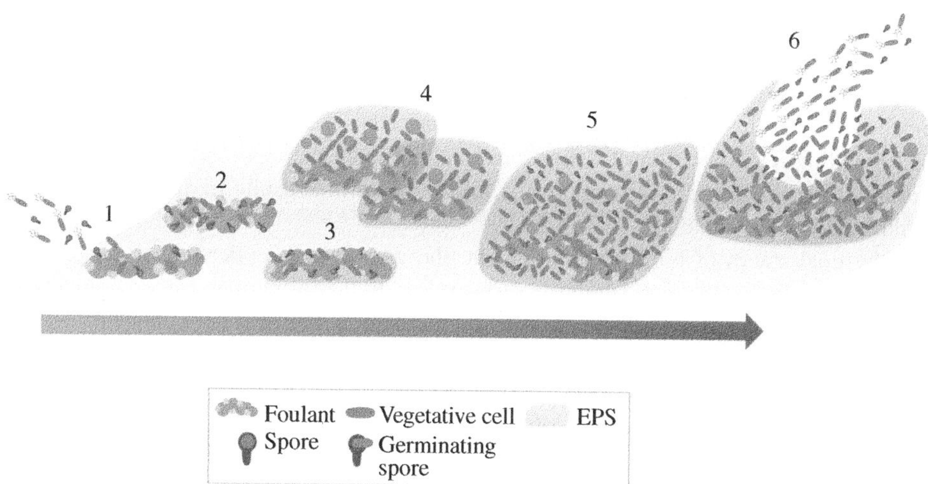

Figure 1.1 Steps involved in biofilm formation over time (arrow) in a dairy processing plant under conditions of flow. (1) Cells and/or spores come into contact with a surface that may be fouled with protein, fat and salts. (2) Cells and spores attach to the fouled surface. (3) Spores germinate and cells grow, beginning to produce EPS. (4) Cells replicate, forming microcolonies enclosed in EPS. (5) Microcolonies increase in size and coalesce, forming complex three-dimensional aggregates of cells and EPS that may contain a variety of niches. (6) Dispersal of cells and spores from the biofilm occurs.

promoting biofilm formation for different bacterial species. Bacteria, such as *Salmonella* spp., are more likely to join a multilayer biofilm in response to nutrient limitation (Gerstel & Romling, 2001), while for *Vibrio cholera*, the presence of glucose and other sugars induces production of a biofilm matrix and multilayer biofilm formation (Kierek & Watnick, 2003).

The second step in biofilm formation requires the cell to form at least a semipermanent association with the surface. This step is frequently referred to as the 'attachment phase'. Many authors have broken this down into a reversible and an irreversible phase, but with increasing knowledge on cell dispersal, the term 'irreversible attachment' is proving to be overstated. In dairy processing plants, there is a wide range of different materials to which bacteria can attach, including 304 and 316 stainless steel, plastic, elastomer (rubber) materials, polyester/polyurethane (conveyor belt materials), epoxy surface coatings and tiles. Bacteria will attach at different rates and strengths to these materials. The ability of bacteria to attach to a surface and the rate at which they attach will, however, change as material (proteins, carbohydrates) from the processing environment comes into contact with the surface and modifies its characteristics. Such so-called 'conditioning films' (see Chapter 3) occur almost as soon as a clean surface comes into contact with a liquid. In addition, the rate of attachment and the ease with which bacteria can be removed from the surface will change as the surface material ages, becomes damaged through mechanical operation or is exposed to cleaning agents and sanitisers.

The effect of surface roughness on the propensity of cells to attach is unclear. Some research reports greater cell attachment on surfaces with high surface roughness, while other research reports that there is no correlation between surface roughness and cell attachment to inert surfaces (Vanhaecke *et al.*, 1990; Flint *et al.*, 2000; Mitik-Dineva *et al.*, 2008, 2009; Truong *et al.*, 2010). While there may be some debate about the influence of surface roughness on attachment, there appears to be general agreement about the importance of using surfaces with minimal cracks and crevices in order to reduce bacterial adherence and biofilm growth and to enhance cleaning effectiveness.

In the next step of biofilm formation, the cells on the surface begin to replicate and produce EPS, which can include polysaccharides, proteins, eDNA and lipids. The production of EPS and the incorporation of extraneous material from the environment, such as food residues (soil) and other microorganisms, into the biofilm, results in an increase in the biofilm's bulk and complexity.

In the final stages of biofilm development, the growth and replication of the primary colonisers (the first cells to attach to the surface) lead to the formation of microcolonies on the surface. These microcolonies independently increase in size over time until they form a series of macrocolonies, which can eventually coalesce to varying degrees, forming complex three-dimensional aggregates of cells and EPS on the surface, variously described as being 'mushroom'- or 'pillar'-like. As the biofilm develops, the presence and metabolic activity of the bacteria within it, coupled with the production of EPS and its associated impact on the diffusion of compounds and gases into, out of and through the biofilm, can lead to the development of a wide variety of microenvironments or niches within the biofilm.

The ultimate structure of the biofilm is dependent on the bacterial species involved in its creation and the chemical and physical characteristics of its environment. Individual macrocolonies may merge together or may remain separated by narrow channels, through which nutrients and other molecules can readily diffuse. The developed biofilm is in a state of flux, where cells within it react to changes in the physical (flow rate, shear) and chemical (nutrient gradients, oxygen concentration) nature of the environment. The variety of conditions occurring within a biofilm can result in the development of phenotypically or genotypically distinct cell populations within it and can ultimately lead to the dispersion or release of cells from the biofilm.

Dispersal from biofilms may be either initiated by the bacteria themselves or mediated by external forces such as fluid shear, abrasion and cleaning. At least three distinct modes of biofilm dispersal have been identified: erosion, sloughing and seeding. Erosion is the continuous release of single cells or small clusters of cells from a biofilm at low levels, owing to either cell replication or an external disturbance to the biofilm. Sloughing is the sudden detachment of large portions of the biofilm, usually during the later stages of its growth, perhaps as conditions with it change or it becomes unstable due to its size. Seeding dispersal is the rapid release of a large number of single cells or small clusters of cells and is always initiated by the bacteria (Kaplan, 2010).

In the 1980s and 90s, interest in biofilms rapidly increased and there were many reports of biofilm formation and development following the generalised steps just described, leading to the proposal of a developmental model of microbial biofilms (O'Toole *et al.*, 2000). This model received wide interest, but, 10 years after it was first proposed, Monds and O'Toole (2009) published a paper expressing concern that evidence in its support had not been forthcoming and that it should not be considered as dogma.

It is known that many, if not all, bacteria are capable of forming or at least living within a biofilm and that living within a biofilm is frequently their normal mode of existence in natural environments (Costerton *et al.*, 1995; Stoodley *et al.*, 2002). As living within a biofilm requires extensive changes in both cell form and function, this strategy entails a significant commitment (Monds & O'Toole, 2009). Once a cell is committed to a biofilm, the spatial stratification within the biofilm can drive an additional physiological differentiation of the population. However, rather than being seen as an indication of the presence of specialised developmental stages, this is increasingly being considered as simply a reflection of the microorganism's response to the development of niches or a microenvironment within the biofilm. In short, it is the ability of bacteria to sense and to respond to their localised environment by regulating gene expression that leads to the development of a sustainable and complex biofilm, rather than an overarching bacterial community-focused goal.

1.4 Biofilm structure

While the structure of a biofilm is ultimately dependent on the species growing within it and the specific physical and chemical conditions in the environment surrounding it, a mature biofilm generally comprises clusters or layers of cells, which form a structure that can vary in thickness from a few micrometres to several millimetres. The cells are surrounded by EPS, which can contain up to 97% water (Zhang *et al.*, 1998). In general, the bacterial cells within a biofilm make up only about 15–20% of its volume, with the remainder being taken up by EPS.

Based on modelling studies, classical porous biofilms containing channels and voids between the mushroom-like outgrowths are predicted to occur under a substrate-transport-limited regime, while compact and dense biofilms are predicted in systems limited by biofilm growth rate and not by the substrate transfer rate. Surface complexity measures, such as roughness and fractal dimension, will increase with increasing transport limitations, while compactness will decrease as the biofilm changes from being dense to being highly porous and open (Picioreanu *et al.*, 1998).

Physical conditions, such as temperature, impact on the species composition (see Chapter 4) and growth rate of bacteria within a biofilm, while in pipelines, fluid flow dynamics can influence biofilm structure. Biofilms grown under laminar flow are reported to be patchy and to consist of aggregates of cells (mushrooms) separated by interstitial voids. Biofilms grown under turbulent flow may also be patchy but are characterised by the occurrence of chains of cells (streamers) that run from the biofilm surface into the bulk fluid phase (Stoodley *et al.*, 1998a). The biofilm as a whole, and the streamers in particular, exhibits viscoelastic properties, which means that it elongates and deforms as flow velocity increases and retracts as velocity decreases (Stoodley *et al.*, 1998b). Recently, it has been shown that the flow of liquid through porous materials, such as industrial filters, can stimulate the formation of streamers, which, over time, can bridge the spaces between surfaces and cause rapid clogging (Drescher *et al.*, 2013).

For many years, it has been known that some bacterial species, growing either as free living cells or within a biofilm, produce or release diffusible signal molecules that increase in concentration as a function of cell numbers. In a process termed 'quorum sensing', bacteria communicate with each other via these signal molecules or autoinducers to regulate their gene expression in response to population density (Miller & Bassler, 2001). The role of quorum sensing in biofilm formation was first reported for biofilms of *Pseudomonas aeruginosa* growing in a flow-through reactor, where it was found that the quorum sensing signal molecule $3OC_{12}$– homoserine lactone (C12) was required for normal biofilm differentiation (Davies *et al.*, 1998). The role of quorum sensing molecules in biofilm formation and differentiation has subsequently received considerable interest. While quorum sensing may not be significant in the structural development of all biofilms, there is evidence that for some species it can be important in events such as the attachment of bacteria to a surface, structural development and maturation and even the control of events leading to the dispersion or release of cells (Davies *et al.*, 1998; Boles & Horswill, 2008; Periasamy *et al.*, 2012; Lv *et al.*, 2014).

1.5 Composition of the EPS

As previously discussed, as cells attach, replicate and grow on a surface they produce EPS. EPS is recognised as playing an important role in the formation and function of biofilms of many species in many different environments. In addition, EPS, which is usually the major component of biofilm matrix, can act as an impermeable or at least semipermeable barrier, limiting the penetration of compounds into and out of the biofilm, and thereby facilitating the establishment of ecological niches within the biofilm and protecting the cells against the actions of antimicrobial compounds.

The composition and structure of components within EPS is varied and complex, being dependent on the bacterial species involved and the environment (Sutherland, 2001; Flemming & Wingender, 2010). EPS compounds that originate from microorganisms include polysaccharides, proteins, lipids and extracellular DNA (eDNA) (Flemming & Wingender, 2010). Polysaccharides have been identified as one of the major components of EPS. However, in many cases, the biochemical properties and functions of polysaccharides remain elusive, due to their complex structures, unique monomer linkages and the fact that their composition and concentration can change over time. Most of the polysaccharides that have been described are long linear or branched molecules, with molecular masses of $0.5–5.0 \times 10^5$ Daltons, and they may be homo- or heteropolysaccharides and either polyanionic (e.g. polysaccharides, such as aliginate or xanthan) or polycatonic compounds (Flemming & Wingender, 2010).

The biofilm matrix can also contain a considerable number of proteins. A wide range of enzymes has been detected within biofilms. Many of these are reported to have bipolymer degrading ability, enabling them to break down complex compounds, such as polysaccharides, proteins, nucleic acids, cellulose and lipids, into nutrients that are more readily available to bacteria. Biopolymer degrading enzymes also play a role in the dispersal of cells from the biofilm. Nonenzymatic proteins in the EPS or biofilm matrix are often involved in the formation and stabilisation of the EPS matrix and are often therefore termed 'structural proteins'. These include the cell surface-associated and extracellular carbohydrate-binding proteins, known as lectins, which form links between the bacterial surface and the EPS (Flemming & Wingender, 2010).

In addition to the obvious role of transferring genetic material between bacteria, via conjugation and DNA transformation, eDNA also appears to play a structural role in maintaining biofilms. The expression of conjugative pili has been shown to stimulate biofilm formation and can stabilise and influence the biofilm structure by forming connections between cells (Ghigo, 2001). The presence of eDNA has been shown to stabilise the young biofilms (Whitchurch *et al.*, 2002). eDNA also has antimicrobial activity and causes cells to lyse by chelating cations that stabilise lipopolysaccharides in the outer membranes of bacterial cells (Flemming & Wingender, 2010).

Lipids, lipopolysaccharides and surfactants can also be found to varying degrees within some EPS, where they are believed to play a role in the initial attachment of the cell to the surface, the development of the biofilm structure and the dispersal of cells from the biofilm (Flemming & Wingender, 2010).

1.6 Composition of the biofilm population

Most biofilms found in nature comprise a range of bacterial species. However, in specialised niches within processing plants, especially in those areas subjected to extremes of temperature, or where the product has been treated to inactivate most microorganisms, it is possible for biofilms dominated by one or a few species to develop. An example of this is in the production of milk powder, where it is possible to find biofilms developing within the evaporators that are dominated by one or two species of thermophilic spore-forming bacteria (Burgess *et al.*, 2010, 2013).

In general, biofilms are very heterogeneous environments characterised by a large degree of chemical, physical and biotic diversity. Variation in diffusion rates into and out of biofilms, as well as in the rates at which compounds are produced or metabolised, can lead to the development of concentration gradients for nutrients, oxygen, ions and signalling molecules. This can result in the creation of microenvironments and biotic diversity, even in monospecies biofilms, as cells adapt to changes in their local environment.

Like any other ecological niche, conditions within biofilms select for cells that are best suited to survive. This means that the resulting population is a reflection of the cells that come into contact with the niche, their ability to grow within the niche and the impact that cell growth and metabolism have on the niche. Based on the diversity of the planktonic population and the selective pressure at the surface and within the developing biofilm, biofilms can comprise one or a small number of species. In most instances, however, it is expected that a biofilm will contain a number of microbial species, with interactions occurring between them. In some cases, such interactions can facilitate the growth and survival of species that may be less suited to survival in a monospecies biofilm under the same environmental conditions (Bremer *et al.*, 2001).

Biotic diversity therefore occurs through a number of mechanisms. In the simplest instance, phenotypic changes take place due to variations in the cell's physiological status, dictated by nutrient or oxygen gradients (Stewart & Franklin, 2008). For example, cells located in the outermost layers of a biofilm that have a ready supply of nutrients and oxygen available can easily grow aerobically. The facultatively anaerobic cells in underlying layers may be oxygen-deprived and so will need to shift to an anaerobic metabolism in order to grow. This can encourage the growth of obligate anaerobic microflora. Cells at deeper layers within the biofilm may be nutrient-limited and have limited growth rates or be metabolically inactive. The response of individual bacterial cells to the local conditions drives phenotypic heterogeneity.

Phenotypic diversity may also arise due to variations in gene expression resulting from differences in transcription initiation or mRNA degradation. So-called 'stochastic gene expression' has been hypothesised to be a cell population's insurance against potential dramatic changes in environmental conditions (Veening *et al.*, 2008).

A third source of phenotypic heterogeneity is genetic mutations. Genetic variation occurring through point mutation, insertion or deletion can potentially increase the phenotypic variability within the biofilm. If such spontaneous mutants confer a significant selective advantage, especially in the presence of a stressor, they will confer a fitness advantage to the mutated cell and its offshoots and promote the survival of the cell population (Plakunov *et al.*, 2010).

Gene transfer within biofilms is enhanced by the close proximity of cells and the ability of the biofilm matrix to trap gene products within the biofilm. Gene transfer occurs within biofilms by two main mechanisms: plasmid conjugation and DNA transformation. In conjugation, direct cell-to-cell contact is required for plasmid transfer. Therefore, while DNA transfer can occur at high rates within a biofilm (Hausner & Wuertz, 1999), the structure of the biofilm and the degree to which cells can move within the biofilm to establish direct contacts will ultimately limit the extent to which conjugation occurs (Molin & Tolker-Nielsen, 2003). DNA transformation occurs when DNA (chromosomal or plasmid) released by one cell is picked up by another. It has been reported that most, if not all, bacteria have the ability to release DNA (Lorenz & Wackernagel, 1994). Cells that have the ability to efficiently take up macromolecular DNA are defined as having developed natural competence. Transformation rates for *Streptococcus mutans* growing within a biofilm have been reported to be 10–600-fold higher compared to the rate in planktonic cultures (Li *et al.*, 2001). Given that the presence of conjugative pili and eDNA, as discussed above, can stabilise biofilms (Whitchurch *et al.*, 2002), it appears that efficient gene transfer is both a consequence of and a contributor to biofilm development (Molin & Tolker-Nielsen, 2003).

1.7 Enhanced resistance of cells within biofilms

A large number of authors have compared the resistance of bacteria within biofilms to their free-living counterparts and declared that the former are far more resistant to a wide range of stressors, including antibiotics, ultraviolet (UV) damage and sanitisers (Costerton *et al.*, 1995; Elasri & Miller, 1999; Langsrud *et al.*, 2003; Bridier *et al.*, 2011). This protection has been postulated to result from a number of factors associated with living within a biofilm, including the binding of EPS to antimicrobial compounds, physical inhibition of the diffusion of antimicrobial compounds by the EPS or chemical reaction of antimicrobial compounds with components of the EPS matrix, all of which decrease the concentration of antimicrobial compounds reaching microorganisms within the biofilm (Thurnheer *et al.*, 2003). For example, chlorine (in a 25 ppm solution), which chemically reacts with organic material, has been shown to only be able to penetrate to a depth of 100 μm into a complex 150–200 μm-thick dairy biofilm (Jang *et al.*, 2006). In addition, chlorine concentrations within a mixed *Pseudomonas aeruginosa* and *Klebseilla pneumoniae* biofilm reached only 20% of the concentration measured in the bulk liquid (De Beer *et al.*, 1994). In contrast, it has been shown that EPS generally does not pose much of a barrier to relatively uncharged molecules, such as the antibiotic rifampin (Zheng & Stewart, 2002).

A direct result of the development of microenvironments within a biofilm is that the physiological state of cells in different parts of the biofilm can be varied. An example of this is the occurrence of so-called 'persister cells': the subpopulation of cells that are not growing (Lewis, 2010). It is postulated that these dormant cells are well suited to survival in stressful environmental conditions, and especially to exposure to antimicrobials, such as antibiotics, which target sites within actively growing cells. Most research on persister cells has focused on their high tolerance to antibiotics, with it being postulated that these cells are not antibiotic-resistant mutants, but rather phenotypic variants that occur stochastically within a clonal population of genetically identical cells (Levin & Rozen, 2006). It is thought that persister cells maintain dormancy due to the overexpression of a broad variety of genes that produce products which induce dormancy if present at high enough levels. Persister cells have been shown to occur at low numbers within stationary phase planktonic cultures and biofilms and it is postulated that such cells may be able to withstand the initial antimicrobial challenge and subsequently grow, reestablishing the population (Lewis, 2008, 2010).

Persister cells aside, the reduced metabolic activity of cells in nutrient-deficient areas within a biofilm may in part account for their increased resistance to antimicrobial agents (Stewart & Olson, 1992; Lisle *et al.*, 1998; Sabev *et al.*, 2006; Soto, 2013). Further, the stress of living in the biofilm (nutrient limitations, cell density triggers, pH changes, oxygen limitations, accumulation of waste products) can induce cells to express stress-responsive genes and to switch to more tolerant phenotypes. For example, in *E. coli*, environmental stress induces a transcriptional regulator that controls the rate at which the alternative sigma factor *RpoS* is produced. This sigma factor can help to prevent DNA damage and its production has been shown to be linked to biofilm formation (Foley *et al.*, 1999).

Biofilm formation may also result in the induction or inhibition of genes, which may specifically or inadvertently, either directly or indirectly, make the cells resistant to the stressor (Sauer *et al.*, 2002; Tremoulet *et al.*, 2002; Schembri *et al.*, 2003; Beaudoin *et al.*, 2012; Zhang *et al.*, 2013).

1.8 Controlling biofilms

Within a processing environment, the renowned difficulty in removing biofilms is caused by a wide variety of factors associated with plant design and operation, as well as the inherent properties of biofilms and the cells within them. Five factors are involved in the development of biofilms in dairy processing plants, namely: the species of microorganisms involved; the type of product being manufactured; the operational conditions (runtime and temperature); the surface material and its condition; and the cleaning and sanitation regimes (chemicals, use and frequency) employed. Given these variables, the factors that can be most easily controlled are the runtime, the cleaning and sanitation regime and, in some cases, the surface materials.

Cleaning and sanitation regimes are required to remove food residues, microorganisms and the cleaning and sanitation agents from food contact surfaces. The effectiveness of a cleaning and sanitation regime is dictated by chemical, thermal and mechanical processes, with combinations of cleaning and sanitation agents, chemical additives (surfactants, wetting agents, chelating agents), the correct temperature and the use of mechanical force (brushing, turbulent flow) being required. It is also essential to have a good understanding of the microorganisms involved – especially whether they are spore-forming or non-spore-forming microorganisms – and of the nature of any fouling material (protein, fat, carbohydrates, mineral salts) associated with the process, which may be incorporated into or cover the biofilm. As the effectiveness of cleaning and sanitation is dependent on a number of factors, it is vital that indicators of cleaning efficacy (microbial numbers, food residue) are monitored on a routine basis.

In dairy processing plants, equipment is normally cleaned-in-place (CIP) by circulating warm or hot cleaning solutions at high velocity (Stewart & Seiberling, 1996), thus satisfying the requirements for chemical, physical and thermal energy input (see Chapters 4 and 12 for more details). A feature of CIP regimes, evident in both industrial- and laboratory-scale systems, is their variable efficiency in eliminating surface-adherent bacteria (Austin & Bergeron, 1995; Faille *et al.*, 2001; Dufour *et al.*, 2004; Bremer *et al.*, 2006). The most important factors influencing the effectiveness of a CIP are: cleaning agent concentration and chemistry; cleaning agent temperature; cleaning time; degree of turbulence of the cleaning solution; and the characteristics of the surface being cleaned. The standard chemicals used in CIP regimes can be formulated to contain compounds, such as surfactants, that improve surface wetting, soil penetration and cleaning properties (Bremer *et al.*, 2006).

As concerns associated with the growth of bacteria within biofilms and their inherent increased resistance to cleaning agents and sanitisers have increased, increasing care has been taken in the design of systems and the specification of materials that will limit biofilm formation and enhance cleaning effectiveness. Dead ends, corners, cracks, crevices, gaskets, valves and joints have long been recognised as being difficult to clean and vulnerable to biofilm formation (Chmielewski & Frank, 2006). It is important to appreciate that any flaws in the design or physical location of equipment that decrease cleaning efficacy will enhance biofilm formation.

1.9 Emerging strategies for biofilm control

It is now well recognised that the removal of microbial cells from surfaces, once they have become attached (biofilms), can be very challenging. For this reason, recent interest has focused on the development of surfaces that either prevent or reduce attachment or contain compounds that are antibacterial and can therefore act against attached cells. It has recently been suggested that antibacterial surfaces should be categorised as being either antibiofouling or bactericidal, depending on the effect that they have on biological systems (Hasan *et al.*, 2013). In a recent review, Hasan *et al.* (2013) defined antibiofouling surfaces as surfaces that resist or prevent cellular attachment due to the presence of an unfavourable surface topography or surface chemistry. They defined bactericidal surfaces as surfaces that disrupt the cell on contact and cause cell death. They also stated that, in some instances, antibacterial surfaces may exhibit both antibiofouling and bactericidal characteristics, giving the example of a surface coated with N,N-dimethyl-2-morpholinone (CB ring), which is capable of inactivating bacteria in a dry environment, and with a zwitterionic carboxybetaine (CB-OH ring), which will resist bacterial attachment in a wet environment (Cao *et al.*, 2012).

Many approaches to the development of antibacterial surfaces involve the immobilisation of an antibacterial agent on the surface to be protected. The classic example of this approach is the historical widespread use of a number of antifouling paints containing either tributyl tin- or copper-based antimicrobial agents in the marine environment. In the food industry, it is important to develop antibacterial surfaces that in themselves will not impact on the safety or quality of the food with which they come into contact. While a number of coatings containing either silver, titanium, hydroxyapatite, antibiotics, quaternary ammonium compounds or fluoride ions (Price *et al.*, 1996; Ding, 2003; Hume *et al.*, 2004; Murata *et al.*, 2007; Zhao *et al.*, 2009) have been explored for their suitability as food contact surfaces, there are safety concerns over the possibility of the compounds being leached from them. In addition, there are a number of other limitations to this approach, including the potential for bacteria to develop resistance, the time it takes for the antibacterial agent to be released from the surface, the low concentration that may result, the potentially short lifetime of the antibacterial functionality and the ability of food components (proteins, lipids) to coat the surfaces, reducing their efficacy (Hasan *et al.*, 2013). Lee *et al.* (2004) proposed an approach to produce permanent, nonleaching antibacterial surfaces by utilising atom-transfer radical polymerisation to modify surfaces with quaternised ammonium groups. This approach is controllable and is reported to present a permanent antibacterial effect, as the surface can be reused without loss of activity (Lee *et al.*, 2004; Yang *et al.*, 2011). While such an approach is believed to potentially have application in the food industry, its commercial applications are still in development (Hasan *et al.*, 2013).

The observation that a number of naturally occurring surfaces in nature, such as insect wings, shark skin and lotus leaves, have the ability to resist fouling by preventing particles, algal spores and bacteria from sticking to them has led researchers to attempt to mimic their activity via microfabrication or nanotechnology (Chung *et al.*, 2007; Anselme *et al.*, 2010; Bazaka *et al.*, 2012). While this field of research is considered to be increasingly promising, with methods to modify the nanotopography of surfaces developing, it seems likely that the degree to which bacterial attachment is inhibited will be species-dependent (Ivanova *et al.*, 2011; Hasan *et al.*, 2013). The impact of surface topography and especially surface roughness on bacterial attachment will be discussed further in Chapter 2 (Section 2.4.3). Further, to be applicable for use in the dairy industry, the antifouling surface will need to be able to work in the presence of not only bacteria but also proteins, fat, sugar and inorganic salts – all compounds which have the potential to attach to and change the chemical and physical nature of a surface.

1.10 Conclusion

It is important to appreciate that microorganisms have been evolving and refining survival strategies for many millions of years. The ability to attach to surfaces and form biofilms is not new, and evidence from the fossil record indicates that microorganisms were living within biofilms at least 500 million years ago (Westall *et al.*, 2001). Over the last 30 years, as our knowledge of the features of biofilms and their way of life has developed, it has become increasingly obvious that the interactions associated with biofilms at the genetic, cellular, population and community level are extremely complex and that the challenge of preventing, controlling or eliminating biofilms is a daunting one.

References

Anselme, K., Davidson, P., Popa, A. M., Giazzon, M., Liley, M. & Ploux, L. 2010. The interaction of cells and bacteria with surfaces structured at the nanometre scale. *Acta Biomatereilia*, **6**, 3824–46.

Austin, J. W. & Bergeron, G. 1995. Development of bacterial biofilms in dairy processing lines. *Journal of Dairy Research*, **62**, 509–19.

Bazaka, K., Jacob, M. V., Crawford, R. J. & Ivanova, E. P. 2012. Efficient surface modification of biomaterial to prevent biofilm formation and the attachment of microorganisms. *Applied Microbiology and Biotechnology*, **95**, 299–311.

Beaudoin, T., Zhang, L., Hinz, A. J., Parr, C. J. & Mah, T. F. 2012. The biofilm-specific antibiotic resistance gene ndvB is important for expression of ethanol oxidation genes in Pseudomonas aeruginosa biofilms. *Journal of Bacteriology*, **194**, 3128–36.

Boles, B. R. & Horswill, A. R. 2008. Agr-mediated dispersal of *Staphylococcus aureus* biofilms. *PLoS Pathogens*, **4**, e1000052.

Bouman, S., Lund, D. B., Driessen, F. M. & Schmidt, D. G. 1892. Growth of thermoresistant streptococci and deposition of milk constituents on plates of heat-exchangers during long operating times. *Journal of Food Protection*, **45**, 806–12.

Bremer, P. J., Monk, I. & Osborne, C. M. 2001. Survival of *Listeria monocytogenes* attached to stainless steel surfaces in the presence or absence of Flavobacterium spp. *Journal of Food Protection*, **64**, 1369–76.

Bremer, P. J., Fillery, S. & Mcquillan, A. J. 2006. Laboratory scale clean-in-place (CIP) studies on the effectiveness of different caustic and acid wash steps on the removal of dairy biofilms. *International Journal of Food Microbiology*, **106**, 254–62.

Bridier, A., Briandet, R., Thomas, V. & Dubois-Brissonnet, F. 2011. Resistance of bacterial biofilms to disinfectants: a review. *Biofouling*, **27**, 1017–32.

Burgess, S. A., Lindsay, D. & Flint, S. H. 2010. Thermophilic bacilli and their importance in dairy processing. *International Journal of Food Microbiology*, **144**, 215–25.

Burgess, S. A., Flint, S. H. & Lindsay, D. 2013. Characterization of thermophilic bacilli from a milk powder processing plant. *Journal of Applied Microbiology*, doi: 10.1111/jam.12366.

Cao, Z., Mi, L., Mendiola, J., Ella-Menye, J. R., Zhang, L., Xue, H. & Jiang, S. 2012. Reversibly switching the function of a surface between attacking and defending against bacteria. *Angewandte Chemie International Edition in English*, **51**, 2602–5.

Chmielewski, R. A. N. & Frank, J. F. 2006. Biofilm formation and control in food processing facilities. *Comprehensive Reviews in Food Science and Food Safety*, **2**, 22–32.

Chung, K. K., Schumacher, J. F., Sampson, E. M., Burne, R. A., Antonelli, P. J. & Brennan, A. B. 2007. Impact of engineered surface microtopography on biofilm formation of *Staphylococcus aureus*. *Biointerphases*, **2**, 89–94.

Costerton, J. W., Lewandowski, Z., Caldwell, D. E., Korber, D. R. & Lappin-Scott, H. M. 1995. Microbial biofilms. *Annual Review of Microbiology*, **49**, 711–45.

Davies, D. G., Parsek, M. R., Pearson, J. P., Iglewski, B. H., Costerton, J. W. & Greenberg, E. P. 1998. The involvement of cell-to-cell signals in the development of a bacterial biofilm. *Science*, **280**, 295–8.

De Beer, D., Srinivasan, R. & Stewart, P. S. 1994. Direct measurement of chlorine penetration into biofilms during disinfection. *Applied and Environmental Microbiology*, **60**, 4339–44.

Ding, S. J. 2003. Properties and immersion behavior of magnetron-sputtered multi-layered hydroxyapatite/titanium composite coatings. *Biomaterials*, **24**, 4233–8.

Drescher, K., Shen, Y., Bassler, B. L. & Stone, H. A. 2013. Biofilm streamers cause catastrophic disruption of flow with consequences for environmental and medical systems. *Proceeding of the National Academy of Sciencies of the United States of America*, **110**, 4345–50.

Dufour, M., Simmonds, R. S. & Bremer, P. J. 2004. Development of a laboratory scale clean-in-place system to test the effectiveness of 'natural' antimicrobials against dairy biofilms. *Journal of Food Protection*, **67**, 1438–43.

Elasri, M. O. & Miller, R. V. 1999. Study of the response of a biofilm bacterial community to UV radiation. *Applied and Environmental Microbiology*, **65**, 2025–31.

Faille, C., Fontaine, F. & Benezech, T. 2001. Potential occurrence of adhering living Bacillus spores in milk product processing lines. *Journal of Applied Microbiology*, **90**, 892–900.

Flemming, H. C. & Wingender, J. 2010. The biofilm matrix. *Nature Reviews Microbiology*, **8**, 623–33.

Flint, S. H., Brooks, J. D. & Bremer, P. J. 2000. Properties of the stainless steel substrate influencing the adhesion of thermo-resistant streptococci. *Journal of Food Engineering*, **43**, 235–42.

Foley, I., Marsh, P., Wellington, E. M., Smith, A. W. & Brown, M. R. 1999. General stress response master regulator rpoS is expressed in human infection: a possible role in chronicity. *Journal of Antimicrobial Chemotherapy*, **43**, 164–5.

Gerstel, U. & Romling, U. 2001. Oxygen tension and nutrient starvation are major signals that regulate agfD promoter activity and expression of the multicellular morphotype in *Salmonella typhimurium*. *Environmental Microbiology*, **3**, 638–48.

Ghigo, J. M. 2001. Natural conjugative plasmids induce bacterial biofilm development. *Nature*, **412**, 442–5.

Hasan, J., Crawford, R. J. & Ivanova, E. P. 2013. Antibacterial surfaces: the quest for a new generation of biomaterials. *Trends in Biotechnology*, **31**, 295–304.

Hausner, M. & Wuertz, S. 1999. High rates of conjugation in bacterial biofilms as determined by quantitative *in situ* analysis. *Applied and Environmental Microbiology*, **65**, 3710–13.

Hume, E. B., Baveja, J., Muir, B., Schubert, T. L., Kumar, N., Kjelleberg, S., Griesser, H. J., Thissen, H., Read, R., Poole-Warren, L. A., Schindhelm, K. & Willcox, M. D. 2004. The control of *Staphylococcus epidermidis* biofilm formation and *in vivo* infection rates by covalently bound furanones. *Biomaterials*, **25**, 5023–30.

Ivanova, E. P., Truong, V. K., Webb, H. K., Baulin, V. A., Wang, J. Y., Mohammodi, N., Wang, F., Fluke, C. & Crawford, R. J. 2011. Differential attraction and repulsion of *Staphylococcus aureus* and *Pseudomonas aeruginosa* on molecularly smooth titanium films. *Scientific Reports*, **1**, 165.

Jamaly, N., Benjouad, A. & Bouksaim, M. 2011. Probiotic potential of *Lactobacillus* strains isolated from known popular traditional Moroccan dairy products. *British Microbiology Research Journal*, **1**, 79–94.

Jang, A., Szabo, J., Hosni, A. A., Coughlin, M. & Bishop, P. L. 2006. Measurement of chlorine dioxide penetration in dairy process pipe biofilms during disinfection. *Applied Microbiology and Biotechnology*, **72**, 368–76.

Kaplan, J. B. 2010. Biofilm dispersal: mechanisms, clinical implications, and potential therapeutic uses. *Journal of Dental Research*, **89**, 205–18.

Karatan, E. & Watnick, P. 2009. Signals, regulatory networks, and materials that build and break bacterial biofilms. *Microbiology and Molecular Biology Reviews*, **73**, 310–47.

Kierek, K. & Watnick, P. I. 2003. Environmental determinants of *Vibrio cholerae* biofilm development. *Applied and Environmental Microbiology*, **69**, 5079–88.

Langsrud, S., Sidhu, M. S., Heir, E. & Holck, A. L. 2003 Bacterial disinfectant resistance – a challenge for the food industry. *International Biodeterioration and Biodegradation*, **51**, 283–90.

Lee, S. B., Koepsel, R. R., Morley, S. W., Matyjaszewski, K., Sun, Y. & Russell, A. J. 2004. Permanent, nonleaching antibacterial surfaces. 1. Synthesis by atom transfer radical polymerization. *Biomacromolecules*, **5**, 877–82.

Levin, B. R. & Rozen, D. E. 2006. Non-inherited antibiotic resistance. *Nature Reviews Microbiology*, **4**, 556–62.

Lewis, K. 2008. Multidrug tolerance of biofilms and persister cells. *Current Topics in Microbiology and Immunology*, **322**, 107–31.

Lewis, K. 2010. Persister cells. *Annual Review of Microbiology*, **64**, 357–72.

Li, Y. H., Lau, P. C., Lee, J. H., Ellen, R. P. & Cvitkovitch, D. G. 2001. Natural genetic transformation of *Streptococcus mutans* growing in biofilms. *Journal of Bacteriology*, **183**, 897–908.

Lisle, J. T., Broadaway, S. C., Prescott, A. M., Pyle, B. H., Fricker, C. & Mcfeters, G. A. 1998. Effects of starvation on physiological activity and chlorine disinfection resistance in *Escherichia coli* O157:H7. *Applied and Environmental Microbiology*, **64**, 4658–62.

Lorenz, M. G. & Wackernagel, W. 1994. Bacterial gene transfer by natural genetic transformation in the environment. *Microbiological reviews*, **58**, 563–602.

Lv, J., Wang, Y., Zhong, C., Li, Y., Hao, W. & Zhu, J. 2014. The effect of quorum sensing and extracellular proteins on the microbial attachment of aerobic granular activated sludge. *Bioresource Technology*, **152**, 53–8.

Miller, M. B. & Bassler, B. L. 2001. Quorum sensing in bacteria. *Annual Review of Microbiology*, **55**, 165–99.

Mitik-Dineva, N., Wang, J., Mocanasu, R. C., Stoddart, P. R., Crawford, R. J. & Ivanova, E. P. 2008. Impact of nano-topography on bacterial attachment. *Biotechnology Journal*, **3**, 536–44.

Mitik-Dineva, N., Wang, J., Truong, V. K., Stoddart, P., Malherbe, F., Crawford, R. J. & Ivanova, E. P. 2009. *Escherichia coli*, *Pseudomonas aeruginosa*, and *Staphylococcus aureus* attachment patterns on glass surfaces with nanoscale roughness. *Current Microbiology*, **58**, 268–73.

Molin, S. & Tolker-Nielsen, T. 2003. Gene transfer occurs with enhanced efficiency in biofilms and induces enhanced stabilisation of the biofilm structure. *Current Opinion in Biotechnology*, **14**, 255–61.

Monds, R. D. & O'Toole, G. A. 2009. The developmental model of microbial biofilms: ten years of a paradigm up for review. *Trends in Microbiology*, **17**, 73–87.

Murata, H., Koepsel, R. R., Matyjaszewski, K. & Russell, A. J. 2007. Permanent, non-leaching antibacterial surface – 2: how high density cationic surfaces kill bacterial cells. *Biomaterials*, **28**, 4870–9.

O'Toole, G., Kaplan, H. B. & Kolter, R. 2000. Biofilm formation as microbial development. *Annual Review of Microbiology*, **54**, 49–79.
Periasamy, S., Joo, H. S., Duong, A. C., Bach, T. H., Tan, V. Y., Chatterjee, S. S., Cheung, G. Y. & Otto, M. 2012. How *Staphylococcus aureus* biofilms develop their characteristic structure. *Proceeding of the National Academy of Sciences of the United States of America*, **109**, 1281–6.
Picioreanu, C., van Loosdrecht, M.C. & Heijnen, J.J. 1998. Mathematical modeling of biofilm structure with a hybrid differential-discrete cellular automaton approach. *Biotechnology & Bioengineering*, **58**, 101–16.
Plakunov, V. K., Strelkova, E. A. & Zhurina, M. V. 2010. Persistence and adaptive mutagenesis in biofilms. *Microbiology*, **79**, 424–34.
Price, J. S., Tencer, A. F., Arm, D. M. & Bohach, G. A. 1996. Controlled release of antibiotics from coated orthopedic implants. *Journal of Biomedical Materials Research*, **30**, 281–6.
Quigley, L., O'Sullivan, O., Stanton, C., Beresford, T. P., Ross, R. P., Fitzgerald, G. F. & Cotter, P. D. 2013. The complex microbiota of raw milk. *FEMS Microbiology Review*, **37**, 664–98.
Sabev, H. A., Robson, G. D. & Handley, P. S. 2006. Influence of starvation, surface attachment and biofilm growth on the biocide susceptibility of the biodeteriogenic yeast *Aureobasidium pullulans*. *Journal of Applied Microbiology*, **101**, 319–30.
Sauer, K., Camper, A. K., Ehrlich, G. D., Costerton, J. W. & Davies, D. G. 2002. *Pseudomonas aeruginosa* displays multiple phenotypes during development as a biofilm. *Journal of Bacteriology*, **184**, 1140–54.
Schembri, M. A., Kjaergaard, K. & Klemm, P. 2003. Global gene expression in *Escherichia coli* biofilms. *Molecular Microbiology*, **48**, 253–67.
Soto, S. M. 2013. Role of efflux pumps in the antibiotic resistance of bacteria embedded in a biofilm. *Virulence*, **4**, 223–9.
Stewart, J. C. & Seiberling, D. A. 1996. Clean in place. *Chemical Engineering*, **102**, 72–9.
Stewart, M. H. & Olson, B. H. 1992. Impact of growth conditions on resistance of *Klebsiella pneumoniae* to chloramines. *Applied and Environmental Microbiology*, **58**, 2649–53.
Stewart, P. S. & Franklin, M. J. 2008. Physiological heterogeneity in biofilms. *Nature Reviews Microbiology*, **6**, 199–210.
Stoodley, P., Dodds, I., Boyle, J. D. & Lappin-Scott, H. M. 1998a. Influence of hydrodynamics and nutrients on biofilm structure. *Journal of Applied Microbiology*, **85** (Suppl. 1), 19S–28S.
Stoodley, P., Lewandowski, Z., Boyle, J. D. & Lappin-Scott, H. M. 1998b. Oscillation characteristics of biofilm streamers in turbulent flowing water as related to drag and pressure drop. *Biotechnology and Bioengineering*, **57**, 536–44.
Stoodley, P., Sauer, K., Davies, D. G. & Costerton, J. W. 2002. Biofilms as complex differentiated communities. *Annual Review of Microbiology*, **56**, 187–209.
Sutherland, I. 2001. Biofilm exopolysaccharides: a strong and sticky framework. *Microbiology*, **147**, 3–9.
Thurnheer, T., Gmur, R., Shapiro, S. & Guggenheim, B. 2003. Mass transport of macromolecules within an *in vitro* model of supragingival plaque. *Applied and Environmental Microbiology*, **69**, 1702–9.
Trémoulet, F., Duché, O., Namane, A., Martinie, B., Labadie, J. C. & The European Listeria Genome Consortium and Labadie, J. C. 2002. Comparison of protein patterns of *Listeria monocytogenes* grown in biofilm or in planktonic mode by proteomic analysis. *FEMS Microbiology Letters*, **210**, 25–31.
Truong, V. K., Lapovok, R., Estrin, Y. S., Rundell, S., Wang, J. Y., Fluke, C. J., Crawford, R. J. & Ivanova, E. P. 2010. The influence of nano-scale surface roughness on bacterial adhesion to ultrafine-grained titanium. *Biomaterials*, **31**, 3674–83.
Vanhaecke, E., Remon, J. P., Moors, M., Raes, F., De Rudder, D. & Van Peteghem, A. 1990. Kinetics of *Pseudomonas aeruginosa* adhesion to 304 and 316-L stainless steel: role of cell surface hydrophobicity. *Applied and Environmental Microbiology*, **56**, 788–95.
Veening, J. W., Smits, W. K. & Kuipers, O. P. 2008. Bistability, epigenetics, and bet-hedging in bacteria. *Annual Review of Microbiology*, **62**, 193–210.

Vert, M., Doi, Y., Hellwich, K.-H., Hess, M., Hodge, P., Kubisa, P., Rinaudo, M. & Schue, F. 2012. Terminology for biorelated polymers and applications (IUPAC Recommendations 2012). *Pure and Applied Chemistry*, **84**, 377–410.

Westall, F., De Wit, M. J., Dann, J., Van Der Gaast, S., De Ronde, C. E. J. & Gerneke, D. 2001. Early Archean fossil bacteria and biofilms in hydrothermally-influenced sediments from the Barberton greenstone belt, South Africa. *Precambrian Research*, **106**, 93–116.

Whitchurch, C. B., Tolker-Nielsen, T., Ragas, P. C. & Mattick, J. S. 2002. Extracellular DNA required for bacterial biofilm formation. *Science*, **295**, 1487.

Yang, W. J., Cai, T., Neoh, K. G., Kang, E. T., Dickinson, G. H., Teo, S. L. & Rittschof, D. 2011. Biomimetic anchors for antifouling and antibacterial polymer brushes on stainless steel. *Langmuir*, **27**, 7065–76.

Zhang, L., Fritsch, M., Hammond, L., Landreville, R., Slatculescu, C., Colavita, A. & Mah, T. F. 2013. Identification of genes involved in *Pseudomonas aeruginosa* biofilm-specific resistance to antibiotics. *PLoS One*, **8**, e61625.

Zhang, X., Bishop, P. L. & Kupferle, M. J. 1998. Measurement of polysaccharides and proteins in biofilm extracellular polymers. *Water Science and Technology*, **37**, 345–8.

Zhao, L., Chu, P. K., Zhang, Y. & Wu, Z. 2009. Antibacterial coatings on titanium implants. *Journal of Biomedical Materials Research Part B: Applied Biomaterials*, **91**, 470–80.

Zheng, Z. & Stewart, P. S. 2002. Penetration of rifampin through *Staphylococcus epidermidis* biofilms. *Antimicrobial Agents and Chemotherapy*, **46**, 900–3.

Zottola, E. A. & Sasahara, K. C. 1994. Microbial biofilms in the food processing industry – should they be a concern? *International Journal of Food Microbiology*, **23**, 125–48.

2 Significance of Bacterial Attachment: A Focus on the Food Industry

Jon Palmer[1], Brent Seale[2] and Steve Flint[1]

[1]Institute of Food, Nutrition and Human Health, Massey University, Palmerston North, New Zealand
[2]School of Applied Sciences, Faculty of Health and Environmental Sciences, Auckland University of Technology, Auckland, New Zealand

2.1 Introduction: The importance of bacterial attachment in biofilm development

The scientific community now acknowledges that in aquatic environments, bacteria exist predominantly in biofilm communities, with 1000–10 000 times more bacteria found attached to a surface in a biofilm community than in a planktonic (suspended) state. Advantages gained by living attached to a surface are thought to include higher concentrations of nutrients close to the surface, promoted genetic exchange, increased resistance to cleaning chemicals and, for a pathogen, increased protection from the host's immune system (Dickson & Koohmarare, 1989). In certain industrial situations, bacterial cell attachment to metallic surfaces may lead to biocorrosion, resulting in damage to pipelines and other important metallic surfaces and entailing millions of dollars in repairs (Beech & Sunner, 2004). In the food industry, it is well known that many pathogenic bacteria are capable of forming biofilms on food contact surfaces, but many other nonpathogenic species also grow within biofilms and cause spoilage issues, resulting in the manufacture of low-quality products. The dominating factor(s) involved in the initial attachment of bacterial cells to surfaces remains elusive, and today it is thought that a multitude of factors are involved, including conditioning films, mass transport, surface charge, hydrophobicity, surface roughness and surface microtopography.

2.2 Conditioning films and bacterial footprints: The importance of conditioning films and bacterial footprints in cell attachment

Organic and inorganic molecules present in liquids are transported to solid–liquid interfaces, either by diffusion or fluid dynamic forces, and accumulate there, forming what is commonly referred to as a 'conditioning film'. As a result, higher concentrations of nutrients are found at surfaces than in the liquid phase (Kumar & Anand, 1998). The greater access to nutrients

Biofilms in the Dairy Industry, First Edition.
Edited by Koon Hoong Teh, Steve Flint, John Brooks and Geoff Knight.

undoubtedly favours bacterial growth at surfaces. The adsorption of organic molecules, such as proteins, to surfaces could also play an important role in bacterial attachment by altering the physicochemical properties of surfaces, such as surface charge and hydrophobicity (Dickson & Koohmarare, 1989; Tang *et al.*, 2011).

Conflicting opinions exist on the importance of conditioning films on initial bacterial attachment. Bernbom *et al.* (2009) reported that water-soluble proteins of animal origin inhibited attachment of a *Pseudomonas* strain to stainless steel. Similarly, Parkar *et al.* (2001) demonstrated that the presence of dilute skim milk (1%) reduced attachment of spores and vegetative cells of thermophilic bacilli to a stainless steel surface. Skim milk was also found to reduce the attachment of *Staphylococcus aureus*, *Listeria monocytogenes* and *Serratia marcescens* to stainless steel (Barnes *et al.*, 1999). Even individual milk components, such as casein and β-lactoglobulin, are reported to reduce attachment of *Listeria monocytogenes* and *Salmonella* Typhimurium to stainless steel (Helke *et al.*, 1993). One reason for the reduced attachment may be that proteins in the bulk fluid phase compete with bacterial cells for binding sites on stainless steel surfaces.

In contrast, Speers and Gilmour (1995) observed that treatment of stainless steel and rubber surfaces with either whey proteins or lactose resulted in an increase in attachment of milk-associated microorganisms. Holah and Gibson (2000), commenting on Johal's (1988) observation that conditioning by meat juices resulted in a reduction in the surface charge of stainless steel, suggested that this was 'enhancing the potential accumulation of bacteria on the surfaces'. Jeong and Frank (1994) suggested that the presence of proteins on a surface favours biofilm formation, as attached proteins could be a source of nutrients for bacteria. In addition, Verran and Whitehead (2006b) reported that a greater number of bacterial cells remained on an inert surface following a cleaning cycle when cells were allowed to attach in the presence of bovine serum albumin (BSA).

The shearing off (removal) of bacterial cells from surfaces can leave behind 'bacterial footprints', which consist of either extracellular polymeric substances (EPS) or cell surface fragments. It is thought that bacterial footprints may play a role in further bacterial cell attachment (Neu, 1992). Dûfrene *et al.* (1996) and Azeredo and Oliveira (2000) reported that EPS enhanced bacteria adhesion to surfaces. However, Gomez-Suarez *et al.* (2002) found that EPS deposited by *Pseudomonas aeruginosa* prevented further bacterial adhesion. One possible explanation for this observation is the production of biosurfactants by bacteria, which can alter the hydrophobicity and surface charge of the bacterial surface. An example of a biosurfactant, produced by *P. aeruginosa*, is the glycolipid rhamnolipid. Sodagari *et al.* (2013) reported that the presence of rhamnolipids in the culture medium inhibited attachment of the Gram-negative bacteria *P. aeruginosa*, *Pseudomonas putida* and *Escherichia coli* and the Gram-positive bacterium *Bacillus subtilis* to both hydrophilic and hydrophobic surfaces. However, rhamnolipids demonstrated a limited ability to remove cells already attached to surfaces. Sodagari *et al.* (2013) also observed that rhamnolipids caused changes in the cell surface hydrophobicity of the *P. aeruginosa*, *P. putida* and *E. coli* strains, although they had no effect on substratum surface properties.

The conflicting observations regarding the importance of conditioning films and bacterial footprints on bacteria attachment may be a reflection of the different surfaces, bacterial strains and experimental conditions employed in studies. This may also be an indication of the diversity of responses displayed by different bacterial groups.

2.3 Bacterial outer surface and attachment

2.3.1 *Role of surface charge in relation to the abiotic surface and bacterial cell*

The cell surface charge is the sum (net) of positive and negative charges on the bacterial cell surface and is usually measured as its zeta-potential, which is calculated from the mobility of bacterial cells in the presence of an electrical field under defined salt concentration and pH. The magnitude of the cell surface charge varies between species and is influenced by cultural conditions (Gilbert *et al.*, 1991; Kim & Frank, 1994) and culture age (Walker *et al.*, 2005), as well as the ionic strength (Dan, 2003) and pH of the suspending medium (Husmark & Ronner, 1990). It is believed that interactions between the surface charges of bacterial cells and the substratum have a strong influence on the attachment of bacterial cells to surfaces.

Most bacteria have a negative zeta-potential at neutral pH (pH 7) (Gilbert *et al.*, 1991; Millsap *et al.*, 1997; Rijnaarts *et al.*, 1999; Lerebour *et al.*, 2004). However, Jucker *et al.* (1996) isolated a strain of *Stenotrophomonas* (*Xanthomonas*) *maltophilia*, with a positive zeta-potential at pH 7, and compared it with a strain of *P. putida*, which had a negative zeta-potential at pH 7. The *S. maltophilia* strain demonstrated high attachment efficiency to glass and Teflon, both of which have a negative surface charge. But as the ionic strength of the suspending medium was increased, a drop in attachment efficiency of *S. maltophilia* was noted, as well as a change to a negative zeta-potential, suggesting that surface charge plays an important role in attachment of *S. maltophilia* to glass and Teflon. Conversely, at high ionic strength, *P. putida* demonstrated an increasing (move towards zero) zeta-potential and a higher attachment efficiency. In another study, Giaouris *et al.* (2005) reported that higher sodium chloride concentrations (10.5%) inhibited the attachment of *Salmonella enterica* to stainless steel coupons. One explanation for these observations, mentioned by Jucker *et al.* (1996) and Van der Wal *et al.* (1997), is that the cell surface charge originates from the dissociation of acidic and basic groups, such as carboxyl, phosphate and amino groups, at the cell surface. In suspensions with a high ionic strength, more ions are available to bind to (or associate with) acidic and basic groups and thus neutralise the cell surface charge.

The pH of the suspending medium also influences cell surface charge by controlling the protonation/deprotonation of acidic and basic groups at the cell surface. At a low pH, the additional hydrogen ions (H^+) present bind to the negatively charged acidic groups, resulting in an increase in the zeta-potential. Conversely, at a high pH, the additional hydroxide ions (OH^-) bind to the positively charged amine groups, resulting in a decrease in the zeta-potential. Bacterial cells and spores have an isoelectric point, which is the pH value at which the zeta-potential is zero (the positive and negative charges are balanced).

Husmark and Ronner (1990) demonstrated that attachment of *Bacillus cereus* spores to surfaces was greatest when the pH of the suspending medium was equal to the isoelectric point of the *B. cereus* spores; in this case, pH 3. In the pH range above the isoelectric point (>pH 4), there was a decrease in spore attachment, which was thought to result from electrostatic repulsion between the surface of the spore and the substratum, because both had a negative charge. Seale *et al.* (2010) also reported that higher levels of attachment occurred in solutions with pH levels close to the isoelectric point of *Geobacillus* spores. Other groups to have reported positive correlations between cell surface charge and attachment include Ukuku and Fett (2002), Dickson and Koohmaraie (1989) and Van Loosdrecht *et al.* (1987a).

There have also been studies in which no relationship was found between cell surface charge and attachment. Flint *et al.* (1997) investigated the attachment of 12 strains of thermoresistant streptococci to stainless steel but did not find a relationship between attachment and surface charge, as measured by separation through anionic and cationic exchange resins. As noted by Flint *et al.* (1997), all of the thermoresistant streptococci strains displayed a negative surface charge at pH 7; this is likely to repel bacterial cells from surfaces such as stainless steel, which also have a negative surface charge. In this case, it is likely that other cell surface characteristics have a more dominant influence on cell attachment. In addition, Gilbert *et al.* (1991) noted that increasing negative charge on the surface of *E. coli* resulted in reduced attachment, demonstrating that attachment cannot be explained by surface charge alone.

2.3.2 *Hydrophobic interactions*

Hydrophobic interactions have widely been suggested to be responsible for the attachment of cells to surfaces (Van Loosdrecht *et al.*, 1987b; Wiencek *et al.*, 1990; Hood & Zottola, 1995). Although the hydrophobic effect has been known for some time, it is difficult to assign it a satisfying definition (Doyle, 2000). Put simply, a hydrophobic molecule would prefer to exist in a hydrophobic environment than in a hydrophilic environment, such as water. Cell surface hydrophobicity is influenced by structures and components found on the bacterial cell surface, such as pili, fimbriae polysaccharides and flagella, which can vary between bacterial strains (Reid *et al.*, 1999) and change throughout the bacterial life cycle.

There are conflicting views on whether hydrophobicity is a strong predictor of cell attachment to surfaces. Van Loosdrecht *et al.* (1987b), Gilbert *et al.* (1991), Peng *et al.* (2001), Iwabuchi *et al.* (2003) and Liu *et al.* (2004) have all suggested that there is a strong correlation ; Van Loosdrecht *et al.* (1987b) went so far as to suggest that surface hydrophobicity is the key factor in determining bacterial attachment to solid surfaces and that surface charge can only become important when surface hydrophobicity is minimal. However, it must be noted that Van Loosdrecht *et al.* (1987a) used polystyrene discs, which are very hydrophobic, to measure cell attachment, thus possibly favouring hydrophobic interactions. On the other hand, Sorongan *et al.* (1991), Parment *et al.* (1992), Flint *et al.* (1997) and Parkar *et al.* (2001) concluded that hydrophobicity had little to no relationship in determining bacterial cell attachment.

One of the key issues with hydrophobicity is determining the best methods by which to measure it. The three most popular are microbial adherence to hydrocarbons, commonly called the MATH test (previously known as bacterial adherence to hydrocarbons, or the BATH test, and first described by Rosenberg *et al.* (1980)), hydrophobic interaction chromatography (HIC) (Smyth *et al.*, 1978) and water contact angle measurements (Van der Mei *et al.*, 1998). In the MATH test, evidence exists that hydrophobicity is not the only interaction taking place between microbial cells and the organic solvent (typically a hydrophobic compound, such as hexadecane). Indeed, the solvents hexadecane and xylene have both been found to break down the cell walls of *S. thermophilus* and *Anoxybacillus* spp. (S.H. Flint, unpublished results).

Ahimou *et al.* (2001), Busscher *et al.* (1995) and Van der Mei *et al.* (1993) have reported that the MATH test can be influenced by electrostatic interactions. Busscher *et al.* (1995) reported that hexadecane, the hydrocarbon most commonly used to measure hydrophobicity, is negatively charged in water, with a zeta-potential of between −80 –and −50mV. Van der Mei *et al.* (1995) concluded that the MATH test should be measured at pH values at which the zeta-potential of the test organism and/or the hydrocarbon is near zero, in order to reduce potential interference from electrostatic interactions. Doyle (2000) suggested the MATH test should be performed either under high ionic strength or at the isoelectric point of the bacterial cells, to again minimise the potential influence of electrostatic interactions.

HIC involves the interaction of microbial cells with a hydrophobic column (e.g. a phenyl-sepharose column). Cells demonstrating high hydrophobicity are retained in the column and cells with low hydrophobicity are eluted. Smyth *et al.* (1978) noted that increasing the ionic concentration – in this case with sodium chloride – affected cell attachment to a HIC column. Wiencek *et al.* (1990) also found that a high ionic strength was required to overcome electrostatic repulsion between bacterial spores and the hydrophobic column. Wiencek *et al.* (1990) used both BATH and HIC methods to measure the relative cell hydrophobicity of bacterial spores and found general agreement between the results obtained with the two methods. Water contact angle measurements describe the tendency of a water droplet to spread across a homogeneous lawn of bacterial cells, usually obtained by filtration of a bacterial cell suspension on to a porous membrane surface (Busscher *et al.*, 1984). As a result, hydrophilic surfaces of high wettability yield low contact angles, while hydrophobic surfaces produce high contact angles. The slow penetration of the liquid drop into the bacterial lawn has led researchers to define the time in which measurements are taken after the addition of the drop on to the bacterial lawn (Gallardo-Moreno *et al.*, 2011).

Hydrophobicity measures the bulk properties of microbial cells at one particular point in time, and can be affected by culture medium, culture age, surface charge and the test method. This could be a principal reason why many studies have struggled to find a correlation between hydrophobicity and attachment. There may also be issues regarding the suitability of attachment assays employed in surface attachment studies. Harimawan *et al.* (2013) employed an atomic force microscope (AFM) to measure the attachment forces between a substratum (stainless steel) and individual bacterial cells and spores of *B. subtilis*. Spores were more hydrophobic than vegetative cells, as measured by contact angle measurements and the MATH test. Measurements obtained with AFM demonstrated spores exhibited greater retraction forces (i.e. adhesive forces to stainless steel) than vegetative cells, which suggests a relationship between hydrophobicity and strength of adhesion to stainless steel surfaces. This approach (i.e. using AFM to measure adhesion forces) may also be useful in studies investigating the influence of, for example, surface charge and the components of conditioning films.

2.3.3 *Role of carbohydrates in attachment*

The role of surface carbohydrates in biofilm structures is well documented, with extracellular polysaccharides described as the main cement holding a biofilm structure together (Sutherland, 2001). However, the involvement of surface polysaccharides in the initial attachment of bacteria to abiotic surfaces is not well understood. Flint *et al.* (1997) reported that surface carbohydrate production by various strains of *S. thermophilus* could not be related to the number of cells attaching to stainless steel. Parkar *et al.* (2001) also found no correlation between the attachment to stainless steel of thermophilic bacilli and the amount of extracellular polysaccharide produced. Allison and Sutherland (1987) compared the attachment of a polysaccharide-producing bacterial strain with that of a nonproducing strain and found no difference between the initial attachment of each. However, the polysaccharide-producing strain formed microcolonies on a surface, while the non-polysaccharide-producing strain remained as single attached cells. In addition, Lehner *et al.* (2005) were unable to relate extracellular polysaccharide production in *Enterobacter sakazakii* (now known as *Cronobacter* spp.) to biofilm formation on glass or PVC plastic.

Evidence supporting the role of surface carbohydrates in initial attachment was reported by Herald and Zottola (1989), who treated cells of *Pseudomonas fragi* with various compounds to disrupt proteins and carbohydrates and concluded that both surface polysaccharides and proteins play a role in attachment of *P. fragi* to stainless steel. It has also been shown that polysaccharide production by cells already attached to a surface leads to irreversible attachment on a variety of surfaces (Donlan, 2002; Romani *et al.*, 2008). There is currently no consensus on whether polysaccharides play a role in the initial attachment of cells to surfaces, but this may be a strain-specific property.

2.3.4 *Teichoic acids, eDNA and cell attachment: Are we missing something?*

Teichoic acids are a major class of surface glycopolymer commonly found in the cell wall of Gram-positive bacteria. They can account for as much as 60% of the cell wall dry weight (Heptinstall *et al.*, 1970; Xia *et al.*, 2010). Most teichoic acids demonstrate zwitterionic properties, due to the negatively charged phosphate groups and positively charged D-alanine residues common in staphylococci. The exact composition of teichoic acids varies from species to species, but generally consists of repeating glycerol phosphate or ribitol phosphate units. Teichoic acids may contain as many as 60 repeats and can extend through and beyond the cell wall (Weidenmaier & Peschel, 2008).

Two types of teichoic acid are produced by most Gram-positive bacteria. One, referred to as 'wall teichoic acid', is attached directly to the peptidoglycan layer, while the other, referred to as 'lipoteichoic acid', is attached to the cytoplasmic membrane. In staphylococci, the wall teichoic acids generally comprise ribitol phosphate units, whereas lipoteichoic acids generally comprise glycerol phosphate units (Vinogradov *et al.*, 2006). The functions of teichoic acids in the bacterial cell wall are not completely understood, but they may be involved in cation binding (Hughes *et al.*, 1973), attachment of cell surface proteins involved in maintaining the Gram-positive cell wall (such as autolysins) (Schlag *et al.*, 2010) and resistance to antimicrobial peptides (Peschel *et al.*, 1999). They may also play a role in cell attachment and biofilm formation, with Gross *et al.* (2001) concluding that the cell surface charge created by cell wall teichoic acids is essential for the initial attachment of bacteria to surfaces. Gross *et al.* (2001) created a mutation in the *dlt* operon of *S. aureus* that mediates the incorporation of D-alanine into teichoic acids, and found cells with this mutation were deficient in their ability to attach to glass and polystyrene surfaces, even though production of polysaccharide intercellular adhesion (PIA) was unaffected. This reduced attachment was attributed to an increased negative charge at the cell surface, caused by the lack of positively charged D-alanine in the cell wall.

Holland *et al.* (2011) also demonstrated the importance of teichoic acids in biofilm development of *Staphylococcus epidermidis* on polystyrene surfaces. Deletion of *tagO*, which encodes for an enzyme responsible for the first stage of teichoic acid production, led to a decrease in attachment and cell surface hydrophobicity and was associated with a decrease in PIA expression. The role of teichoic acids in cell attachment and biofilm development in other Gram-positive bacteria known to produce biofilms in the food industry is practically unexplored and little is known about the how teichoic acids affect cell surface properties and attachment to surfaces.

Extracellular DNA (eDNA) found within the biofilm matrix was initially considered to be residual DNA left over from lysed cells. However, it is now clear that eDNA is an integral part of the biofilm matrix (Flemming & Wingender, 2010). Early studies by Whitchurch *et al.* (2002) showed that biofilm development by *P. aeruginosa* was disrupted by the addition of DNase I into the suspending medium. More recently, Harmsen *et al.* (2010) demonstrated that eDNA was important for initial cell attachment and biofilm formation by *L. monocytogenes* on both polystyrene and glass surfaces. Vilain *et al.* (2009) found that attachment and biofilm formation by *B. cereus* were enhanced when eDNA was present on the surface. Finally, Das *et al.* (2010) found that treatment of staphylococci and streptococci with DNase I resulted in reduced attachment to hydrophilic and hydrophobic surfaces, suggesting surface DNA is also involved in the initial attachment process. Das *et al.* (2010) also observed that eDNA was more important for attachment of Gram-positive bacteria to hydrophobic than to hydrophilic surfaces. On a cautionary note, the role of eDNA should be considered carefully in studies in which removal of eDNA has been achieved using DNAses, as commercial DNAses often contain trace amounts of host proteases, which may contribute to biofilm disruption (Marti *et al.*, 2010).

Several questions still remain about the structure and composition of eDNA. For example, is the primary sequence of eDNA distinguishable from genomic DNA? Steinberger and Holden (2005) concluded that cellular DNA and eDNA appeared identical in *P. aeruginosa* biofilms. On the other hand, Bockelmann *et al.* (2006) reported that eDNA produced in biofilms by a γ proteobacteria strain, and analysed by random amplified polymorphic DNA (RAPD), had some noticeable differences from genomic DNA. Another question that still needs to be answered is whether eDNA is actively excreted into and maintained within the biofilm matrix or whether the presence of eDNA is solely a result of cell lysis.The exact role of teichoic acids and eDNA in bacterial attachment to surfaces and biofilm development is still in its infancy, and many questions remain about how teichoic acids and eDNA interact with other bacterial surface components and influence bacterial cell attachment.

2.4 Role of the abiotic surface in attachment

2.4.1 *Are all abiotic surfaces created even?*

Modern food processing equipment is fabricated using a wide variety of materials, including stainless steels, elastomers, polyester, polyurethane, polytetrafluoroethylene (PTFE) and even rubber, often used as O-rings when joining sections of stainless steel pipes. Stainless steel is the most common material used for food contact surfaces, because it is easy to fabricate and is durable, chemically and physiologically inert, corrosion resistant and easy to clean (Holah & Gibson, 2000).

Some studies have found that bacteria attach in higher numbers to hydrophobic surfaces, such as PTFE, than to hydrophilic surfaces, such as glass and metals (Pasmore *et al.*, 2002; Teixeira *et al.*, 2005). Others, such as Marouani-Gardi *et al.* (2009), have found little difference between biofilm formation by *E. coli* O157:H7 on stainless steel (hydrophilic) versus polyurethane (hydrophobic) surfaces.

The hydrophobicity rating of stainless steel surfaces is a topic with limited published work, with surfaces often only described as simply 'hydrophilic' or 'hydrophobic' in nature (Carpentier & Cerf, 1993). Descriptions of the hydrophobic/hydrophilic nature of stainless steel vary within the literature. Brugnoni *et al.* (2007) and Teixeira *et al.* (2005) described stainless steel as a hydrophobic surface, while Li and Logan (2004) listed all metal oxides used in their study as hydrophobic. However, Lejeune (2003), Planchon *et al.* (2007) and Lerebour *et al.* (2004) described stainless steel surfaces as hydrophilic in nature and Boulangé-Petermann (1996) commented that all metal surfaces are hydrophilic compared with polymers. The hydrophilic/hydrophobic nature of stainless steel appears to be relative: it is hydrophobic when compared with glass (a very hydrophilic material), but hydrophilic when compared with PTFE (a very hydrophobic surface). The hydrophobicity also depends on the grade and surface finish of the stainless steel, with electropolished 316 stainless steel appearing more hydrophilic than 316 stainless steel with a 2B finish. The confusion and contradiction relating to the hydrophilic/hydrophobic nature of stainless steel will most likely remain until there is an agreement on the most appropriate method by which to measure surface hydrophobicity.

The surface charge of an abiotic surface is also likely to play an important role in bacterial attachment. Fukuzaki *et al.* (1995) reported that the zeta-potential of stainless steel particles at pH 7 was weakly negative, with the stainless steel having an isoelectric point between pH 4.0 and 4.5. Bren *et al.* (2004) proposed that hydroxyl groups of surface oxides can interact with H^+ and OH^- groups according to the following reaction:

$$\mathrm{MeOH}_2^+ \xleftrightarrow{\mathrm{H+}} \mathrm{MeOH} + H^+ \xleftrightarrow{\mathrm{H+}} \mathrm{MeO}^- + 2H^+ \tag{2.1}$$

The relative levels of oxide groups that are protonated (positively charged), neutral or dissociated (negatively charged) are very dependent upon the pH of the overlying medium. Thus, in a low-pH medium, the dominant group would be $MeOH_2^+$, but at neutral or high pH values, MeOH or MeO^- groups might dominate. Different metal oxides may also have slightly different pKa and pKb values. Thus, altering the relative levels of metal oxides at the surface may lead to changes in the surface charge. This was demonstrated by Takehara and Fukuzaki (2002), who observed that stainless steel treated with nitric acid and ozone at 300 °C contained different ratios of chromium and iron oxides at the surface. The surfaces also had different relative adsorption curves for H^+ and OH^- titrations, suggesting that the surface treatment can play an important role in determining the surface charge of stainless steel.

However, it is very difficult to make broad assumptions that certain abiotic surfaces are more prone to biofilm formation in all food environments, as abiotic surfaces are largely influenced by environmental conditions and the bacterial species or strains that may be present. Some manufacturing operations, such as milk powder production, provide highly selection conditions, resulting in a single-species biofilm (Scott *et al.*, 2007). In this industrial situation, it may be possible to use surface modification not to stop biofilm formation, but to limit biofilm formation and extend production run times.

2.4.2 *Surface modification and ion impregnation of stainless steel to reduce cell attachment*

The modification of stainless steel surfaces to prevent or reduce bacterial attachment and biofilm formation is a challenging issue. There has been some success in creating antifouling paints for ship hulls (Srinivasan & Swain, 2007), but these paints generally leach toxic compounds into the environment, making them unsuitable for the food industry. The use of poylethylene glycol (PEG)-based compounds has also been proposed; these block microbial and protein attachment to surfaces, an approach termed a 'molecular brush' (Kingshott *et al.*, 2003). Roosjen *et al.* (2003) achieved reductions in attachment by 2 log for Gram-negative and Gram-positive bacteria by coating glass surfaces with PEG. However, when Wei *et al.* (2003) coated stainless steel with PEG it prevented adsorption of β-lactoglobulin but had no effect on attachment of strains of *Listeria* and *Pseudomonas*.

It would be surprising if the application of an antiadhesive compounds such as PEG to stainless steel surfaces saw use in industrial situations. In real environments, once surfaces become coated with organic material and bacteria start to attach, changes to the surface properties lead to failure of the antiadhesive surface.

Zhao *et al.* (2008) found that stainless steel implanted with N^+, O^+ or SiF_3^+ ions had a lower surface energy, higher contact angle and lower surface roughness compared with native stainless steel, with SiF_3^+ having the lowest surface energy. Attachment of *S. epidermidis* and *S. aureus*, over a period of 24 hours, was approximately 0.5 log lower for stainless steel implanted with SiF_3^+ than for native stainless steel. Unfortunately, only Gram-positive bacteria were tested in this study, so it was not possible to assess the effect SiF_3^+ might have on Gram-negative bacteria. Pereni *et al.* (2006) reported Ni-P-PTFE- and silicone-coated stainless steel had a lower surface energy than native stainless steel and reduced attachment of a *P. aeruginosa* strain by approximately 0.5 logs. The significance of a 0.5 log reduction in attachment within the food processing environment is debatable. The most likely benefit obtained from this approach will come from a reduced attachment strength between biofilms (bacterial cells and EPS) and the surface, which may result in biofilms that are easier to remove during cleaning.

2.4.3 *Surface roughness and microtopography*

Stainless steel can be manufactured with a number of surface finishes, which differ in surface characteristics such as surface roughness and microtopography. Both roughness and microtopography may contribute to cell attachment and biofilm formation.

There is no consensus on whether surface roughness is an important factor influencing cell attachment. Some groups have observed greater cell attachment on surfaces with higher

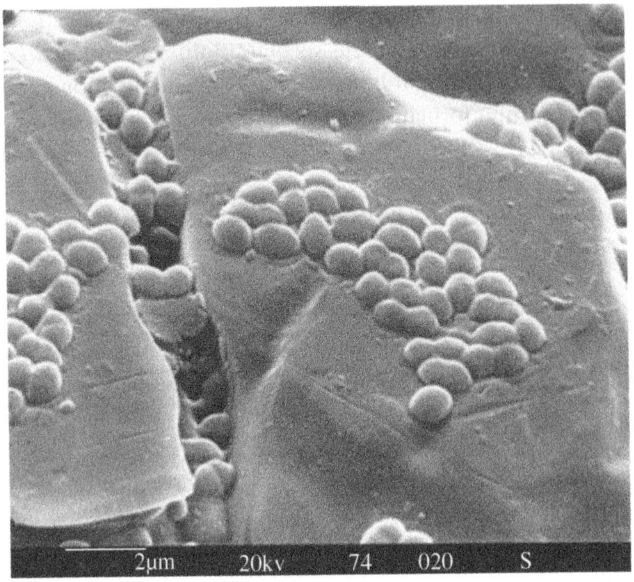

Figure 2.1 Adhesion of *Streptococcus thermophilus* cells to 304-grade stainless steel.

levels of surface roughness (Pedersen, 1990; Lecleroq-Perlat & Lalande, 1994), but others have observed no correlation between the two (Mafu *et al.*, 1990; Vanhaecke *et al.*, 1990; Flint *et al.*, 2000). The lack of consensus here may be a result of the degree of surface roughness considered in each of these studies.

Surface topography may also play a part in cell attachment to surfaces (Kumar & Anand, 1998). Surface topography can include features such as the parallel grooves obtained from polishing stainless steel with silicon carbide and the cracks and scratches that have been observed on stainless steel using scanning electron microscopy (SEM) (Zoltai *et al.*, 1981) and AFM (Arnold & Bailey, 2000). Verran and Whitehead (2006a) found that cell attachment was greater on surfaces with surface features (scratches and pits) that were of similar size to microbial cells, compared with surface features that were much larger than microbial cells. Several groups have observed through SEM that bacteria are able to attach within the surface cavities of stainless steel surfaces (Zoltai *et al.*, 1981; Verran & Whitehead, 2006a).

Verran *et al.* (2001) and Jullien *et al.* (2002) suggested that surface topography had little effect on the total number of bacterial cells attaching, but might protect cells from removal during cleaning and thus allow biofilm regrowth to occur more rapidly. Flint *et al.* (2000) also commented that surface topography around the critical size close to the diameter of the bacterial cells might entrap bacteria on the stainless steel surface, thus providing cells with some degree of protection from cleaning agents (Figure 2.1).

Nature has also played a role in advancing our understanding of how to control biofilm growth in industrial situations. An example is the development of Sharklet technology, which is an engineered surface based upon the microtopography of shark skin. Chung *et al.* (2007) reported that biofilm formation of *S. aureus* was disrupted over a 21-day test period and that biofilm coverage was significantly reduced. More recently, Graham *et al.* (2013) tested bacterial attachment to silicone-based polydimethylsiloxane (PDMS) surfaces displaying an array of topographical features, including rib features and holes with various spacing. All of the PDMS surfaces with engineered topographical features demonstrated reduced cell attachment compared to smooth PDMS surfaces.

2.5 *Staphylococcus* and attachment, an example: Surface proteins implicated in cell attachment to abiotic surfaces

The frequency of isolation of staphylococci from food contact surfaces in food processing plants varies a great deal. Sharma and Anand (2002) reported that for a dairy processing plant in India, 16% of all bacteria isolated were identified as *S. aureus*. In the United Kingdom, *Staphylococcus* was the second most common genus, after *Pseudomonas*, amongst bacteria isolated from biofilms within food factories (Gibson *et al.*, 1999). However, in Italy, Normanno *et al.* (2005) reported that of 1515 food contact surfaces tested, only 1.6% were positive for coagulase-positive staphylococci. The contamination of ready-to-eat foods by staphylococci is generally thought to occur through the raw materials or the hands of the people making the food. It is generally believed that between 10 and 40% of the population are carriers of enterotoxogenic *S. aureus* (Soriano *et al.*, 2002), and, as a consequence, much of the blame for coagulase-positive staphylococci contamination of foods comes from poor-quality raw materials or poor hygiene standards.

Studies have found that many, but not all, staphylococci isolated from cases of bovine mastitis (Darwish and Asfour, 2013) and dairy farms (Tremblay *et al.*, 2013; Lee *et al.*, 2014) are strong biofilm producers on a range of surfaces, including stainless steel (Heilmann *et al.*, 1997), milk-coated stainless steel (Hamadi *et al.*, 2014) and polystyrene (Cucarella *et al.*, 2001). Although the study of staphylococci biofilms tend to be mainly based on medical strains and medically relevant surfaces, biofilm formation by staphylococci on materials used in food processing and dairy plants is gaining more attention.

Key aspects of the initial attachment of staphylococci to solid surfaces are thought to include surface hydrophobicity (Hogt *et al.*, 1983, 1986), surface proteins (Timmerman *et al.*, 1991; Veenstra *et al.*, 1996; Heilmann *et al.*, 1997; Cucarella *et al.*, 2001; Knobloch *et al.*, 2001; Geoghegan *et al.*, 2010) and teichoic acid structure (Gross *et al.*, 2001). Several surface proteins have been implicated in the ability of staphylococci to attach to surfaces. Cucarella *et al.* (2001) identified two mutants of *S. aureus*, through the use of the transposon Tn917, which demonstrated significantly lower attachment to surfaces. Both mutants had the Tn917 transposon inserted at the same locus on the chromosome of the bacteria. This locus encoded a cell wall-associated protein of 2276 amino acids, with a size of 254 kDa, termed 'biofilm-associated protein' (BAP) (Arrizubieta *et al.*, 2004). All isolates of *S. aureus* harbouring the BAP gene showed high levels of attachment to inert surfaces and were strong biofilm producers. Tormo *et al.* (2005) reported that strong biofilm-producing strains from the species *S. epidermidis*, *Staphylococcus chromogenes*, *Staphylococcus xylosus*, *Staphylococcus simulans* and *Staphylococcus hyicus* all produced a BAP-like protein with an amino acid sequence similarity to BAP of greater than 80%, suggesting that the BAP surface protein is an important protein in the attachment of staphylococci to surfaces.

Other groups have also described the isolation of mutants unable to attach to solid surfaces or unable to form a biofilm due to the loss of a surface protein. Heilmann *et al.* (1996) isolated a transposon-insertion mutant of *S. epidermidis* unable to attach to polystyrene. In comparison with the wild type, the mutant lacks five cell surface-associated proteins with masses of 120, 60, 52, 45 and 38 kDa. Restoration of the 60 kDa protein by complementation studies demonstrated that only the 60 kDa protein was required for initial attachment to polystyrene. Also noted was a decrease in the hydrophobicity of the mutant compared with the wild-type strain and the more pronounced ability of the mutant to attach to a hydrophilic surface, in this case glass. Heilmann *et al.* (1996) suggested that the increase in attachment to glass may be a result of the mutant lacking the five surface proteins, allowing hydrophilic surface structures to become unmasked and thus making the cell surface more hydrophilic. This in turn increases the likelihood of hydrophilic/hydrophilic interaction between the mutant bacterial cell surface and the glass surface.

Further analysis by Heilmann *et al.* (1997) showed that the 60 kDa adhesion protein appeared to be a fragment of a much larger protein bearing sequence homology to an autolysin (AtlE) found in *S. aureus*. Heilmann *et al.* (1997) proposed that the 60 and 52 kDa protein fragments were produced by cleavage of the 120 kDa protein. This is similar to an AtlE found in *S. aureus*, which is composed of two lytic active domains of 60 and 52 kDa in size. The ability of the 60 kDa adhesion protein to bind to both polystyrene surfaces and plasma protein-coated surfaces suggests that it is a multifunctional surface protein that allows cells to attach to solid surfaces and host cell surfaces.

Veenstra *et al.* (1996) identified a 280 kDa surface protein from *S. epidermidis*, subsequently named SSP1 (*Staphylococcus* surface protein), and through the use of immunogold labelling and examination by electron microscopy suggested that it was located on fimbriae-like structures on the cell surface. Proteolytic cleavage of SSP1 by trypsin resulted in the production of SSP2, a 250 kDa product, as demonstrated by SDS-PAGE. The proteolytic cleavage of cells with SSP1 on the surface coincided with the loss of adhesive function and increased concentration of SSP2, suggesting the conversion of SSP1 to SSP2. Veenstra *et al.* (1996) suggested that the bacterial cell may be able to control its own phenotype between high and low adhesion states through the proteolytic cleavage of SSP1 to SSP2.

The question of the distribution of the previously mentioned surface proteins among staphylococci remains open: do *Staphylococcus* isolates possess all three surface proteins or do some isolates only have one or even none? Tremblay *et al.* (2013) concluded that the presence of intracellular polysaccharide (*icaA*) and/or the BAP surface protein was associated with a greater ability to form biofilms. However, initial attachment was not investigated. If some *Staphylococcus* isolates possess all three surface proteins associated with attachment, then does each protein have a specific affinity with a particular surface (e.g. hydrophobic or hydrophilic surfaces) or are they all generic in terms of overall surface affinity? Do other bacteria associated with attachment to solid surfaces also possess multiple surface proteins, as *Staphylococcus* strains appears to, or do they have a smaller or possibly larger repertoire of surface proteins that can be called upon to help in initial attachment of cells?

References

Ahimou, F., Poquot, M., Thonart, P. & Rouxhet, P. G. 2001. Influence of electrical properties on the evaluation of the surface hydrophobicity of *Bacillus subtilus*. *Journal of Microbiological Methods*, **45**, 119–26.

Allison, D. A. & Sutherland, I. W. 1987. The role of exopolysaccharides in adhesion of freshwater bacteria. *Journal of General Microbiology*, **133**, 1319–27.

Arnold, J. W. & Bailey, G. W. 2000. Surface finishes on stainless steel reduce bacterial attachment and early biofilm formation: scanning electron and atomic force microscopy study. *Poultry Science*, **79**, 1839–45.

Arrizubieta, M. J., Toledo-Arana, A., Amorena, B., Penadés, J. R. & Lasa, I. 2004. Calcium inhibits BAP-dependent multicellular behavior in *Staphylococcus aureus*. *Journal of Bacteriology*, **186**, 7490–8.

Azeredo, J. & Oliveira, R. 2000. The role of exopolymers in the attachment of *Sphingomonas paucimobilis Biofouling*, **16**, 59–67.

Barnes, L-M., Lo, M. F., Adams, M. R. & Chamberlain, H. H. L. 1999. Effect of milk proteins on adhesion of bacteria to stainless steel surfaces. *Applied and Environmental Microbiology*, **65**, 4543–8.

Beech, I. B. & Summer, J. 2004. Biocorrosion: towards understanding interactions between biofilms and metals. *Current Opinion in Biotechnology*, **15**, 181–6.

Bernbom, N., Ng, Y. Y., Jørgensen, R. L., Arpanaei, A., Meyer, R. L., Kingshott, P., Vejborg, R. M., Klemm, P. & Gram. L. 2009. Adhesion of food-borne bacteria to stainless steel is reduced by food conditioning films. *Journal of Applied Microbiology*, **106**, 1268–79.

Bockelmann, U., Janke, A., Kuhn, R., Neu, T. R., Wecke, J., Lawrence J. R. & Szewzyk, U. 2006. Bacterial extracellular DNA forming a defined network-like structure. *FEMS Microbiology Letters*, **262**, 31–8.

Boulangé-Petermann, L. 1996. Processes of bioadhesion on stainless steel surfaces and cleanability: a review with special reference to the food industry. *Biofouling*, **10**, 275–300.

Bren, L., English, L., Fogarty, J., Policoro, R., Zsidi, A., Vance, J., Drelich, J., Istphanous, N. & Rohly, K. 2004. Hydrophilic/electron acceptor surface properties of metallic biomaterials and their effect on osteoblast activity. *Journal of Science and Technology*, **18**, 1711–22.

Brugnoni, L. I., Lozano, J. E. & Cubitto, M. A. 2007. Potential of yeast isolated from apple juice to adhere to stainless steel surfaces in the apple juice processing industry. *Food Research International*, **40**, 332–40.

Busscher, H. J., Weerkamp, A. H., Van der Mei, H. C., Van Pelt, A. W. J., Jong, H. P. & Arends, J. 1984. Measurement of the surface free energy of bacterial cell surfaces and its relevance for adhesion. *Applied and Environmental Microbiology*, **48**, 980–3.

Busscher, H. J., Van De Belt-Gritter, B. & Van der Mei, H. C. 1995. Implications of microbial adhesion to hydrocarbons for evaluating cell surface hydrophobicity 1. Zeta potentials of hydrocarbon droplets. *Colloids and Surfaces B: Biointerfaces*, **5**, 111–16.

Carpentier, B. & Cerf, O. 1993. Biofilms and their consequences with particular reference to hygiene in the food industry. *Journal of Applied Bacteriology*, **75**, 499–511.

Chung K. K., Sahumacher, J. F., Sampson, E. M., Antonelli, P.J. & Brennan, A. B. 2007. Impact of engineered surface microtopography on biofilm formation of *Staphylococcus aureus*. *Biointerphases*, **2**, 89–94.

Cucarella, C., Solano, C., Valle, J., Amorena, B., Lasa, I. & Penades, J. R. 2001. BAP a *Staphylococcus aureus* surface protein involved in biofilm formation. *Journal of Bacteriology*, **183**, 2888–96.

Dan, N. 2003. The effect of charge regulation on cell adhesion to substrates: salt-induced repulsion. *Journal of Colloid Interface Science*, **27**, 41–7.

Darwish, S. F. & Asfour, H. A. E. 2013. Investigation of biofilm forming ability in staphylococci causing bovine mastitis using phenotypic and genotypic assays. *The Scientific World Journal*, **2013**, 1–9.

Das, T., Sharma, P. K., Busscher, H. J., Van der Mei, H.C. & Krom, B.P. 2010. Role of extracellular DNA in initial bacterial adhesion and surface aggregation. *Applied and Environmental Microbiology*, **76**, 3405–8.

Dickson, J. S. & Koohmarare, M. 1989. Cell surface charge characteristics and their relationship to bacterial attachment to meat surfaces *Applied and Environmental Microbiology*, **55**, 832–6.

Donlan, R. M. 2002. Microbial life on surfaces. *Emerging Infectious Diseases*, **8**, 881–90.

Doyle, R. J. 2000. Contribution of the hydrophobic effect to microbial infection. *Microbes Infection*, **2**, 391–400.

Dûfrene, Y. F., Boonaert, C. J. P. & Rouxhet, P. G. 1996. Adhesion of *Azospirillum brasilense*: role of proteins at the cell-support interface. *Colloids and Surfaces B–Biointerfaces*, **7**, 113–28.

Flemming, H. & Wingender, J. 2010. The biofilm matrix. *Nature Reviews Microbiology*, **8**, 623–33.

Flint, S. H., Brooks, J. D. & Bremer, P. J. 1997. The influence of cell surface properties of thermophilic streptococci on attachment to stainless steel. *Journal of Applied Microbiology*, **83**, 508–17.

Flint, S. H., Brooks, J. D. & Bremer, P. J. 2000. Properties of the stainless steel substrate influencing the adhesion of thermo-resistant streptococci. *Journal of Food Engineering*, **43**, 235–42.

Fukuzaki, S., Urano, H. & Hagata, K. 1995. Adsorption of pectin onto stainless steel surfaces: role of electrostatic interactions. *Journal of Japanese Society Food Science Technol-Nippon Shokuhin Kagaku Kogaku Kaishi*, **12**, 700–8.

Gallardo-Moreno, A. M., Navarro-Pérez, M. L., Vadillo-Rodriguez, V., Bruque, J. M. & González-Martin, M. L. 2011. Insights into bacterial contact angles: Difficulties in defining hydrophobicity and surface Gibbs energy. *Colloids and Surfaces B: Biointerfaces*, **88**, 373–80.

Geoghegan, J. A., Corrigan, R. M., Gruszka, D. T., Speziale, P., O'Gara, J. P., Potts, J. R., Foster, T. J. 2010. Role of surface protein SasG in biofilm formation *by Staphylococcus aureus. Journal of Bacteriology*, **192**, 5663–73.

Giaouris, E., Chorianopoulos, N. & Nychas, G.-J. E. 2005. Effect of temperature, pH, and water activity on biofilm formation by *Salmonella enterica* Enteritidis PT4 on stainless steel surfaces as indicated by the bead vortexing method and conductance measurements. *Journal of Food Protection*, **68**, 2149–54.

Gibson, H., Taylor, J. H., Hall, K. H. & Holah, J. T. 1999. Effectiveness of cleaning techniques used in the food industry in terms of removal of bacterial biofilms. *Journal of Applied Microbiology*, **87**, 41–8.

Gilbert, P., Evans, D. J., Evans, E., Duguid, I. G. & Brown, M. R. W. 1991. Surface characteristics and adhesion of *E. coli* and *Staphylococcus epidermidis. Journal of Applied Bacteriology*, **71**, 72–7.

Gomez-Suarez, C., Pasma, J., Van der Borden, A. J., Wingender, J., Flemming, H., Busscher, H. & Van der Mei, H. 2002. Influence of extracellular polymeric substances on deposition and redeposition of *Pseudomonas aeruginosa* to surfaces. *Microbiology-SGM*, **148**, 1161–9.

Graham, M. V., Mosier, A. P., Kiehl, T. R., Kaloyeros, A. E. & Cady, N. C. 2013. Development of antifouling surfaces to reduce bacterial attachment. *Soft Matter*, **9**, 6235–44.

Gross, M., Cramton, S. E., Götz, F. & Peschel, A. 2001. Key role of teichoic acid net charge in *Staphylococcus aureus* colonization of artificial surfaces. *Infection and Immunity*, **69**, 3423–6.

Hamadi, F., Asserne, F., Elabed, S., Bensouda, S., Mabrouki, M. & Latrache, H. 2014. Adhesion of *Staphylococcus aureus* on stainless steel treated with three types of milk. *Food Control*, **38**, 104–8.

Harimawan, A., Zhong, S., Chwee-Teck, L. & Yin-Peng, T. 2013. Adhesion of *Bacillus subtilis* spores and vegetative cells onto stainless steel – DLVO theories and AFM spectroscopy. *Colloids and Surfaces B: Biointerfaces*, **405**, 233–41.

Harmsen, M., Lappann, M., Knøchel, S. & Molin, S. 2010. Role of extracellular DNA during biofilm formation by *Listeria monocytogenes*. *Applied and Environmental Microbiology*, **76**, 2271–9.

Heilmann, C., Gerke, C., Perdreau-Remington, F. & Götz, F. 1996. Characterization of Tn917 insertion mutants of *Staphylococcus epidermidis* affected in biofilm formation. *Infection and Immunity*, **64**, 277–82.
Heilmann, C., Hussan, M., Peters, G. & Götz, F. 1997. Evidence for autolysin mediated primary attachment of *Staphylococcus epidermidis* to a polystyrene surface. *Molecular Microbiology*, **24**, 1013–24.
Helke, D. M., Somers, E. B. & Wong, A. C. L. 1993. Attachment of *Listeria monocytogenes* and *Salmonella typhimurium* to stainless steel and buna-N in the presence of milk and milk components. *Journal of Food Protection*, **56**, 479–84.
Heptinstall, S., Archibald, A. R. & Baddiley, J. 1970. Teichoic acids and membrane function in bacteria. *Nature*, **225**, 519–21.
Herald, P. J. & Zottola, E. A. 1989. Effect of various agents upon the attachment of *Pseudomonas fragi* to stainless steel. *Journal of Food Science*, **54**, 461–4.
Hogt, A. H., Dakart, J., Vries, J. A. & Feijen, J. 1983. Adhesion of coagulase negative *Staphylococcus* to biomaterials. *Journal of General Microbiology*, **129**, 2959–68.
Hogt, A. H., Dankart, J., Hulstaert, C. E. & Feijen, J. 1986. Cell surface characteristics of coagulase negative *Staphylococcus* and their adherence to fluorinated poly(ethylenepropylene). *Infection and Immunity*, **51**, 294–301.
Holah, J. & Gibson, H. 2000. Food industry biofilms. In: Allison, D., Gilbert, J. P., Lappin-Scott, H. & Wilson, M. (eds) Community Structure and Co-operation in Biofilms. Cambridge University Press, Cambridge, pp. 211–35.
Holland. L. M., Conlon, B. & O'Gara, J. P. 2011. Mutation of *tag*O reveals an essential role for wall teichoic acids in *Staphylococcus epidermidis* biofilm development. *Microbiology-SGM*, **157**, 408–18.
Hood, S. K. & Zottola, E. A. 1995. Biofilms in food processing. *Food Control*, **6**, 9–18.
Hughes, A. H., Hancock, I. C. & Baddiley, J. 1973. The function of teichoic acids in cation control in bacterial membranes. *Biochemical Journal*, **132**, 83–93.
Husmark, U. & Ronner, U. 1990. Forces involved in adhesion of *Bacillus* spores to solid surfaces under different environmental conditions. *Journal of Applied Bacteriology*, **69**, 557–62.
Iwabuchi, N., Sunairi, M., Anzai, H., Morisaki, H. & Nakajima, M. 2003. Relationships among colony morphotypes, cell surface properties and bacterial adhesion to substrata in *Rodococcus*. *Colloids and Surfaces B: Biointerfaces*, **30**, 51–60.
Jeong, D. K. & Frank, J. F. 1994. Growth of *Listeria monocytogenes* at 21 °C in biofilms with microorganisms isolated from meat and dairy environments. *Lebensm-Wiss Technology*, **27**, 415–24.
Johal, S. 1988. Bacterial adhesion to processing surfaces in the meat industry. PhD thesis, University of Surrey, UK.
Jucker, B. A., Harms, H. & Zehnder, A. J. B. 1996. Adhesion of the positively charged bacterium *Stenotrophomanas (Xanthomonas) maltophilia* 70401 to glass and teflon. *Journal of Bacteriology*, **178**, 5472–9.
Jullien, C., Benezech, T., Carpentier, B., Lebret, V. & Faille, C. 2002. Identification of surface characteristics relevant to the hygienic status of stainless steel for the food industry. *Journal of Food Engineering*, **56**, 77–87.
Kim, K. Y. & Frank, J. F. 1994. Effect of nutrients on biofilm formation by *Listeria monocytogenes* on stainless steel. *Journal of Food Protection*, **58**, 246–51.
Kingshott, P., Wei, J., Baggee-Ravn, D., Gadegaard, N. & Gram, L. 2003. Covalent attachment of poly(ethylene glycol) to surfaces, critical for reducing bacterial adhesion. *Langmuir*, **19**, 6912–21.
Knobloch, J. K. M., Bartscht, K., Sabottke, A., Rohde, H., Feucht, H. & Mack, D. 2001. Biofilm formation by *Staphylococcus epidermidis* depends on functional RsbU. An activator of the zyB operon: differential activation mechanisms due to ethanol and salt stress. *Journal of Bacteriology*, **183**, 2624–33.
Kumar, C. G. & Anand, S. K. 1998. Significance of microbial biofilms in food industry: a review. *International Journal of Food Microbiology*, **42**, 9–27.

Lecleroq-Perlat, M-N. & Lalande, M. 1994. Cleanability in relation to surface chemical composition and surface finishing of some materials commonly used in food industries. *Journal of Food Engineering*, **23**, 501–17.

Lee, S. H. L., Mangolin, B. L. C., Goncalves, J. L., Neeff, D. V., Silva, M. P., Cruz, A. G. & Oliveira, C. A. F. 2014. Biofilm-producing ability of *Staphylococcus aureus* isolates from Brazilian dairy farms. *Journal of Dairy Science*, **97**, 1812–16.

Lehner, A., Riedel, K., Eberl, L., Breeuwer, P., Diep, B. & Stephan, R. 2005. Biofilm formation, extracellular polysaccharide production and cell to cell signalling in various *Enterobacter sakazakii* strains: aspects promoting environmental persistence. *Journal of Food Protection*, **68**, 2287–94.

Lejeune, P. 2003. Contamination of abiotic surfaces: what a colonizing bacterium sees and how to blur it. *Trends in Microbiology*, **11**, 179–84.

Lerebour, G., Cupferman, S. & Bellon-Fontaine, M. N. 2004. Adhesion of *Staphylococcus aureus* and *Staphylococcus epidermidis* to Episkin® reconstructed epidermis model and to an inert 304 stainless steel substrate. *Journal of Applied Microbiology*, **97**, 7–16.

Li, B. & Logan, B. E. 2004. Bacterial adhesion to glass and metal-oxide surfaces. *Colloids and Surfaces B: Biointerfaces*, **36**, 81–90.

Liu, Y., Yang, S., Li, Y., Xu, H., Qin, L. & Tay, J. 2004. The influence of cell and substratum surface hydrophobicities on microbial attachment. *Journal of Bacteriology*, **110**, 251–6.

Mafu, A. A., Roy, D., Foulet, J. & Magny, P. 1990. Attachment of *Listeria monocytogenes* to stainless steel, glass, polypropylene and rubber surfaces after short contact times. *Journal of Food Protection*, **53**, 742–6.

Marouani-Gardi, N., Chassaing, D. & Carpentier, B. 2009. Comparative evaluation of biofilm formation and tolerance to a chemical shock of pathogenic and nonpathogenic *Escherichia coli* O157:H7 strains. *Journal of Food Protection*, **72**, 157–64.

Marti, M., Trotonda, M. P., Tormo-Mas, M. A., Vergara-Irigaray, M., Cheung, A. L., Lasa, I. & Penades, J. R. 2010. Extracellular proteases inhibit protein-dependent biofilm formation in *Staphylococcus aureus*. *Microbes and Infection*, **12**, 55–64.

Millsap, K. W., Reid, G., Van der Mei, H.C. & Bussher, H. 1997. Cluster analysis of genotypically characterized *Lactobacillus* species based on physicochemical cell surface properties and their relationship with adhesion to hexadacane. *Canadian Journal of Microbiology*, **43**, 284–91.

Neu, T. R. 1992. Microbial footprints and the general ability of microorganisms to label interfaces. *Canadian Journal of Microbiology*, **38**, 1005–8.

Normanno, G., Firinu, A., Virgilio, S., Mula, G., Dambrosio, A., Poggiu, A., Decastelli, L., Mioni, R., Scuota, S., Bolzoni, G., Di Giannatale, E., Salinetti, A. P., La Salandra, G., Bartoli, M., Zuccon, F., Pirino, T., Sias, S., Parisi, A., Quaglia, N. C. & Celano, G. V. 2005. Coagulase-postive staphylococci and *Staphylococcus aureus* in foods products marketed in Italy. *International Journal of Food Microbiology*, **98**, 73–9.

Parkar, S. G., Flint, S. H., Palmer, J. S. & Brooks, J. D. 2001. Factors influencing attachment of thermophilic bacilli to stainless steel. *Journal of Applied Microbiology*, **90**, 901–8.

Parment, P. A., Svanborg-Eden, C., Chaknis, M. J., Sawant, A. D., Hagber, G. L., Wilson, L. A. & Adhearn, D. G. 1992. Hemagglutination (Fimbriae) and hydrophobicity in adherence of *Serratia marcescens* to urinary tract epithelium and contact lenses. *Current Microbiology*, **25**, 113–18.

Pasmore, M., Todd, P., Pfirfer, B., Rhodes, M. & Boeman, C. N. 2002. Effect of polymer surface properties on the reversibility of attachment of *Pseudomonas aeruginosa* in the early stages of biofilm development. *Biofouling*, **18**, 65–71.

Pedersen, K. 1990. Biofilm development on stainless steel and PVC surfaces in drinking water. *Water Research*, **24**, 239–43.

Peng, J. S., Tsai, W. C. & Chou, C. C. 2001. Surface characteristics of *Bacillus cereus* and its adhesion to stainless steel. *International Journal of Food Microbiology*, **65**, 105–11.

Pereni, C. I., Zhao, Q., Liu, Y. & Abel, E. 2006. Surface free energy effect on bacterial retention. *Colloids and Surfaces B: Biointerfaces*, **48**, 143–7.

Peschel, A., Otto, M., Jack, R.W., Kalbacher, H., Jung, G. & Götz, F. 1999. Inactivation of the dlt operon in *Staphylococcus aureus* confers sensitivity to defensins protegrins and other antimicrobial peptides. *Journal of Biological Chemistry*, **274**, 8405–10.

Planchon, S., Gaillard-Martinie, B., Leroy, S., Bellon-Fontaine, M. N., Fada, S. & Talon, R. 2007. Surface properties and behaviour on abiotic surfaces of *Staphylococcus carnosus*, a genetically homogeneous species. *Food Microbiology*, **24**, 44–51.

Reid, G., Bialowska-Hobrzanska, H., Van der Mei, H. C. & Busscher, H. J. 1999. Correlation between genetic, physic-chemical surface characteristics and adhesion of four strains of *Lactobacillus*. *Colloids and Surfaces B: Biointerfaces*, **13**, 75–81.

Rijnaarts, H. H. M., Norde, W., Lyklema, J. & Zehnder, A. J. B. 1999. DLVO and steric contributions to bacterial deposition in media of different ionic strengths. *Journal of Colloid Interface Science*, **14**, 179–95.

Romani, A. M., Fund, K., Artigas, J., Schwartz, T., Sabater, S. & Obst, U. 2008. Relevance of polymeric matrix enzymes during biofilm formation. *Microbial Ecology*, **56**, 427–36.

Roosjen, A., Kaper, H.J., Van der Mei, H.C., Norde, W. & Busscher, H. J. 2003. Inhibition of adhesion of yeasts and bacteria by poly(ethylene oxide)-brushes on glass in a parallel plate flow chamber. *Microbiology*, **149**, 3239–46.

Rosenberg, M., Gutnick, D. & Rosenberg, E. 1980. Adherence of bacteria to hydrocarbons: a simple method for measuring cell-surface hydrophobicity. *FEMS Microbiology Letters*, **9**, 29–33.

Schlag, M., Biswas, R., Krismar, B., Kohler, T., Zoll, S., Yu, W., Schwarz, H., Peschel, A. & Götz, F. 2010. Role of staphylococcal wall teichoic acid in targeting the major autolysin Atl. *Molecular Microbiology*, **75**, 864–73.

Scott, S. A., Brooks, J. D., Rakonjac, J., Walker, K. M. R. & Flint, S. H. 2007. The formation of thermophilic spores during the manufacture of whole milk powder. *International Journal of Dairy Technology*, **60**, 109–17.

Seale, R. B., Bremer, P. J., Flint, S. H. & Mcquillen, A. J. 2010. Characterization of spore surfaces from a *Geobacillus* sp. isolate by pH dependence of surface charge and infrared spectra. *Journal of Applied Microbiology*, **109**, 1339–48.

Sharma, M. & Anand, S. K. 2002. Characterization of constitutive microflora of biofilms in dairy processing lines. *Food Microbiology*, **19**, 627–36.

Smyth, C. J., Jonsson, P., Olsson, E., Soderland, O., Rosengren, J., Hjerten, A. & Adstrom, T. 1978. Differences in hydrophobic surface characteristics of porcine enteropathogenic *E. coli* with or without K88 antigen as revealed by HIC. *Infection and Immunity*, **22**, 462–72.

Sodagari, M., Wang, H., Newby, Z. & Ju, L. K. 2013. Effect of rhamnoilpids on initial attachment of bacteria on glass and octadecyltrichorosilane-modified glass. *Colloids and Surfaces B: Biointerfaces*, **103**, 121–8.

Soriano, J. M., Font, G., Molto, J. G. & Manes, J. 2002. Enterotoxigenic staphylococci and their toxins in restaurant foods. *Trends in Food Science and Technology*, **13**, 60–7.

Sorongon, M. L., Bloodgood, R. A. & Burchard, R. P. (1991) Hydrophobicity adhesion and surface exposed proteins of gliding bacteria. *Applied Environmental Microbiology*, **57**, 3193–9.

Speers, J. G. S. & Gilmour, A. 1985. The influence of milk and milk components on the attachment of bacteria to farm dairy equipment surfaces. *Journal of Applied Bacteriology*, **59**, 325–32.

Srinivasan, R. & Swain, G. 2007. Managing the use of copper based antifouling paints. *Environmental Management*, **39**, 423–41.

Steinberger, R. E. & Holden, P. A. 2005. Extracellular DNA in single- and multiple-species unsaturated biofilms. *Applied and Environmental Microbiology*, **71**, 5404–10.

Sutherland, I. W. 2001. Biofilm exopolysaccharides: a strong and sticky framework. *Microbiology*, **147**, 3–9.

Takehara, A. & Fukuzaki, S. 2002. Effect of the surface charge of stainless steel on adsorption behaviour of pectin. *Biocontrol Science*, **7**, 9–15.

Tang, L., Pillai, S., Revsbech, N. P., Schramm, A., Bischoff, C. & Meyer R. L. 2011. Biofilm retention on surfaces with variable roughness and hydrophobicity. *Biofouling*, **27**, 111–21.

Teixeira, P., Lopes, Z., Azeredo, J., Oliveira, R. & Vieira, M. J. 2005. Physico-chemical surface characterization of a bacterial population isolated from a milking machine. *Food Microbiology*, **22**, 247–51.

Timmerman, C. P., Fleer, A., Besnier, L., Degraff, L., Cremers, F. & Verhoef, J. 1991. Characterisation of a proteinaceous adhesion of *Staphylococcus epidermidis* which mediates attachment to polystyrene. *Infection and Immunity*, **59**, 4187–92.

Tormo, M. A., Knecht, E., Götz, F., Lasa, N. & Penadés, J. R. 2005. Bap-dependent biofilm formation by pathogenic species of *Staphylococcus*: evidence of horizontal gene transfer? *Microbiology*, **151**, 2465–75.

Tremblay, Y., Lamarche, D., Chever, P., Haine, D., Messier, S. & Jacques, M. 2013. Characterisation of the ability of coagulase-negative staphylococci isolated from the milk of Canadian farms to form biofilms. *Journal of Dairy Science*, **96**, 234–46.

Ukuku, D. O. & Fett, W. F. 2002. Relationship of cell surface change and hydrophobicity to strength of attachment of bacterial to cantaloupe rind. *Journal of Food Protection*, **65**, 1093–9.

Van der Mei, H. C., De Vries, J. & Buscher, H. J. 1993. Hydrophobic and electrostatic cell surface properties of thermophilic dairy *streptococci*. *Applied and Environmental Microbiology*, **59**, 4305–12.

Van der Mei, H. C., Van De Belt-Gritter, B. & Busscher, H. J. 1995. Implications of microbial adhesion to hydrocarbons for evaluating cell surface hydrophobicity 2. Adhesion mechanisms. *Colloids and Surfaces B: Biointerfaces*, **5**, 117–26.

Van der Mei, H. C., Bos, R. & Busscher, H. J. 1998. A reference guide to microbial cell surface hydrophobicity based on contact angles. *Colloids and Surfaces B: Biointerfaces*, **11**, 213–21.

Van der Wal, A., Norde, W., Zehnder, A. J. B. & Lyklema, J. 1997. Determination of the total charge in the cell walls of gram-positive bacteria. *Colloids Surfaces B–Biointerfaces*, **9**, 81–100.

Van Loosdrecht, M. C. M., Lyklema, J., Norde, W., Schroa, G. & Zehnder, A. J. B. 1987. Electrophoretic mobility and hydrophobicity as a measure to predict the initial steps of bacterial adhesion. *Applied and Environmental Microbiology*, **53**, 1898–901.

Van Lossdrecht, M. C. M, Lyklema, J., Norde, W., Schroa, G. & Zehnder, A. J. B. 1987. The role of bacterial cell wall hydrophobicity in adhesion. *Applied and Environmental Microbiology*, **53**, 1893–7.

Vanhaecke, E., Remon, J-P. Mears, M., Roes, F., Rudder, D. D. & Van-Peteghem, A. 1990. Kinetics of *Pseudomonas aeruginosa* adhesion to 304 and 316-l stainless steel role of cell surface hydrophobicity. *Applied and Environmental Microbiology*, **56**, 788–95.

Veenstra, G. C., Cremers, F. F. M., Van Dijk, H. & Fleer, H. 1996. Ultrastructural organization and regulation of biomaterial adhesion of *Staphylococcus epidermidis*. *Journal of Bacteriology*, **178**, 537–41.

Verran, J. & Whitehead, K. A. 2006a. The effect of surface topography on the retention of microorganisms. *Food and Bioproducts Processing*, **84**, 253–9.

Verran, J. & Whitehead, K. A. 2006b. Assessment of organic materials and microbial components on hygiene surfaces. *Food and Bioproducts Processing*, **84**, 260–4.

Verran, J., Rowe, D. L. & Boyd, R. D. 2001. The effect of nanometer dimension topographical features on the hygienic status of stainless steel. *Journal of Food Protection*, **64**, 1183–7.

Vilain, S., Pretorius, J. M., Theron, J. & Brozel, V. S. 2009. DNA as an adhesion: *Bacillus cereus* requires extracellular DNA to form biofilms. *Applied and Environmental Microbiology*, **75**, 2861–8.

Vinogradov, E., Sadovskaya, I., Li, J. & Jabbouri, S. 2006. Structural elucidation of extracellular and cell wall teichoic acids of *Staphylococcus aureus* MN8m, a biofilm forming strain. *Carbohydrate Research*, **342**, 738–43.

Walker, S. L., Hill, J. E., Redman, J. A. & Elimelech, M. 2005. Influence of the growth phase on adhesion kinetics of *Escherichia coli* D12g. *Applied and Environmental Microbiology*, **71**, 3093–9.

Wei, J., Ravn, D. B., Gram, L. & Kingshott, P. 2003. Stainless steel modified with poly(glycol) can prevent protein adsorption but not bacterial adhesion. *Colloids and Surfaces B: Biointerfaces*, **32**, 275–91.
Weidenmaier, C. & Peschel, A. 2008. Teichoic acids and related cell wall glycopolymers in Gram positive physiology and host interactions *Nature Reviews Microbiology*, **6**, 276–86.
Whitchurch, C. B., Tolker-Nielsen, T., Ragas, P. C. & Mattick, J. S. 2002. Extracellular DNA required for bacterial biofilm formation. *Science*, **295**, 1487.
Wiencek, M. K., Klapes, A. N. & Foegeding, P. M. 1990. Hydrophobicity of *Bacillus* and *Clostridium* spores. *Applied and Environmental Microbiology*, **56**, 2600–5.
Xia, G., Kohler, T. & Peschel, A. 2010. The wall teichoic acid and lipoteichoic acid polymers of *Staphylococcus aureus. International Journal of Medical Microbiology*, **300**, 148–54.
Zhao, Q., Liu, Y., Wang, C., Wang, S., Peng, N. & Jeynes, C. 2008. Reduction of bacterial adhesion on ion-implanted stainless steel surfaces. *Medical Engineering and Physics*, **30**, 341–9.
Zoltai, P. T., Zoltola, E. A. & Mckay, L. 1981. Scanning electron microscopy of microbial attachment to milk contact surface. *Journal of Food Protection*, **44**, 204–8.

3 The Effect of Milk Composition on the Development of Biofilms

Steve Flint[1], Norshhaidah Mohd Jamaludin[2], Ben Somerton[1], Jon Palmer[1] and John Brooks[3]

[1]Institute of Food, Nutrition and Human Health, Massey University, Palmerston North, New Zealand
[2]Institute of Fundamental Sciences, Massey University, Palmerston North, New Zealand
[3]School of Applied Sciences, Faculty of Health and Environmental Sciences, Auckland University of Technology, Auckland, New Zealand

3.1 Introduction

Biofilms are a biologically complex structure of microorganisms associated with solid surfaces (Kumar & Anand, 1998). Generally, biofilm formation begins when microorganisms transition from an aqueous phase and attach to a solid surface (Costerton *et al.*, 1987). Often, such a surface is conditioned with nutrients that enable the microorganism to grow and develop (Costerton *et al.*, 1987), which may promote interspecies and intraspecies interactions (Teh *et al.*, 2012).

Biofilm formation was first discussed by Zobell (1943), who investigated the attachment of soil bacteria using a buried-slide method. Costerton *et al.* (1987) described a biofilm as a functional group of microorganisms embedded on to a surface alongside resultant extracellular polymeric substances (EPS). Once deposited, such microorganisms grow and multiply rapidly to form a colony of cells large enough to trap nutrients and debris (Kumar & Anand, 1998).

Conditioning is the accumulation of molecules at the solid–liquid interface on a food contact surface, leading to a higher concentration of nutrients than at the overlying aqueous phase (Kumar & Anand, 1998). According to Kumar and Anand (1998), conditioning occurs when nutrients such as proteins and lipids (e.g. from milk) become adsorbed on to a surface via a diffusion process. It has been shown that the rate of transport and the duration of adsorption on to the surface play an integral part in this process (Characklis, 1981). Kumar and Anand (1998) further indicated that the turbulent flow of the liquid also plays a key part in enhancing the transition of the bacteria.

The process of microbial attachment is generally considered to involve two phases: a reversible phase and an irreversible phase. The initial, reversible phase is associated with

Biofilms in the Dairy Industry, First Edition.
Edited by Koon Hoong Teh, Steve Flint, John Brooks and Geoff Knight.

weak interaction via van der Waals bonds and electrostatic and hydrophobic interactions (Ortega *et al.*, 2008). The later, irreversible phase consists of stronger attachment, leading to the formation of biofilm via anchoring by appendages and/or excretion of EPS. The process of conditioning and biofilm development is affected by the type of surface, the type of microorganism and the available nutrients (Speers & Gilmour, 1985; Herald & Zottola, 1988; Austin & Bergeron, 1995). The presence of milk in dairy plants provides sustenance for microorganisms, encouraging their growth and propagation in a biofilm (Speers & Gilmour, 1985). Since milk is frequently in contact with stainless steel surfaces during dairy production, it has the potential to influence biofilm production (Barnes *et al.*, 1999). Other factors, such as the type of surface and the type of microbial strain, will also influence the dynamics of the adherence of microorganisms. Understanding how changes in the composition of milk influence the biofilm formation process may help identify potential measures by which to control biofilm production in dairy plants.

3.2 Milk composition

Milks consist of water, specific proteins, easily digested fats, lactose, minerals and vitamins. All milks have a nearly neutral pH (Jensen, 1995). Proteins include casein, serum albumin and whey proteins (lactalbumin and lactoglobulin) (Swaisgood, 1995). Casein micelles are the largest structures in the fluid portion of the milk, making up 80% of the total protein content (Jensen, 1995). The typical concentration of protein is 3.3% w/v. Fats consist mostly of triacylglycerol (TAG) molecules, which contain over 400 types of fatty acid (FA) (Rudd, 2013). The fat component is secreted in the form of a fat globule and is surrounded by a membrane called a milk lipid globule membrane (MLGM) (Jensen, 1995). The typical concentration of fat in milk is 3.4% w/v. The carbohydrate portion consists mostly of lactose, followed by small amounts of glucose and galactose (Newburg *et al.*, 1995). Minerals include sodium, potassium, chloride, calcium, magnesium, citrate, phosphate, and sulphate (Atkinson *et al.*, 1995).

The composition of a given cow's milk is influenced by several factors, including its lactation phase, breed and diet (Chandan & Kilara, 2010).

A study had demonstrated the influence of milk composition on the adherence of 15 different bacterial cultures isolated from soiled milking equipment to glass, rubber and stainless steel surfaces. Lactose and non-casein protein caused a slight increase in the number of bacteria that attached to all three surfaces (Speers & Gilmour, 1985), due to the synthesis of a polymer essential to cell adherence (similar to the attachment of oral streptococci to tooth surfaces in plaque development; Mukasa & Slade, 1973). In a separate study, it was shown that positively charged ions such as sodium, calcium, magnesium and cationic surfactants can encourage bacterial attachment to a filter membrane by binding to and neutralising negatively charged surfaces (Bellona & Drewes, 2005). A third study found that ferrous ions promote the attachment of *Listeria monocytogenes* and *Serratia marcesens* to clean stainless steel by shielding the negative surface charge on bacteria and steel (Barnes *et al.*, 1999). Recent studies have shown enhanced biofilm formation in the presence of divalent cations, and have demonstrated that changes in the monovalent to divalent cation ratio have the potential to influence the biofilm formation of thermophilic bacilli (Somerton *et al.*, 2012, 2013).

3.3 Influence of organic molecules (protein and lipid) on the development of biofilms in the dairy industry

Kirtley and McGuire (1989) showed that the development of a protein conditioning film may influence the formation of a biofilm in a dairy plant. Speers and Gilmour (1985) investigated the effect of milk and various milk components on the attachment of different milk microflora to a variety of surface types. The highest number of attached bacteria occurred in the presence of non-casein protein (lactoglobulin and α-lactalbumin). However, casein showed little effect on bacterial attachment to the surface. In contrast, Meadows (1971) observed that bacteria immersed in casein or gelatin suspensions attached in large numbers. This could be due to the different strains and surfaces used in the two studies. According to Speers and Gilmour (1985), the viscosity of the non-casein protein may have caused an accumulation of bacteria and aided in the attachment to the surface, stabilising the formation of biofilms. However, since non-casein and casein protein possess comparable viscosity, this reason for this is vague and unclear.

Some studies have attempted to elucidate the influence of lipids on the formation of biofilms (Maxcy, 1973; Pasvolsky *et al.*, 2014). Maxcy (1973) demonstrated that all samples with phospholipid presoiling produced a yellow film on stainless steel surfaces. This yellow layer, associated with the phospholipid fraction, also resulted in high bacterial counts on the yellow surface. Maxcy (1973) concluded that the accumulation of high bacterial counts on the equipment was attributable to inadequate sanitation of the equipment and a build-up of the fatty materials. The dipolar nature of the phospholipid promoted the adsorption of bacteria to stainless steel and caused subsequent interaction with other soil components. Pasvolsky *et al.* (2014) examined the influence of fatty acids – particularly butyric acid – in milk on the production of 'floc' biofilm (bundle) among *Bacillus* species. *Bacillus* species can survive pasteurisation due to their ability to form heat-resistant spores, and may survive in dairy farm equipment, forming robust biofilms. Pasvolsky *et al.* (2014) hypothesised that bacteria grown in milk were attracted to fat molecules. To support this hypothesis, they experimented with *Bacillus subtilis* 3610 grown in lysogeny broth with and without butyric acid at different concentrations (0.01–0.02%). Their study showed that butyric acid triggers the formation of the floc bundles and that the formation of floc increases with the concentration of butyric acid. They proposed that butyric acid served as a 'stress signal' for the bacteria, enhancing biofilm formation and thus protecting the bacteria from the toxic effect of butyric acid. Further investigations showed that butyric acid triggered the upregulation of the *tapA* gene, which was responsible for the stimulation of biofilm formation (Winkelman *et al.*, 2013; Pasvolsky *et al.*, 2014). Teh *et al.* (2013) showed that lipolytic enzymes released by heat-resistant bacteria originating from raw milk caused lipolysis of milk fat, which produced butyric acid. Perhaps butyric acid present in dairy plants stimulates the biofilm formation of bacteria.

3.4 Protein and lipid molecules reduce attachment of bacteria to surfaces

Not all milk components aid the attachment of bacteria to surfaces. A study conducted by Barnes *et al.* (1999) on the effect of milk proteins on the attachment of bacteria to clean stainless steel showed that skim milk reduced the attachment of *Staphylococcus aureus*,

L. monocytogenes and *Serratia marcesens*. The individual milk components α-casein, β-casein and κ-casein also decreased the attachment of *S. aureus* and *L. monocytogenes* to the stainless steel, as compared to untreated stainless steel. In addition, Barnes *et al.* (1999) investigated the potential role of surface roughness in the attachment of bacteria, comparing stainless steel 2B with No. 8 mirror-finished stainless steel. However, no significant difference was observed in bacterial attachment between the two surfaces, although the reduction in bacterial attachment caused by the addition of milk components was still observed.

Barnes *et al.* (2001) carried out a further study to investigate the effect of milk proteins on bacterial attachment to stainless steel. This study showed that *S. aureus* and *L. monocytogenes* attached at the lowest extent in the presence of κ-casein, and the authors suggested that the protein might prevent bacterial attachment due to its hydrophilic regions. Bacterial attachment was greatest in the presence of α-lactalbumin, and this was believed to be due to α-lactalbumin covering less of the surface, allowing a greater surface area of stainless steel to be exposed for bacterial attachment. Barnes also showed that, as the milk dilution increased, the coverage of the surface by milk proteins also decreased, as measured by x-ray photoelectron spectroscopy, resulting in an increased extent of attachment of bacteria to the surface. Barnes *et al.* (2001) also speculated that, as the concentration of protein on the surface lowers, the adsorbed protein molecules may orientate themselves in such a way as to reduce the steric hindrance as bacteria approach the attachment surface. Speers and Gilmour (1985) noted that the presence of whole milk, fat and casein did not cause any significant increase in bacterial attachment in the presence of milk components, except for casein protein and lactose. Fletcher (1976), meanwhile, showed that different types of protein influence bacterial attachment differently, highlighting that proteins such as bovine serum albumin (BSA), gelatin, fibrinogen and pepsin inhibited the attachment of bacteria to an exposed surface, while casein protein, after adsorbtion to the surface, had little effect on bacterial attachment. However, Fletcher (1976) also stated that the extent of bacterial attachment depended on the type of surface and the species of bacteria involved. Both Fletcher (1976) and Speers and Gilmour (1985) agree that the competition between various components present in the milk might be the reason behind the reduction of bacteria attachment. Speers and Gilmour (1985) also suggested that milk fat globules and naturally occurring antibodies in milk might cause the inhibition of bacterial attachment. Brooks and Seaman (1973) proposed that certain protein molecules modify the ionic condition of the surface, forming a macromolecular 'scaffolding' to which the bacteria are unable to attach (Maroudas, 1975).

In a study investigating the impact of dairy lipids on bacterial adherence to stainless steel surfaces, Dat *et al.* (2014) reported that skim milk-, buttermilk- and butter serum-conditioned surfaces reduced bacterial attachment of *Lactococcus lactis, Leuconstoc cremoris* and *Lactobacillus casei*. Bacterial attachment was lower in buttermilk and butter serum relative to skim milk. Skim milk contains less fat (0.2%) than butter serum (3.1%) or buttermilk (1.8%). Dat *et al.* (2014) proposed that the different compositions created different surface roughness after conditioning, changing the bacterial attachment behaviour. Conditioning the surfaces of some areas of dairy plants with particular protein molecules and lipids might reduce bacterial attachment and thus, perhaps, reduce biofilm production on milk processing equipment. Barnes *et al.* (1999) highlighted that a pretreatment with macromolecules possessing particular properties could provide a temporary solution in tackling the biofouling problem in the food industry. Similarly, Dat *et al.* (2014) stated that surface conditioning with milk byproducts such as buttermilk, butter serum and skim milk might provide an initial control for bacterial attachment. This is supported by a study conducted by Busscher *et al.* (1998), who showed that the daily consumption of buttermilk reduced biofilm formation on a silicone rubber voice prosthesis over a period of at least 8 days.

3.5 Effect of ions on the development of biofilms of thermophilic bacilli

Observations made in New Zealand milk powder manufacturing plants have indicated that, during the processing of milk formulations high in sodium and low in calcium and magnesium ions, biofilm formation and contamination by thermophilic bacilli, predominantly consisting of *Geobacillus* spp. and *Anoxybacillus flavithermus*, is markedly abated (Somerton *et al.*, 2012). As it is perceived that biofilms in the manufacturing lines of milk powder manufacturing plants act as the main reservoir of thermophilic bacilli, the influence on biofilm growth of these bacteria appears a likely explanation. A range of free sodium (Na^+), potassium (K^+), calcium (Ca^{2+}) and magnesium (Mg^{2+}) ion concentrations and ratios was tested on *Geobacillus* spp. and *A. flavithermus* throughout biofilm formation, which involved the transition of planktonic cells to an irreversibly attached form and the subsequent establishment of a biofilm. Somerton *et al.* (2012) aimed to increase our understanding of the observed decrease in thermophile counts in final milk powder products with high monovalent to divalent cation ratios, and to obtain insights of practical significance.

Three mechanisms for the effect of Na^+, K^+, Ca^{2+} and Mg^{2+} on *Geobacillus* spp. and *A. flavithermus* biofilm formation were proposed:

1. Their effect on cation homeostasis and their requirement as a nutrient source.
2. Their direct electrostatic effect on cohesive forces among bacterial cells, the stainless steel attachment substrate and extracellular matrix polymers.
3. Their effect on the physiology and metabolism of bacteria, which may indirectly influence the attachment and cohesive forces of a biofilm.

The effects of Na^+, K^+, Ca^{2+} and Mg^{2+} on planktonic *Geobacillus* spp. and *A. flavithermus* were investigated in order to gain insights into the effect of cations on the bacteria prior to their transition to a surface-attached form. It was hypothesised that if cations influenced *Geobacillus* spp. and *A. flavithermus* in the planktonic form, this might subsequently influence their ability to transition from a planktonic to a stainless steel-attached form. It was found that the response of *Geobacillus* spp. and *A. flavithermus* to Ca^{2+} and Mg^{2+} was predominantly responsible for an increase in the optical density of the planktonic cultures (Somerton *et al.*, 2012). It was concluded that the optical density of the cultures depended on surface protein production, rather than differences in total viable cell counts, spore counts, cell size, cell aggregation or the production of surface polysaccharide. This is a novel finding, as usually the optical density of planktonic bacterial cultures is proportional to cell and spore numbers (Rippey & Watkins, 1992; Griffiths *et al.*, 2011). Also, it was proposed that Ca^{2+} and Mg^{2+} stimulated the production of surface protein by *Geobacillus* spp. and *A. flavithermus*, which increases the metabolic diversity of the bacteria, increases their interaction with the environment and may enhance their ability to attach to a substrate (Somerton *et al.*, 2013). These findings indicated that the cations had a physiological effect on planktonic *Geobacillus* spp. and *A. flavithermus*, and, conversely, the electrostatic effects of the cations had no apparent influence on culture optical density. These findings are of fundamental significance to our knowledge of the effect of cations on *Geobacillus* spp. and *A. flavithermus* metabolism, physiology and biofilm formation.

In addition, it was found that when a cation was supplemented alone, high Na^+, K^+ or Ca^{2+} concentrations of between 63 and 250 mM significantly decreased the optical density of *Geobacillus* spp. cultures. It was proposed that the high individual cation concentrations imbalanced cation homeostasis of the *Geobacillus* spp., which inhibited their metabolism and growth. This is an example of an effect of cations on the homeostasis of the *Geobacillus* spp. Furthermore, Mg^{2+} protected the *Geobacillus* spp. strains from inhibitory concentrations of Na^+, K^+ or Ca^{2+} (63–250 mM). These results have a practical significance as they indicate that growth of *Geobacillus* spp. in a milk formulation with a high monovalent to divalent cation ratio (i.e. high Na^+ and low Mg^{2+} concentrations) may be inhibitory. In addition, they have a fundamental significance, as they indicate that cations at high monovalent to divalent cation ratios inhibit the growth and metabolism of bacteria by imbalancing cation homeostasis.

Overall, the results obtained from investigations of the effect of cations on planktonic *Geobacillus* spp. and *A. flavithermus* indicate that the divalent cations Ca^{2+} and Mg^{2+} promote growth and physiologically prime bacteria for biofilm formation, and that high concentrations of the monovalent cations Na^+ and K^+ inhibit the growth of *Geobacillus* spp. These findings have a practical significance as they indicate that *Geobacillus* spp. growth and biofilm formation may be inhibited in a milk formulation with a high monovalent to divalent cation ratio.

The effect of different Na^+, K^+, Ca^{2+} and Mg^{2+} concentrations and monovalent to divalent cation ratios on both the transition of planktonic *Geobacillus* spp. and *A. flavithermus* to an irreversibly attached form on stainless steel and the subsequent establishment of a biofilm was investigated by Somerton *et al.* (2013). The same authors also investigated the effect of preconditioning planktonic *Geobacillus* spp. and *A. flavithermus* with different cation concentrations and monovalent to divalent cation ratios prior to attachment and biofilm formation. They found that the attachment and biofilm formation of *Geobacillus* spp. and *A. flavithermus* were not altered when the ionic strength of the growth medium ranged between 2 and 125 mM, or when monovalent to divalent cation ratios of 2 : 1 and 10 : 1 were compared. This indicated that the electrostatic effects of the cations did not influence the transition of planktonic *Geobacillus* spp. and *A. flavithermus* to a stainless steel-attached form or the proliferation of the bacteria in an established biofilm. Preconditioning *Geobacillus* spp. and *A. flavithermus* with cations often increased subsequent attachment of the bacteria relative to unconditioned bacteria. This indicated that the bacteria physiologically responded to the cations during preconditioning, subsequently increasing their ability to attach to stainless steel. For example, the *Geobacillus* spp. and *A. flavithermus* may have responded to the cations by upregulating the expression of surface-exposed polymers that assist attachment. These findings indicate that the transition of *Geobacillus* spp. and *A. flavithermus* from milk formulations to stainless steel product contact surfaces in milk powder manufacturing plants is predominantly mediated by bacterial physiological factors (e.g. surface-exposed adhesins), rather than the direct electrostatic effect of cations surrounding bacteria.

Interestingly, biofilm formation after 6 hours by *Geobacillus* sp. F75 tended to decrease as the monovalent to divalent cation ratio of milk formulations increased. This demonstrated the potential for *Geobacillus* spp. biofilm formation to be inhibited in milk formulations with high monovalent to divalent cation ratios during milk powder manufacture.

MALDI-TOF mass spectroscopy was used to investigate the influence of Na^+, K^+, Ca^{2+} and Mg^{2+} on protein expression by *Geobacillus* sp. F75 grown in a biofilm. Protein expression was investigated in order to gain insights into the influence of the cations on the physiology of *Geobacillus* spp., so as to test the hypothesis that in the presence of different cations leads to different physiologies (Somerton *et al.*, 2013).

In cultures supplemented with 2 mM Mg^{2+}, 16 *Geobacillus* sp. F75 proteins were not expressed (they were speculated to be downregulated) and one protein was expressed (i.e. upregulated) compared to cultures that were not supplemented with cations or cultures supplemented with all cations (Na^+, K^+, Ca^{2+} and Mg^{2+}). This finding has a fundamental significance, as it indicates that Mg^{2+} influences the physiology of *Geobacillus* spp. during biofilm formation.

Five of the downregulated proteins were identified as having functions involved in sporulation, so it was proposed that Mg^{2+} prevents sporulation and thereby promotes the cell division and metabolism of *Geobacillus* spp. in a biofilm. This finding provides evidence to suggest that in milk formulations with high monovalent to divalent cation ratios (which have low Mg^{2+} concentrations), *Geobacillus* spp. will have a tendency to opt for sporulation as opposed to cell division and growth. Thus, this finding has a practical significance as it indicates that the proliferation of *Geobacillus* spp. biofilms in the processing of milk formulations with high monovalent to divalent cation ratios may be abated, lowering the thermophillic bacilli cell counts in the final milk powder products thereby derived.

In order to further investigate the observation that biofilm formation 6 hours after attachment by *Geobacillus* sp. F75 was inhibited in a milk formulation with a high monovalent to divalent cation ratio, up to 18 hours' biofilm formation by three *Geobacillus* spp. isolates and three *A. flavithermus* isolates was investigated in milk formulations with varied Na^+, Ca^{2+} and Mg^{2+} concentrations and monovalent to divalent cation ratios (Somerton *et al.*, 2013). This study was conducted for three reasons: first, to investigate the prevalence of the inhibition of biofilm formation by isolates from the *Geobacillus* spp. and *A. flavithermus* groups in milk formulations with high monovalent to divalent cation ratios; second, to investigate the bacterial isolates' potential to inhibit biofilm formation across the entire 18-hour duration of operation of a typical milk powder manufacturing plant; and third, to characterise the role of high Na^+, low Ca^{2+} and low Mg^{2+} concentrations in the inhibition of biofilm formation of the bacterial isolates in milk formulations with high monovalent to divalent cation ratios.

Biofilm formation by all three *Geobacillus* spp. isolates was inhibited for up to 18 hours in a milk formulation with a high monovalent to divalent cation ratio, whereas biofilm formation by all three *A. flavithermus* isolates was similar for between 6 and 18 hours in a milk formulation with a high monovalent to divalent cation ratio as compared to that in a milk formulation with an average monovalent to divalent cation ratio. These results demonstrated that biofilm formation by *Geobacillus* spp. in the manufacturing lines of milk powder manufacturing plants is markedly compromised throughout the processing duration of milk formulations with high monovalent to divalent cation ratios. This has a practical significance, as, given that a substantial proportion of thermophillic bacilli that may contaminate milk powders belong to the *Geobacillus* spp. group and *Geobacillus* spp. spores have a greater tolerance to high temperatures than *A. flavithermus* spores (Scott *et al.*, 2007), it is proposed that milk powders derived from milk formulations with high monovalent to divalent cation ratios have the potential to record markedly decreased thermophilic bacilli counts and, as a consequence, have a superior quality and may fetch higher selling prices.

It was concluded that high Na^+, low Ca^{2+} and low Mg^{2+} concentrations were collectively required to maximally inhibit *Geobacillus* spp. biofilm formation. When a milk formulation with a high monovalent to divalent cation ratio was supplemented with either Ca^{2+} or Mg^{2+}, the increased Ca^{2+} and Mg^{2+} concentrations protected the *Geobacillus* spp. isolates from the toxic effect of the high Na^+ concentration. This finding is similar to results observed in planktonic experiments, where it was observed that Mg^{2+} protected *Geobacillus* spp. from toxic concentrations of Na^+, K^+ or Ca^{2+} (Somerton *et al.*, 2012). These findings have a fundamental and practical significance, as they show that Mg^{2+} has a protective effect against toxic Na^+ concentrations and that a high monovalent to divalent cation ratio can inhibit *Geobacillus* spp. in both a planktonic and a biofilm form.

It was proposed to be unlikely that the electrostatic effects of the cations in a milk formulation with a high monovalent to divalent cation ratio were responsible for the inhibition of biofilm formation by the *Geobacillus* spp. isolates. The electrostatic effects of cations with surface-exposed polymers and the extracellular matrices of bacteria are similar for all types of bacteria (Neuhaus & Baddiley, 2003). Given that biofilm formation by the *Geobacillus* spp. isolates, but not the *A. flavithermus* isolates, was inhibited in the milk formulation with a high monovalent to divalent cation ratio, this finding has a fundamental significance, as it is proposed that the predominant mechanism influencing the inhibition of the *Geobacillus* spp. was not an electrostatic effect.

It is proposed that the dominant mechanism influencing the inhibition of *Geobacillus* spp. was either an imbalance of cation homeostasis or a physiological response of the bacteria to the high Na^+, low Ca^{2+} and low Mg^{2+} concentrations. These findings are of a fundamental significance, as they provide insights into how cations, at high monovalent to divalent cation ratios, inhibit *Geobacillus* spp. biofilm formation. High Na^+ concentrations may either have a toxic effect or cause a physiological response compromising the growth of the *Geobacillus* spp. in a biofilm. Low Ca^{2+} and Mg^{2+} concentrations may deprive the *Geobacillus* spp. of sufficient Ca^{2+} and Mg^{2+} for growth and metabolism or cause the bacteria to elicit a physiological response that decreases the growth of the *Geobacillus* spp. in a biofilm.

In contrast to *Geobacillus* spp., biofilm formation by *A. flavithermus* is not affected by the high monovalent to divalent cation ratios present in some milk formulations. *A. flavithermus* may have a greater tolerance than *Geobacillus* spp. to the Na^+ concentrations present in the milk formulations studied, or may have a greater capacity to acquire Ca^{2+} and Mg^{2+}. This suggests that *A. flavithermus* growth and biofilm formation are not inhibited in manufacturing lines during the manufacture of milk powders with high monovalent to divalent cation ratios.

Figure 3.1 shows the effect of sodium ions and calcium ions on the biofilm formation of a *Geobacillus* species. Adding sodium ions to a standard milk formulation (Figure 3.1b,c) shows some inhibition of biofilm formation. Adding calcium ions to a milk formulation lacking in calcium increases the formation of biofilm.

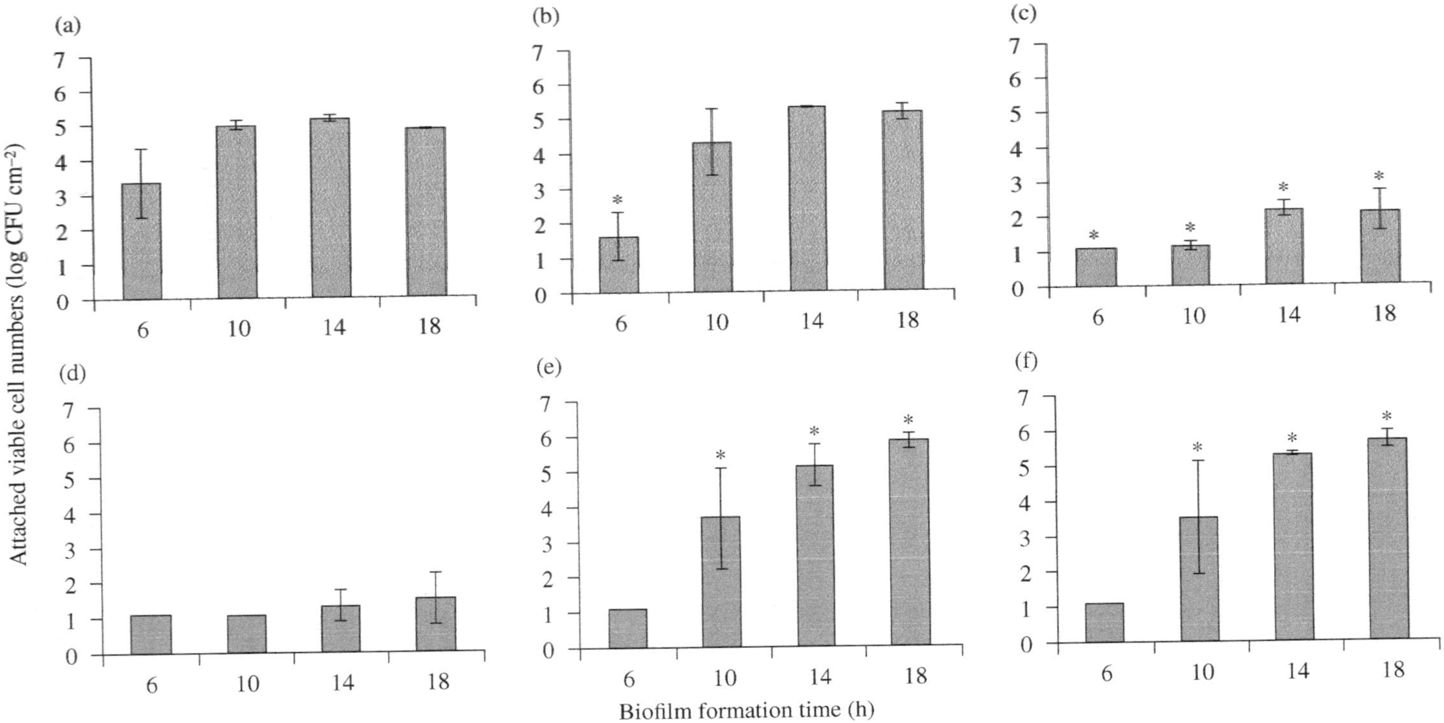

Figure 3.1 Biofilm formation, after 6–18 hours of incubation at 55 °C, by viable *Geobacillus* spp. 183 cells (log CFU/cm^2) on stainless steel coupons completely submerged in (a) a standard milk formulation, (b) the same milk formulation supplemented with 50 mM NaCl or (c) the same milk formulation supplemented with 100 mM NaCl, or in (d) a reduced-calcium milk formulation (2), (e) the same milk formulation supplemented with 2 mM $CaCl_2$ or (f) the same milk formulation supplemented with 2 mM $MgCl_2$. Experiments were repeated as triplicates and error bars represent ±1 standard deviation (σn^{-1}). Asterisk denotes a significant difference ($P \leq 0.05$) between cation-supplemented and unsupplemented milk formulations for the respective milk formulation and time point.

If the Na^+ concentration of a milk formulation is to be increased and the Ca^{2+} and Mg^{2+} concentrations of a milk formulation are to be reduced, the cation concentrations should be manipulated as early in the manufacturing process as possible. This will have the benefit of preventing biofilm formation of *Geobacillus* spp. during milk powder manufacture. This has the potential to lower the thermophilic bacilli count in a milk powder product, which can increase the quality and selling price of the product.

Further investigations of the prevalence of *Geobacillus* spp. strains that are inhibited when grown in milk formulations with high monovalent to divalent cation ratios will more accurately and conclusively determine the extent of growth inhibition of the *Geobacillus* spp. group in milk formulations with a high monovalent to divalent cation ratio. Furthermore, if the growth inhibition of *Geobacillus* spp. strains in milk formulations with high monovalent to divalent cation ratios proves to be widespread, a Na^+ toxicity test might be developed which could be used to differentiate between the *Geobacillus* spp. and *A. flavithermus* groups.

It would be useful to investigate the possibility that *Geobacillus* spp. biofilm formation is inhibited for longer than 18 hours (i.e. for up to 30 hours) when grown in milk formulations with high monovalent to divalent cation ratios. If it is found that many *Geobacillus* spp. strains are inhibited for up to 30 hours then the manufacturing run time could be extended. This would decrease manufacturing costs associated with cleaning regimes, such as the cost of cleaning chemicals and loss of production time.

Further investigations are needed to clarify whether it is low Ca^{2+}, low Mg^{2+} or both low Ca^{2+} and Mg^{2+} which inhibits *Geobacillus* spp. biofilm formation in milk formulations with a high monovalent to divalent cation ratio. If it is found that only one of the two cation concentrations needs to be low to inhibit *Geobacillus* spp. biofilm formation, then only one will have to be targeted when developing strategies by which to decrease such formation during milk powder manufacture. Identifying the minimum inhibitory concentration of Na^+ and the maximum inhibitory concentrations of Ca^{2+} and Mg^{2+} which inhibit *Geobacillus* spp. biofilm formation would also provide target concentrations when developing such strategies, while identifying the molecular mechanisms used by *Geobacillus* spp. and *A. flavithermus* to detect, monitor and respond to fluctuations in external cation concentrations would provide insights into the molecular mechanisms which liberate the effects of cations on cation homeostasis and into the physiology of *Geobacillus* spp. and *A. flavithermus*. Identification of the physiological factors which either make *Geobacillus* spp. susceptible to high monovalent to divalent cation ratios in milk formulations or enable *A. flavithermus* to tolerate high monovalent to divalent cation ratios in milk formulations would help our understanding of the observations made by Somerton *et al.* (2013).

Investigations into the potential for Na^+ to competitively exclude Ca^{2+} and Mg^{2+} from the cell envelope would provide further insights into the molecular mechanisms that liberate the effects of cations on cation homeostasis and into the physiology of *Geobacillus* spp. and *A. flavithermus*. Identification and characterisation of the proteins upregulated by *Geobacillus* spp. and *A. flavithermus* in response to Ca^{2+} and Mg^{2+} in planktonic culture would provide some insight into the involvement of the surface-exposed proteins in attachment and biofilm formation. Identification and characterisation of the physiological factors, such as adhesion, that assist attachment by *Geobacillus* spp. and *A. flavithermus* would help our understanding of the attachment mechanism used by thermophilic bacilli and aid in the development of strategies by which to negate their attachment to stainless steel in the manufacturing lines of milk powder manufacturing plants.

Investigating the effect of Mg^{2+} on the number of spores in *Geobacillus* sp. biofilms might support results obtained from protein expression experiments, which indicate that Mg^{2+} induces a downregulation of sporulation protein expression, and thus of sporulation.

Finally, investigations into the effect of a range of cation concentrations and monovalent to divalent cation ratios on *Geobacillus* spp. and *A. flavithermus* biofilm formation in a continuous-flow reactor would create circumstances more closely aligned to those found in a milk powder manufacturing plant. Under such circumstances, the direct electrostatic effects of cations on bacterial attachment and biofilm formation might be more influential.

3.6 Conclusion

Biofilm formation may be enhanced or reduced through simple changes in milk composition, including the content of both organic and inorganic compounds. Our understanding of these effects is by no means complete, but already we have enough information to suggest that biofilm formation in a dairy manufacturing environment might be controlled by manipulation of the composition of the milk or milk product.

References

Atkinson, S., Alston-Mills, B., Lonnerdal, B. & Neville, M. C. 1995. Major minerals and ionic constituents of human and bovine milks. In: Jensen, R. G. (ed.) Handbook of Milk Composition. Academic Press, Waltham, MA.

Austin, J. W. & Bergeron, G. 1995. Development of bacterial biofilms in dairy processing lines. *Journal of Dairy Research*, **62**, 509–19.

Barnes, L.-M., Lo, M. F., Adams, M. R. & Chamberlain, A. H. L. 1999. Effect of milk proteins on adhesion of bacteria to stainless steel surfaces. *Applied and Environmental Microbiology*, **65**, 4543–8.

Barnes, L.-M., Adams, M. R., Watts, J. F., Zhdan, P. A. & Chamberlain, A. H. L. 2001. Correlated XPS, AFM, and bacterial adhesion studies on milk and milk proteins adherent to stainless steel. *Biofouling*, **17**, 1–22.

Bellona, C. & Drewes, J. E. 2005. The role of membrane surface charge and solute physico-chemical properties in the rejection of organic acids by NF membranes. *Journal of Membrane Science*, **249**, 227–34.

Brooks, D. E. & Seaman, G. V. F. 1973. The effect of neutral polymers on the electrokinetic potential of cells and other charged particles:1.Models for zeta potential increase. *Journal of Colloid and Interface Science*, **43**, 670–86.

Busscher, H. J., Bruinsma, G., Van Weissenbruch, R., Leunisse, C., Van Der Mei, H. C., Dijk, F. & Albers, F. W. J. 1998. The effect of buttermilk consumption on biofilm formation on silicone rubber voice prostheses in an artificial throat. *European Archives of Oto-Rhino-Laryngology*, **225**, 410–13.

Chandan, R. C. & Kilara, A. 2010. Dairy Ingredients for Food Processing. John Wiley & Sons, Chichester.

Characklis, W. G. 1981. Fouling biofilm development: a process analysis. *Biotechnology and Bioengineering*, **23**, 1923–60.

Costerton, J. W., Cheng, K., Geesey, G. G., Ladd, T. I., Nickel, J. C., Dasgupta, M. & Marrie, T. J. 1987. Bacterial biofilms in nature and disease. *Annual Reviews in Microbiology*, **41**, 435–64.

Dat, N. M., Manh, L. D., Hamanaka, D., Hung, D. V., Tanaka, F. & Uchino, T. 2014. Surface conditioning of stainless steel coupons with skim milk, buttermilk, and butter serum solutions and its effect on bacterial adherence. *Food Control*, **42**, 94–100.

Fletcher, M. 1976. The effects of proteins on bacterial attachment to polystyrene. *Journal of General Microbiology*, **94**, 400–4.

Griffiths, M. J., Garcin, C., Van Hille, R. P. & Harrison, S. T. L. 2011. Interference by pigment in the estimation of microalgal biomass concentration by optical density. *Journal of Microbiological Methods*, **85**,119–23.

Herald, P. J. & Zottola, E. A. 1988. Attachment of *Listeria monocytogenes* to stainless steel surfaces at various temperatures and pH values. *Journal of Food Science*, **53**, 1549–62.

Jensen, R. G. 1995. Handbook of Milk Composition. Academic Press, Waltham, MA.

Kirtley, S. A. & McGuire. 1989. On differences in surface constitution of dairy product contact materials. *Journal of Dairy Science*, **72**, 1748–53.

Kumar, C. G. & Anand, S. 1998. Significance of microbial biofilms in food industry: a review. *International Journal of Food Microbiology*, **42**, 9–27.

Maroudas, N. 1975. Adhesion and spreading of cells on charged surfaces. *Journal of Theoretical Biology*, **49**, 417–24.

Maxcy, R. 1973. Nature and cause of yellow film occurring on dairy equipment. *Journal of Dairy Science*, **56**, 164–7.

Meadows, P. S. 1971. The attachment of bacteria to solid surfaces. *Archiv für Mikrobiologie*, **148**, 57–62.

Mukasa, H. & Slade, H. D. 1973. Mechanism of adherence of *Streptococcus mutans* to smooth surfaces. I. Roles of insoluble dextran-levan synthetase enzymes and cell wall polysaccharide antigen in plaque formation. *Infection and Immunity*, **8**, 555–62.

Newburg, D. S., Neubauser, Suzanne H. 1995. Carbohydrates in milks: analysis, quantities, and significance. In: Jensen, R. G. (ed.) Handbook of Milk Composition. Academic Press, Waltham, MA.

Neuhaus, F. C. & Baddiley. 2003. A continuum of anionic charge: Structures and funcations of D-alanyl-teichoic acids in Gram-positive bacteria. *Microbiology and Molecular Biology Reviews*, **67**, 686–723.

Ortega, M. P., Hagiwara, T., Watanabe, H. & Sakiyama, T. 2008. Factors affecting adhesion of *Staphylococcus epidermidis* to stainless steel surface. *Japan Journal of Food Engineering*, **9**, 251–9.

Pasvolsky, R., Zakin, V., Ostrova, I. & Shemesh, M. 2014. Butyric acid released during milk lipolysis triggers biofilm formation of *Bacillus* species. *International Journal of Food Microbiology*, **181**, 19–27.

Rippey, S. R. & Watkins, W. D. 1992. Comparative rates of disinfection of microbial indicator organisms in chlorinated sewage effluents. *Water Science and Technology*, **26**, 2185–9.

Rudd, L. B. 2013. Milk Fat: Composition, Nutritional Value and Health Implications. Nova Science Publisher, Hauppauge, NY.

Scott, S. A., Brooks, J. D., Rakonjac, J., Walker, K. M. R. & Flint, S. H. 2007. The formation of thermophilic spores during the manufacture of whole milk powder. *International Journal of Dairy Technology*, **60**, 109–17.

Somerton, B., Palmer, J., Brooks, J., Smolinski, E., Lindsay, D. & Flint, S. 2012. Influence of cations on growth of thermophilic *Geobacillus* spp. and *Anoxybacillus flavithermus* in planktonic culture. *Applied and Environmental Microbiology*,**78**, 2477–81.

Somerton, B., Flint, S., Palmer, J., Brooks, J. & Lindsay, D. 2013. Preconditioning with cations increases the attachment of *Anoxybacillus flavithermus* and *Geobacillus* species to stainless steel. *Applied and Environmental Microbiology*, **79**, 4186–90.

Speers, J. & Gilmour, A. 1985. The influence of milk and milk components on the attachment of bacteria to farm dairy equipment surfaces. *Journal of Applied Microbiology*, **59**, 325–32.

Swaisgood, H. E. 1995. Protein and amino acid compositions of bovine milk. In: Jensen, R. G. (ed.) Handbook of Milk Composition. Academic Press, Waltham, MA.

Teh, K. H., Flint, S., Palmer, J., Andrewes, P., Bremer, P. & Lindsay, D. 2012. Proteolysis produced within biofilms of bacterial isolates from raw milk tankers. *International Journal of Food Microbiology*, **157**, 28–34.

Teh, K. H., Lindsay, D., Palmer, J., Andrewes, P., Bremer, P. & Flint, S. 2013. Lipolysis within single culture and co-culture biofilms of dairy origin. *International Journal of Food Microbiology*, **163**, 129–35.

Winkelman, J. T., Bree, A. C., Bate, A. R., Eichenberger, P., Gourse, R. L. & Kearns, D. B. 2013. RemA is a DNA-binding protein that activates biofilm matrix gene expression in *Bacillus subtilis*. *Molecular Microbiology*, **88**, 984–97.

Zobell, C. E. 1943. The effect of solid surfaces upon bacterial activity. *Journal of Bacteriology*, **46**, 39–56.

4 Overview of the Problems Resulting from Biofilm Contamination in the Dairy Industry

Brent Seale[1], Phil Bremer[2], Steve Flint[3], John Brooks[1] and Jon Palmer[3]

[1]School of Applied Sciences, Faculty of Health and Environmental Sciences, Auckland University of Technology, Auckland, New Zealand
[2]Department of Food Science, University of Otago, Dunedin, New Zealand
[3]Institute of Food, Nutrition and Human Health, Massey University, Palmerston North, New Zealand

4.1 Introduction

Milk is an excellent medium for the growth of microorganisms. While bacteria have traditionally been studied in terms of their planktonic growth, recently, studies have focused on bacteria adhered to and growing on surfaces and the problems they can cause in milk products. This chapter will discuss how biofilms can affect food safety, spoilage and processing efficiency within dairy manufacturing plants.

4.2 Microbiological flora associated with dairy manufacturing

Within a dairy manufacturing plant, a wide range of temperatures and conditions are used in the production of diverse products, such as butter, cheese, liquid milk and milk powder. These conditions can provide unique ecological niches for the growth of a range of organisms, including psychrophilic (cold-loving), psychrotrophic (cold-tolerant), mesophilic, thermoduric (heat-tolerant) and thermophilic (heat-loving) organisms. This section discusses each of these types of organism in some detail.

4.2.1 *Psychrotrophs*

Bacterial spoilage of milk caused by psychrotrophic microorganisms results in significant economic losses for the dairy industry (Randolph, 2006), particularly since many dairy products are stored and transported at low temperatures in order to limit the growth of microorganisms. Facultative psychrotrophs can persist and potentially grow within a temperature range of

Biofilms in the Dairy Industry, First Edition.
Edited by Koon Hoong Teh, Steve Flint, John Brooks and Geoff Knight.

0–40 °C. They typically occur at higher numbers in milk than do obligate psychrophiles, which will grow only between 0 and 15 °C. Psychrotrophic bacteria are ubiquitous and can be found in a number of environments, including soil, water and vegetation (Cousins, 1982). They consist of both Gram-positive and Gram-negative bacteria. The majority of psychrotrophs belong to the genera *Pseudomonas*, *Aeromonas*, *Acinetobacter*, *Serratia*, *Alcaligenes*, *Achromobacter*, *Enterobacter*, *Flavobacteriumm*, *Klebsiella*, *Bacillus*, *Arthrobacter*, *Clostridium*, *Lactobacillus*, *Listeria*, *Staphylococcus*, *Corynebacterium*, *Microbacterium* and *Micrococcus*. During chilled storage, the predominant microflora in milk shifts from being Gram-positive to predominantly Gram-negative.

Pseudomonas are the predominant Gram-negative bacteria found in chilled milk, owing to their ability to grow within the bulk phase and on the surface of refrigerated milk containers. Many *Pseudomonas* spp. produce heat-stable extracellular lipases, proteases and lecithinases that can cause spoilage of milk. While the pseudomonads that created them are destroyed by pasteurisation, these heat-stable enzymes can remain active post-pasteurisation and cause spoilage, as discussed in further detail in Section 4.4.

The growth of *Paenibacillus* spp. in milk at refrigerated temperatures has recently been identified as a concern in the dairy industry (Ivy *et al.*, 2012). Isolates from *Bacillus* spp. represent the predominant bacteria found early in the shelf life of liquid pasteurised milk products, as they can survive pasteurisation. However, in the absence of post-pasteurisation contamination caused by Pseudomonads, *Paenibacillus* spp. can become the predominant microorganism and reduce the shelf life of the product (Ranieri & Boor, 2010). Isolates of *Paenibacillus* are typically found in raw milk in very low numbers and remain in low numbers in post-pasteurised milk, but are able to grow at the low temperatures in refrigerated liquid milk. High numbers of *Paenibacillus* spp. can cause off-flavours and curdling of milk (Huck *et al.*, 2007; Ranieri *et al.*, 2009; De Jonghe *et al.*, 2010). Yegorenkova *et al.* (2011) have shown that *Paenibacillus polymyxa* is able to form biofilms on abiotic surfaces, but it remains unknown whether biofilms produced by dairy isolates of *Paenibacillus* are responsible for the spoilage of refrigerated milk (Yegorenkova *et al.*, 2011).

4.2.2 *Mesophiles*

Mesophiles are a group of bacteria that grow between 10 and 45 °C. Like psychrotrophs, they are ubiquitous, being found in soil, water, vegetation and animals. A number of pathogens that are of concern for the dairy industry are mesophiles. These include *Campylobacter* spp., *Salmonella*, *Staphylococcus aureus*, *Bacillus cereus* and *Cronobacter sakazakii* (van Acker *et al.*, 2001; Oliver *et al.*, 2005). They can be controlled in a dairy environment through refrigeration, pasteurisation and implementation of Good Manufacturing Practices (GMP) in a manufacturing plant. These organisms will be discussed in further detail in Section 4.3.

4.2.3 *Thermodurics*

Thermoduric microorganisms consist of mesophiles and thermophiles that are able to survive temperatures considerably higher than those under which they are able to grow. Specifically, thermoduric organisms are able to survive, but not grow, at pasteurisation temperatures.

Thermoduric bacteria include species of *Micrococcus, Streptococcus, Lactobacillus* and *Bacillus. Streptococcus thermophilus* is able to survive the pasteurisation process and attach to surfaces downstream. The adhered cells are considerably more resistant to sanitisers and heat than planktonic cells and are able to form biofilms in post-pasteurisation processing lines. More information regarding *S. thermophilus* will be discussed in Chapter 6.

4.2.4 *Thermophiles*

Thermophiles are organisms that grow at between 45 and 70 °C. *Geobacillus* spp., *Anoxybacillus flavithermus* and *Bacillus licheniformis* are thermophiles that are commonly found in milk. These organisms are typically associated with soil and compost and are believed to be introduced into raw milk in low numbers (<10 CFU/ml) during milking (McGuiggan *et al.*, 2002). Like thermodurics, these organisms can survive pasteurisation, and, if conditions such as water activity and temperature are suitable, can attach, grow and form biofilms. While not pathogenic, their presence in milk can lead to the production of enzymes that break down the milk proteins, creating off-flavours in the final product. These organisms also have the ability to form endospores that enable them to survive even greater extremes of temperature, low water activity and chemical activity, which means they can be extremely difficult to eradicate from a dairy manufacturing plant. More details regarding thermophilic bacilli, their spores and their impact on the dairy industry is found in Chapter 7.

4.3 Effects of biofilms on food safety

Food safety is a serious concern for the food processing industry. The monitoring and control of pathogens in processing lines is crucial in order to avoid contamination of products and prevent potentially unsafe or low-quality products being sold to consumers. Over the last 30 years, the importance of biofilms in the contamination of foods has gained prominence (Shi & Zhu, 2009). In dairy manufacturing plants, biofilms can be divided into two categories:

1. **Process biofilms:** Biofilms that are unique to a dairy manufacturing plant and form on surfaces that are in direct contact with the flowing product.
2. **General biofilms:** Biofilms that form in the general food processing environment.

4.3.1 Bacillus cereus

Biofilms created by *Bacillus* spp. are a serious problem in food processing environments, owing to the potential of *Bacillus* spp. to cause illness or spoil food products (Faille *et al.*, 2014). Among the various species of bacilli, *B. cereus* is the major concern in a large range of food processing environments, due to the ability of some strains to cause gastrointestinal illness (Rusul, 1995). *Bacillus cereus* is a Gram-positive, spore-forming, facultative aerobic rod-shaped bacterium. It is ubiquitous in the environment and can be found in soil, plants and even within the intestinal tract of insects and mammals (Granum & Lund, 1997; Pirttijarvi *et al.*,

1999; Stenfors Arnesen *et al.*, 2008). Vegetative cells are able to produce enterotoxins, which can result in either diarrhoeal or emetic food poisoning (Granum, 1994). Diarrhoeal food poisoning resulting from toxicoinfection can occur if the vegetative cells produce complex enterotoxin while growing in the small intestine. Foodborne intoxication due to an emetic toxin can result if the vegetative cells have the opportunity to grow up to large numbers in food.

There have been a number of studies on the growth of *B. cereus* in pasteurised milk products (Becker *et al.*, 1994; Larsen & Jørgensen, 1997; Notermans *et al.*, 1997; Faille *et al.*, 2001). The ability of *B. cereus* to form spores undoubtedly plays an important role in its ability to become established within dairy manufacturing plant processing lines as a process biofilm (Rönner & Husmark, 1992; Shaheen *et al.*, 2010). Spores are able to survive pasteurisation and adhere to surfaces downstream in a processing line. Spores of *B. cereus* are particularly interesting because they possess an exosporium, which has a greater surface hydrophobicity than the spores of other bacilli (Ronner *et al.*, 1990; Husmark & Ronner, 1992). The increased surface hydrophobicity of these spores enables them to adhere in higher numbers to stainless steel than most other microorganisms (Husmark & Ronner, 1990; Faille *et al.*, 2002; Tauveron *et al.*, 2006). If conditions are favourable, these spores can germinate and the vegetative cells can grow, multiply and form a biofilm.

Vegetative cells of *B. cereus* are able to form biofilms under static or flow conditions (Wijman *et al.*, 2007) and can produce extracellular polymeric substances (Karunakaran & Biggs, 2011) and DNA (Vilain *et al.*, 2009), which play a role in biofilm formation and structure. Several studies have also shown that sporulation can occur within established biofilms of *B. cereus* (Storgårds *et al.*, 2006; Faille *et al.*, 2014); these spores can subsequently leave the biofilm and contaminate other parts of the processing line or the product being processed.

4.3.2 Listeria monocytogenes

Listeria monocytogenes has been recognised for many years as an important foodborne pathogen (Schlech *et al.*, 1983). It is associated with a number of foodborne disease outbreaks and has a relatively high mortality rate, particularly among immune-compromised or elderly people and pregnant mothers and their unborn children (Farber & Peterkin, 1991). *L. monocytogenes* is a facultative psychrotroph that can grow and replicate at refrigerated temperatures (Rosso *et al.*, 1996), creating a serious problem for many food industries, and in particular the dairy industry, where refrigeration is the main hurdle in many products to the growth of microorganisms. *L. monocytogenes* is believed to be associated with general biofilms growing in the processing environment. A previous study observed biofilms of *L. monocytogenes* in milking equipment, such as milk meters (Latorre *et al.*, 2010). The growth of *L. monocytogenes* as biofilms is dependent upon the growth of other isolates within a dairy or meat processing plant (Jeong & Frank, 1994a,b). It has been reported that species of *Pseudomonas*, *Bacillus* and *Streptococcus* can reduce the number of *L. monocytogenes* that adhere to a surface, indicating competition for adhesion sites. In contrast, other studies have reported that the presence of other bacteria promotes the establishment of pathogenic bacteria such as *L. monocytogenes* through the formation of multispecies biofilms and enhances its resistance to sanitisers (Sashara and Zottola, 1993; Bremer *et al.*, 2001, 2002). A study by Borucki *et al.* (2003) showed that persistent strains in bulk milk samples increased biofilm formation capability compared with nonpersistent strains. However, no relationship was observed between the ability of certain strains to form biofilms and their virulence.

4.3.3 Cronobacter sakazakii

Cronobacter sakazakii (previously known as *Enterobacter sakazakii*) is a concern for food manufacturers globally, as its presence in food products is a serious threat to infants and can cause a number of illness, such as necrotising enterocolitis (van Acker *et al.*, 2001), bacteraemia (Muytjens *et al.*, 1988) and a rare form of infant meningitis (Bar-Oz *et al.*, 2001). It has been isolated from powdered infant milk formula (Simmons *et al.*, 1989). Two separate studies have found that *C. sakazakii* is able to adhere to and form biofilms on a number of different surfaces associated with dairy manufacturing plants (Iversen *et al.*, 2004; Kim *et al.*, 2006). It also has the ability to survive spray drying, desiccation and osmotic stress (Osaili & Forsythe, 2009). Studies using genotyping methods to track isolates of *C. sakazakii* in dairy manufacturing plants have reported that the microorganism can be found external to the processing lines and not associated with raw milk (Craven *et al.*, 2010; Jacobs *et al.*, 2011; Sonbol *et al.*, 2013). A study by Jacobs *et al.* (2011) reported that *C. sakazakii* isolates present in textile filters at the top of the spray dryer were of the same genotype as isolates found in the final product. These studies indicate that the source of the contamination lies not within the processing lines but in external sources entering near the final stages of manufacture. However, more research is required in order to determine whether biofilms in these areas are the source of contamination.

4.4 Effects of biofilms on spoilage

Spoilage caused by the growth of microorganisms and their production of enzymes such as lipases and proteases is a serious concern for the dairy industry. Spoilage can result from the production of heat-stable enzymes by bacteria in raw milk holding tanks prior to pasteurisation, from microorganisms that survive pasteurisation or from post-pasteurisation contaminating microorganisms that actively grow and produce enzymes. This section will examine the effects of biofilms on the spoilage of dairy products.

The production of spoilage enzymes by dairy biofilms has until recently been an unrecognised source of degradation of dairy products (Teh *et al.*, 2014a). Teh *et al.* (2012, 2013) have recently shown that the production of spoilage enzymes is generally higher in biofilms than in enzymes produced by planktonic cells. This raises the possibility that these spoilage enzymes can be secreted from dairy biofilms during milk transportation, handling and processing and might end up in the final dairy product. Heat-stable enzymes that are produced by psychrotrophic biofilms upstream of pasteurisation could remain active post-pasteurisation and spoil the final product, as only trace amounts of an enzyme are required to cause damage (Shah, 1994).

The increased enzymatic activity within biofilms can be explained by their different metabolic activity and physiology compared with planktonic cells (Oosthuizen *et al.*, 2001). Proteolysis has been found to be strain-, temperature- and growth mode-dependent

(Teh *et al.*, 2014a). For example, proteolysis and growth were observed at 37 °C (and only at 37 °C) by the biofilm cells of *Pseudomonas fluorescens* but not by the corresponding planktonic cells. This could be linked with the ability of cells in biofilms to grow at higher temperatures than their planktonic counterparts (Rogers *et al.*, 1994; Nilsson *et al.*, 2011). The accumulation of enzymes within biofilms may also aid in the survival of the bacteria in a dairy environment. In another study the hydrolysis of tributyrin within a *Staphylococcus aureus* biofilm was greater when the biofilm was grown under a low-nutrient environment (Teh *et al.*, 2013). This was postulated to be due to a stress response, with an accumulation of enzymes within the biofilm acting as a survival mechanism (Budhani & Struthers, 1998; Thomason *et al.*, 2012). Furthermore, butyric acid released from the lipolysis of milk has been shown to promote biofilm formation by *Bacillus subtilis* (Pasvolsky *et al.*, 2014).

Recent research has examined the effect of multispecies biofilms on the production of spoilage enzymes. Isolates of *B. licheniformis* and *Pseudomonas fragi*, growing as a multispecies biofilm, displayed greater proteolysis than their corresponding single-culture biofilms (Teh *et al.*, 2012). This observation was in agreement with a previous study, which found that the production of amyloyltic enzymes increased in co-culture biofilms of *Bacillus amyloliquefaciens* and *Zymomonas mobilis* compared with either single-culture biofilm (Abate *et al.*, 1999). The effect of nutrient availability on lipolysis in co-culture biofilms was inconclusive, perhaps as a result of the complexities of microbial interaction. In general, mutualistic interactions were observed where the amount of lipolysis was increased in all of the co-culture biofilms of *Streptococcus uberis* C05 when grown in a nutrient-rich environment (Teh *et al.*, 2013). The bacterial strain, the co-culture combinations and the availability of nutrients can all influence the lipolytic effect of a biofilm.

In addition to psychrotrophic bacteria, mesophilic and thermophilic bacteria found in raw milk may also be entrapped within biofilms, and the populations within these biofilms may shift to favour the growth of mesophilic and thermophilic bacteria in a dairy manufacturing plant. Bacterial populations within multispecies biofilms can shift due to environmental factors and microbial interactions (Martiny *et al.*, 2003; Elias & Banin, 2012). Thermophilic bacteria, such as *Geobacillus stearothermophilus*, are common contaminants in milk powder manufacturing and are known to produce heat-stable enzymes (Chopra & Mathur, 1985; Burgess *et al.*, 2010). During heat treatment, the heat-stable enzymes produced within these biofilms may be secreted into the heat-treated products, thereby shortening their shelf life.

Another study by Teh *et al.* (2014a,b) demonstrated that proteolysis can occur in ultra-high-temperature (UHT) milk that was previously exposed to biofilms for 10 hours in an *in vitro* model simulating the transportation of raw milk by a milk tanker. The effects of the enzymes produced by bacteria within a biofilm formed on an *in vitro* model of a milk tanker with three different microbial loads (10^3, 10^5 and 10^7 CFU/ml), comprising *P. fluorescens*, *Serratia liquefaciens* and *S. aureus*, were examined. The degradation of the UHT milk exposed to a slightly contaminated vessel (10^3 CFU/ml) was observed only at 40 °C within 6 months of storage, while degradation after exposure to a moderately contaminated vessel (10^5 CFU/ml) was observed at both 30 and 40 °C. Milk exposed to a highly contaminated vessel (10^7 CFU/ml) was extensively degraded, and the effect was observed immediately when the milk was heat treated, resulting in the coagulation of the milk. The degradation of milk was most likely caused by the presence of heat-stable proteases, as the milk was previously subjected to heat treatment (141 °C for 15 seconds). This is in agreement with other studies, in which heat-stable proteases were able to retain their activity and affect the quality of UHT product during storage (Champagne *et al.*, 1994; Shah, 1994; Celestino *et al.*, 1997). This study demonstrated that the presence of multispecies biofilms on the internal surfaces of a milk tanker during raw milk transportation may have detrimental effects on the quality of manufactured products, as a result of enzyme secretion (Teh *et al.*, 2014b).

Nonstarter lactic acid bacteria (NSLAB) can contribute to the development of undesirable flavours and undesirable appearance in cheese. The majority of NSLABs are *Lactobacillus* spp., but *Pediococcus* and *Leuconostoc* spp. can also be present (Peterson & Marshall, 1990). The source of NSLABs in cheese is believed to be post-pasteurisation contamination from biofilms in the equipment (Austin & Bergeron, 1995; Somers *et al.*, 2001; Agarwal *et al.*, 2006). A study by Somers *et al.* (2001) demonstrated that biofilms formed during the cheese-making process can potentially survive cleaning, resulting in the contamination of subsequent batches of cheese. All NSLABs can generate problems during cheese processing, and modification of cleaning regimes to target biofilms and reduce the chances of contamination is required (Somers *et al.*, 2001; Agarwal *et al.*, 2006).

The presence of proteolytic activity by *Lactobacillus* spp. can contribute to desirable flavours in Cheddar cheese, but can also cause an increase in bitter peptides that results in undesirable flavours (Driessen *et al.*, 1984; Arihara & Luchansky, 2000). NSLABs can also cause gas formation and calcium lactate crystallisation, resulting in a white haze on the cheese (Agarwal *et al.*, 2006). However, not all NSLABs create undesirable effects. Ragusano cheese is created from brine-salted raw milk without the addition of a starter culture. Lactic acid is produced by bacteria naturally occurring in the milk and growing on the surface of traditional wooden vats called tinas (Licitra *et al.*, 2007). It has also been reported that biofilms that naturally reside on the surfaces of wooden shelves can inhibit the growth of *L. monocytogenes* on the surface of cheese by 2 orders of magnitude after 12 days of incubation at 15 °C (Mariani *et al.*, 2011).

The research presented in this section highlights how enzymes produced by bacterial cells within biofilms can result in the spoilage of final dairy products, even following milk treatment steps such as pasteurisation and UHT. Biofilms can form on the internal surface of a milk tanker regardless of the quality of the raw milk. When milk tankers are not adequately cleaned, biofilm growth and enzyme secretion can damage even good-quality milk. Raw milk that is extensively degraded causes relatively little damage to a dairy company, as the milk will be either rejected or diverted to less critical products. Milk that is contaminated with a low level of heat-stable enzymes, however, is potentially damaging to a dairy company's products, its financial return and its reputation, as there is a risk the damage will go undetected until the product is consumed.

4.5 Effects of biofilms on processing efficiency

The presence of established biofilms in milk processing lines can have a serious effect on processing efficiency in a dairy manufacturing plant. The adherence of denatured proteins to a surface can promote bacterial adhesion and growth. The build-up of proteins and biofilms can restrict the flow of products, reduce thermal transfer through stainless steel and promote corrosion. In order to remove these build-ups, longer and more intensive cleaning regimes are required, which result in financial loss due to increased downtime between product runs and use of greater volumes of cleaning chemicals. The metabolic activity of established biofilms can also cause biocorrosion of the underlying stainless steel. This section will discuss the effects of biofilm growth on processing efficiency in a dairy manufacturing plant.

4.5.1 *Effects of fouling and biofilms on heat transfer and flow rates*

Many dairy manufacturing plant processes require heating of milk products flowing at very high flow rates (1.5 m/s). At temperatures greater than 65 °C, whey proteins begin to denature and aggregate, which can result in faster adherence to surfaces compared with the native state (Belmar-Beiny & Fryer, 1992). These adhered proteins create fouling layers that can restrict fluid and flow and reduce thermal conductivity through stainless steel surfaces (Yoon & Lund, 1989) (Figure 4.1). These fouling layers can also alter the characteristics of the stainless steel surfaces, resulting in a higher number of bacteria adhering to them (De Jong, 1997). A study by Flint *et al.* (2001) found that vegetative cells and spores of *G. stearothermophilus* adhered in 10–100 times greater numbers to a fouled stainless steel surface than to a clean surface. Fouling and biofilms can result in decreased production run times, increased product losses and increased cleaning times as attempts are made to control the fouling problem. A decrease in heat-transfer coefficients and fluid flow caused by the build-up of deposits requires the use of increased energy to maintain specific temperatures and flow rates (Russell, 1993; De Jong, 1997).

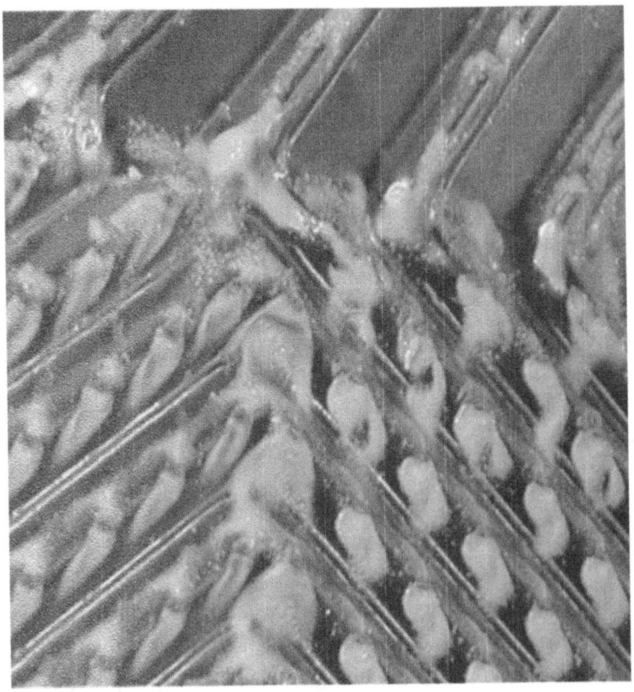

Figure 4.1 Accumulation of fouling and biofilm on the plates of a plate heat exchanger, which will result in reduced heat transfer and restrict the flow of product.

Figure 4.2 Severely blocked ultrafiltration membrane with biofilm visible on the edge of the cartridge.

Fouling and biofilms can become a serious issue for dairy processing membranes after prolonged use (Tang *et al.*, 2009a,b, 2010; Anand *et al.*, 2014) (Figure 4.2). Membranes are used to remove bacteria from skim milk, to concentrate casein micelles and to recover serum proteins from whey. Research by Tang *et al.* (2009b) showed that the predominant organisms causing biofilm production on membranes are of the genera *Bacillus*, *Pseudomonas* and *Klebsiella*. Fouling can cause a severe flux decline and affect the quality of the final product, resulting in higher production costs due to an increase in cleaning frequency and replacement of membranes (Tang *et al.*, 2010). The effects of biofilms on product quality and processing efficiency will be discussed in further detail in Chapter 8.

4.5.2 *Cleaning*

Cleaning-in-place (CIP) regimes are generally used in dairy processing plants. A typical CIP regime consists of five steps, with the rinse and circulation times depending on pipe length and the equipment being cleaned. The typical steps in a CIP regime, as defined by Stewart and Seiberling (1996), are:

1. **Pre-rinse:** The purpose of this step is to remove as much 'loose' soil as possible prior to the formulated alkaline wash. Removal of most of the organic fat, carbohydrate or proteinaceous soil is generally accomplished with ambient- or warm-temperature water.
2. **Alkaline wash:** This step uses heated (70–80 °C), recirculated, formulated solutions. Since relatively long contact times are required, recirculation of cleaning solution is essential for economical operation.
3. **Post-rinse with water:** This step normally occurs at ambient temperature. Its purpose is to rinse away most of the alkaline cleaner. This solution is sometimes recovered for the pre-rinse in the next CIP cleaning program.

4. **Acid rinse:** This step occurs at ambient or heated (55–80 °C) temperatures, using recirculated acid solutions. Its purpose is two-fold: (i) to neutralise and remove residual alkaline cleaner, which would otherwise form films on equipment that cannot readily be removed by a simple post-rinse with water; and (ii) to remove mineral deposits.
5. **Post-rinse:** This step uses water or recirculated sanitising rinse. It is used to apply a bactericidal agent to all cleaned surfaces. The post-rinse is sometimes heated to permit faster drying of equipment.

CIP regimes were designed to remove foulant and bacterial growth from the food contact surfaces within dairy manufacturing plants. A feature of CIP regimes, evident in both industrial- and laboratory-scale systems, is their variable effectiveness in eliminating surface-adherent bacteria (Austin & Bergeron, 1995; Faille *et al.*, 2001; Dufour *et al.*, 2004; Bremer *et al.*, 2006). The most important factors influencing the effectiveness of the CIP are the cleaning time, the cleaning agent temperature, the cleaning agent concentration and chemistry, the degree of turbulence of the cleaning solution and the characteristics of the surface being cleaned. The standard chemicals used in CIP regimes can be formulated to contain compounds such as surfactants that improve their surface wetting, soil penetration and cleaning properties (Bremer *et al.*, 2006).

Bacterial contaminants, such as *S. thermophilus*, *Salmonella* spp. and *L. monocytogenes*, have a greater resistance to heat and sanitisers in the presence of organic material when growing on surfaces compared with planktonic cells (Frank & Koffi, 1990; Flint *et al.*, 2002). The increased resistance is associated with the amount of growth, biofilm structure and the potential change in the physiology of the adhered cells (Frank & Koffi, 1990; Dhir & Dodd, 1995; Joseph *et al.*, 2001).

The efficacy of CIP is also dependent on the washing temperatures used. For example, a study by Latorre *et al.* (2010) showed that high bacterial cell counts were detected in dairy farms with low washing temperatures (47–53 °C). This may have been caused by an old or incorrect setting of the heating systems (Bava *et al.*, 2011). The same thing was observed in a study by Elmoslemany *et al.* (2009), in which bacterial spores were found attached to stainless steel surfaces in dairy manufacturing plants following cleaning. Furthermore, a reduction in the caustic concentration and temperature can reduce the efficiency of CIP by approximately 2 log in mixed-culture biofilms (Bremer *et al.*, 2006).

4.5.3 *Corrosion*

Corrosion of a metal surface results from physicochemical interactions between the surface and its environment, in which electrons are transferred from the metal to an external electron acceptor (Beech, 2004). This causes a release of metal ions from the surface, leading to its deterioration (Figure 4.3). This process can occur through oxidation or reduction reactions. In aerobic environments, corrosion of metal occurs through the reduction of water, while in anaerobic environments it occurs through the production of hydrogen (Borenstein, 1994). However, the rate at which these reactions occurs is determined by a number of factors, including corrosion products, metal type, micro-organisms and the chemical composition of the aqueous environment (Borenstein, 1994).

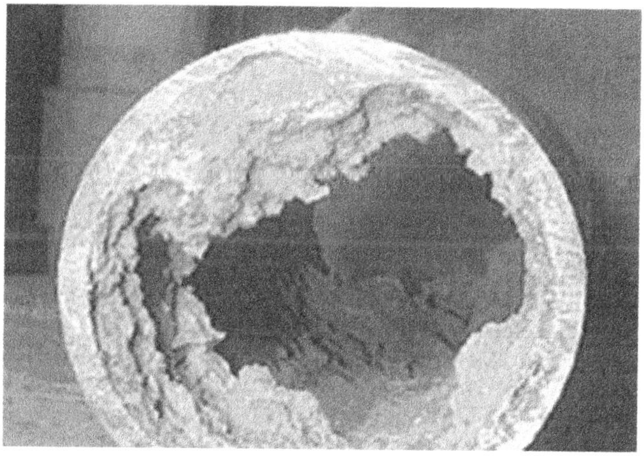

Figure 4.3 Biofilm build-up in a pipe used to transport waste, showing signs of corrosion of the steel surface.

A number of studies have looked at microbiologically influenced corrosion, which is defined as the initiation or aggravation of corrosion due to microbial activity on a surface (Hamilton, 1991; Zuo, 2007).

The dairy industry uses 304- and 316-grade stainless steel for metal surfaces and machinery, because they are durable, corrosion-resistant (resistant to phosphoric acid) and easy to clean. Typically, 304-grade stainless steel is used for refrigerated storage tanks, pasteurisers, maturation tanks and cheese racks, while 316-grade stainless steel is used for pasteurisers, plate and tubular heat exchangers, packing machinery and ultrafiltration equipment. Grade 316 contains of 2–3% molybdenum, which improves its resistance to chlorides.

There is a lack of knowledge at present of the effects of biofilms on the corrosion of metal surfaces within a dairy manufacturing plant. One study found that dairy microflora such as *Bacillus*, *Pseudomonas*, *Micrococcus*, *Niesseria*, *Streptococcus* and *Lactobacillus* could cause pitting on stainless steel surfaces in dairy effluent (Babu *et al.*, 2006). This corrosion was caused by oxygen reduction and fermentation processes, which converted sulphate and iron into ferrous sulphide, which acted as a cathode to the parent metal's anode. However, concentrations of sulphate in raw and pasteurised milk should be extremely low. Currently, not much is known about the occurrence of corrosion due to biofilms present in product pipelines.

4.6 Conclusion

In conclusion, biofilm contamination in a dairy manufacturing plant can have serious effects on the quality and safety of dairy products. Biofilms can also affect processing efficiency by reducing flow and heat transfer rates within processing pipelines. Established biofilms are difficult to remove with CIP and are resistant to sanitising agents. They can also promote corrosion of stainless steel surfaces, mainly in waste systems.

References

Abate, C. M., Castro, G. R., Siñeriz, F. & Callieri, D. A. S. 1999. Production of amylolytic enzymes by *Bacillus amyloliquefaciens* in pure culture and in co-culture with *Zymomonas mobilis*. *Biotechnology Letters*, **21**, 249–52.

Agarwal, S., Sharma, K., Swanson, B. G., Yüksel, G. & Clark, S. 2006. Nonstarter lactic acid bacteria biofilms and calcium lactate crystals in Cheddar cheese. *Journal of Dairy Science*, **89**, 1452–66.

Anand, S., Singh, D., Avadhanula, M. & Marka, S. 2014. Development and control of bacterial biofilms on dairy processing membranes. *Comprehensive Reviews in Food Science and Food Safety*, **13**, 18–33.

Arihara, K. & Luchansky, J. B. 2000. Dairy Lactobacilli. In: Hui, Y. H. & Khachatourians, G. G. (eds) Food Biotechnology: Microorganisms. VCH, New York, NY.

Austin, J. W. & Bergeron, G. 1995. Development of bacterial biofilms in dairy processing lines. *Journal of Dairy Research*, **62**, 509–19.

Babu, B. R., Maruthamuthu, S., Rajasekar, A., Muthukumar, N. & Palaniswamy, N. 2006. Microbiologically influenced corrosion in dairy effluent. *International Journal of Environmental Science and Technology*, **3**, 159–66.

Bar-Oz, B., Preminger, A., Peleg, O., Block, C. & Arad, I. 2001. *Enterobacter sakazakii* infection in the newborn. *Acta Paediatrica*, **90**, 356–8.

Bava, L., Zucali, M., Sandrucci, A., Brasca, M., Vanoni, L., Zanini, L. & Tamburini, A. 2011. Effect of cleaning procedure and hygienic condition of milking equipment on bacterial count of bulk tank milk. *Journal of Dairy Research*, **78**, 211–19.

Becker, H., Schaller, G., Von Wiese, W. & Terplan, G. 1994. *Bacillus cereus* in infant foods and dried milk products. *International Journal of Food Microbiology*, **23**, 1–15.

Beech, I. B. 2004. Corrosion of technical materials in the presence of biofilms – current understanding and state-of-the art methods of study. *International Biodeterioration and Biodegradation*, **53**, 177–83.

Belmar-Beiny, M. T. & Fryer, P. J. 1992. Initial stages of fouling from whey protein solutions. 3rd UK Heat Transfer Conference, Birmingham, UK, 1021–7.

Borenstein, S. 1994. Microbiologically Influenced Corrosion Handbook. Elsevier, Amsterdam.

Borucki, M. K., Peppin, J. D., White, D., Loge, F. & Call, D. R. 2003. Variation in biofilm formation among strains of *Listeria monocytogenes*. *Applied and Environmental Microbiology*, **69**, 7336–42.

Bremer, P. J., Monk, I. & Osborne, C. M. 2001. Survival of *Listeria monocytogenes* attached to stainless steel surfaces in the presence or absence of *Flavobacterium* spp. *Journal of Food Protection*, **64**, 1369–76.

Bremer, P. J., Monk, I. & Butler, R. 2002. Inactivation of *Listeria monocytogenes/Flavobacterium* spp. biofilms using chlorine: impact of substrate, pH, time and concentration. *Letters in Applied Microbiology*, **35**, 321–5.

Bremer, P., Fillery, S. & Mcquillan, A. J. 2006. Laboratory scale Clean-In-Place (CIP) studies on the effectiveness of different caustic and acid wash steps on the removal of dairy biofilms. *International Journal of Food Microbiology*, **106**, 254–62.

Budhani, R. K. & Struthers, J. K. 1998. Interaction of *Streptococcus pneumoniae* and *Moraxella catarrhalis*: Investigation of the indirect pathogenic role of β-lactamase-producing moraxellae by use of a continuous-culture biofilm system. *Antimicrobial Agents and Chemotherapy*, **42**, 2521–6.

Burgess, S. A., Lindsay, D. & Flint, S. H. 2010. Thermophilic bacilli and their importance in dairy processing. *International Journal of Food Microbiology*, **144**, 215–25.

Celestino, E. L., Iyer, M. & Roginski, H. 1997. Reconstituted UHT-treated milk: Effects of raw milk, powder quality and storage conditions of UHT milk on its physico-chemical attributes and flavour. *International Dairy Journal*, **7**, 129–40.

Champagne, C. P., Laing, R. R., Roy, D., Mafu, A. A., Griffiths, M. W. & White, C. 1994. Psychrotrophs in dairy products: their effects and their control. *Critical Reviews in Food Science & Nutrition*, **34**, 1–30.

Chopra, A. K. & Mathur, D. K. 1985. Purification and characterization of heat-stable proteases from *Bacillus stearothermophilus* RM-67. *Journal of Dairy Science*, **68**, 3202–3211.

Cousins, M. A. 1982. Presence and activity of psychrotropic microorganisms in milk and dairy products: a review. *Journal of Food Protection*, **45**, 172–207.

Craven, H. M., Mcauley, C. M., Duffy, L. L. & Fegan, N. 2010. Distribution, prevalence and persistence of *Cronobacter* (*Enterobacter sakazakii*) in the nonprocessing and processing environments of five milk powder factories. *Journal of Applied Microbiology*, **109**, 1044–52.

De Jong, P. 1997. Impact and control of fouling in milk processing. *Trends in Food Science and Technology*, **8**, 401–5.

De Jonghe, V., Coorevits, A., De Block, J., Van Coillie, E., Grijspeerdt, K., Herman, L., De Vos, P. & Heyndrickx, M. 2010. Toxinogenic and spoilage potential of aerobic spore-formers isolated from raw milk. *International Journal of Food Microbiology*, **136**, 318–25.

Dhir, V. K. & Dodd, C. E. R. 1995. Susceptibility of suspended and surface-attached *Salmonella enteritidis* to biocides and elevated temperatures. *Applied and Environmental Microbiology*, **61**, 1731–8.

Driessen, F. M., Vries, J. D. & Kingma, F. 1984. Adhesion and growth of thermoresistant streptococci on stainless steel during heat treatment of milk. *Journal of Food Protection*, **11**, 836–901.

Dufour, M., Simmonds, R. S. & Bremer, P. J. 2004. Development of a laboratory scale clean-in-place system to test the effectiveness of 'natural' antimicrobials against dairy biofilms. *Journal of Food Protection*, **67**, 1438–43.

Elias, S. & Banin, E. 2012. Multi-species biofilms: Living with friendly neighbors. *FEMS Microbiology Reviews*, **36**, 990–1004.

Elmoslemany, A. M., Keefe, G. P., Dohoo, I. R. & Jayarao, B. M. 2009. Risk factors for bacteriological quality of bulk tank milk in Prince Edward Island dairy herds. Part 1: Overall risk factors. *Journal of Dairy Science*, **92**, 2634–43.

Faille, C., Fontaine, F. & Benezech, T. 2001. Potential occurrence of adhering living Bacillus spores in milk product processing lines. *Journal of Applied Microbiology*, **90**, 892–900.

Faille, C., Jullien, C., Fontaine, F., Bellon-Fontaine, M. N., Slomianny, C. & Benezech, T. 2002. Adhesion of *Bacillus* spores and *Escherichia coli* cells to inert surfaces: role of surface hydrophobicity. *Canadian Journal of Microbiology*, **48**, 728–38.

Faille, C., Bénézech, T., Midelet-Bourdin, G., Lequette, Y., Clarisse, M., Ronse, G., Ronse, A. & Slomianny, C. 2014. Sporulation of *Bacillus* spp. within biofilms: a potential source of contamination in food processing environments. *Food Microbiology*, **40**, 64–74.

Farber, J. M. & Peterkin, P. I. 1991. *Listeria monocytogenes*, a food-borne pathogen. *Microbiological Reviews*, **55**, 476–511.

Flint, S., Palmer, J., Bloemen, K., Brooks, J. & Crawford, R. 2001. The growth of *Bacillus stearothermophilus* on stainless steel. *Journal of Applied Microbiology*, **90**, 151–7.

Flint, S., Brooks, J., Bremer, P., Walker, K. & Hausman, E. 2002. The resistance to heat of thermoresistant streptococci attached to stainless steel in the presence of milk. *Journal of Industrial Microbiology and Biotechnology*, **28**, 134–6.

Frank, J. F. & Koffi, R. A. 1990. Surface-adherent growth of *Listeria monocytogenes* is associated with increased resistance to surfactant sanitizers and heat. *Journal of Food Protection*, **53**, 550–4.

Granum, P. E. 1994. *Bacillus cereus* and its toxins. *Journal of Applied Microbiology*, **76**, 61S–6S.

Granum, P. E. & Lund, T. 1997. *Bacillus cereus* and its food poisoning toxins. *FEMS Microbiology Letters*, **157**, 223–8.

Hamilton, W. A. 1991. Sulphate-reducing bacteria and their role in biocorrosion. In: Flemming, H.-C. & Geesey, G. G. (eds) Biofouling and Biocorrosion in Industrial Water Systems. Springer, Berlin.

Huck, J. R., Woodcock, N. H., Ralyea, R. D. & Boor, K. J. 2007. Molecular subtyping and characterization of psychrotolerant endospore-forming bacteria in two New York State fluid milk processing systems. *Journal of Food Protection*, **70**, 2354–64.

Husmark, U. & Ronner, U. 1990. Forces involved in adhesion of *Bacillus cereus* spores to solid surfaces under different environmental conditions. *Journal of Applied Microbiology*, **69**, 557–62.

Husmark, U. & Ronner, U. 1992. The influence of hydrophobic, electrostatic and morphological properties on the adhesion of *Bacillus* spores. *Biofouling*, **5**, 335–44.

Iversen, C., Lane, M. & Forsythe, S. J. 2004. The growth profile, thermotolerance and biofilm formation of *Enterobacter sakazakii* grown in infant formula milk. *Letters in Applied Microbiology*, **38**, 378–82.

Ivy, R. A., Ranieri, M. L., Martin, N. H., Den Bakker, H. C., Xavier, B. M., Wiedmann, M. & Boor, K. J. 2012. Identification and characterization of psychrotolerant sporeformers associated with fluid milk production and processing. *Applied and Environmental Microbiology*, **78**, 1853–64.

Jacobs, C., Braun, P. & Hammer, P. 2011. Reservoir and routes of transmission of *Enterobacter sakazakii* (*Cronobacter* spp.) in a milk powder-producing plant. *Journal of Dairy Science*, **94**, 3801–10.

Jeong, D. K. & Frank, J. F. 1994a. Growth of *Listeria monocytogenes* at 10 °C in biofilms with microorganisms isolated from meat and dairy processing environments. *Journal of Food Protection*, **57**, 576–86.

Jeong, D. K. & Frank, J. F. 1994b. Growth of *Listeria monocytogenes* at 21 °C in biofilms with microorganisms isolated from meat and dairy processing environments. *LWT-Food Science and Technology*, **27**, 415–24.

Joseph, B., Otta, S. K., Karunasagar, I. & Karunasagar, I. 2001. Biofilm formation by *Salmonella* spp. on food contact surfaces and their sensitivity to sanitizers. *International Journal of Food Microbiology*, **64**, 367–72.

Karunakaran, E. & Biggs, C. A. 2011. Mechanisms of *Bacillus cereus* biofilm formation: an investigation of the physicochemical characteristics of cell surfaces and extracellular proteins. *Applied Microbiology and Biotechnology*, **89**, 1161–75.

Kim, H., Ryu, J.-H. & Beuchat, L. R. 2006. Attachment of and biofilm formation by *Enterobacter sakazakii* on stainless steel and enteral feeding tubes. *Applied and Environmental Microbiology*, **72**, 5846–56.

Larsen, H. D. & Jørgensen, K. 1997. The occurrence of *Bacillus cereus* in Danish pasteurized milk. *International Journal of Food Microbiology*, **34**, 179–86.

Latorre, A. A., Van Kessel, J. S., Karns, J. S., Zurakowski, M. J., Pradhan, A. K., Boor, K. J., Jayarao, B. M., Houser, B. A., Daugherty, C. S. & Schukken, Y. H. 2010. Biofilm in milking equipment on a dairy farm as a potential source of bulk tank milk contamination with *Listeria monocytogenes*. *Journal of Dairy Science*, **93**, 2792–802.

Licitra, G., Ogier, J. C., Parayre, S., Pediliggieri, C., Carnemolla, T. M., Falentin, H., Madec, M. N., Carpino, S. & Lortal, S. 2007. Variability of bacterial biofilms of the 'tina' wood vats used in the Ragusano cheese-making process. *Applied and Environmental Microbiology*, **73**, 6980–7.

Mariani, C., Oulahal, N., Chamba, J. F., Dubois-Brissonnet, F., Notz, E. & Briandet, R. 2011. Inhibition of *Listeria monocytogenes* by resident biofilms present on wooden shelves used for cheese ripening. *Food Control*, **22**, 1357–62.

Martiny, A. C., Jørgensen, T. M., Albrechtsen, H.-J., Arvin, E. & Molin, S. 2003. Long-term succession of structure and diversity of a biofilm formed in a model drinking water distribution system. *Applied and Environmental Microbiology*, **69**, 6899–907.

Mcguiggan, J. T. M., Mccleery, D. R., Hannan, A. & Gilmour, A. 2002. Aerobic spore-forming bacteria in bulk raw milk: factors influencing the numbers of psychrotrophic, mesophilic and thermophilic *Bacillus* spores. *International Journal of Dairy Technology*, **55**, 100–7.

Muytjens, H. L., Roelofs-Willemse, H. & Jaspar, G. H. 1988. Quality of powdered substitutes for breast milk with regard to members of the family Enterobacteriaceae. *Journal of Clinical Microbiology*, **26**, 743–6.

Nilsson, R. E., Ross, T. & Bowman, J. P. 2011. Variability in biofilm production by *Listeria monocytogenes* correlated to strain origin and growth conditions. *International Journal of Food Microbiology*, **150**, 14–24.

Notermans, S., Dufrenne, J., Teunis, P., Beumer, R., Te Giffel, M. & Peeters Weem, P. 1997. A risk assessment study of *Bacillus cereus* present in pasteurized milk. *Food Microbiology*, **14**, 143–51.

Oliver, S. P., Jayarao, B. M. & Almeida, R. A. 2005. Foodborne pathogens in milk and the dairy farm environment: food safety and public health implications. *Foodbourne Pathogens and Disease*, **2**, 115–29.

Oosthuizen, M. C., Steyn, B., Lindsay, D., Brözel, V. S. & Holy, A. 2001. Novel method for the proteomic investigation of a dairy-associated *Bacillus cereus* biofilm. *FEMS Microbiology Letters*, **194**, 47–51.

Osaili, T. & Forsythe, S. 2009. Desiccation resistance and persistence of *Cronobacter* species in infant formula. *International Journal of Food Microbiology*, **136**, 214–20.

Pasvolsky, R., Zakin, V., Ostrova, I. & Shemesh, M. 2014. Butyric acid released during milk lipolysis triggers biofilm formation of *Bacillus* species. *International Journal of Food Microbiology*, **181**, 19–27.

Peterson, S. D. & Marshall, R. T. 1990. Nonstarter lactobacilli in Cheddar cheese: a review. *Journal of Dairy Science*, **73**, 1395–410.

Pirttijarvi, T. S. M., Andersson, M. A., Scoging, A. C. & Salkinoja-Salonen, M. S. 1999. Evaluation of methods for recognising strains of the *Bacillus cereus* group with food poisoning potential among industrial and environmental contaminants. *Systematic and Applied Microbiology*, **22**, 133–44.

Randolph, H. 2006. Identifying spoilage bacteria and potential shelf-life problems. *Dairy Foods*, **107**, 67.

Ranieri, M. L. & Boor, K. J. 2010. Tracking and eliminating sporeformers in dairy systems. *Australian Journal of Dairy Technology*, **65**, 74–80.

Ranieri, M. L., Huck, J. R., Sonnen, M., Barbano, D. M. & Boor, K. J. 2009. High temperature, short time pasteurisation temperatures inversely affect bacterial numbers during refrigerated storage of pasteurized fluid milk. *Journal of Dairy Science*, **92**, 4823–32.

Rogers, J., Dowsett, A. B., Dennis, P. J., Lee, J. V. & Keevil, C. W. 1994. Influence of temperature and plumbing material selection on biofilm formation and growth of *Legionella pneumophila* in a model potable water system containing complex microbial flora. *Applied and Environmental Microbiology*, **60**, 1585–92.

Rönner, U. & Husmark, U. 1992. Adhesion of *Bacillus cereus* spores – a hazard to the dairy industry. In: Melo, L. F. (ed.) Biofilms – Science and Technology. Kluwer Academic Publishers, The Netherlands.

Ronner, U., Husmark, U. & Henriksson, A. 1990. Adhesion of *bacillus* spores in relation to hydrophobicity. *Journal of Applied Bacteriology*, **69**, 550–6.

Rosso, L., Bajard, S., Flandrois, J. P., Lahellec, T. C., Fournaud, J. & Veit, P. 1996. Differential growth of *Listeria monocytogenes* at 4 and 8C: Consequences for the shelf life of chilled products. *Journal of Food Protection*, **59**, 944–9.

Russell, P. 1993. The formation of biofilms. *Milk Industry London*, **95**, 10–11.

Rusul, G. 1995. Prevalence of *Bacillus cereus* in selected foods and detection of enterotoxin using TECRA-VIA and BCET-RPLA. *International Journal of Food Microbiology*, **25**, 131–9.

Sashara, K. C. & Zottola, E. A. 1993. Biofilm formation by *Listeria monocytogenes* utilizes a primary colonizing microorganism in flowing systems. *Journal of Food Protection*, **56**, 1022–8.

Schlech, W. F., Lavigne, P. M., Bortolussi, R. A., Allen, A. C. & Haldane, E. V. 1983. Epidemic listeriosis: evidence for transmission by food [*Listeria monocytogenes*]. *New England Journal of Medicine*, **308**, 203–6.

Shah, N. P. 1994. Psychrotrophs in milk: a review. *Milchwissenschaft*, **49**, 432–7.

Shaheen, R., Svensson, B., Andersson, M. A., Christiansson, A. & Salkinoja-Salonen, M. 2010. Persistence strategies of *Bacillus cereus* spores isolated from dairy silo tanks. *Food Microbiology*, **27**, 347–55.

Shi, X. & Zhu, X. 2009. Biofilm formation and food safety in food industries. *Trends in Food Science & Technology*, **20**, 407–13.

Simmons, B. P., Gelfand, M. S., Haas, M., Metts, L. & Ferguson, J. 1989. *Enterobacter sakazakii* infections in neonates associated with intrinsic contamination of a powdered infant formula. *Infection Control and Hospital Epidemiology: The Official Journal of the Society of Hospital Epidemiologists of America*, **10**, 398–401.

Somers, E. B., Johnson, M. E. & Wong, A. C. L. 2001. Biofilm formation and contamination of cheese by nonstarter lactic acid bacteria in the dairy environment. *Journal of Dairy Science*, **84**, 1926–36.

Sonbol, H., Joseph, S., Mcauley, C. M., Craven, H. M. & Forsythe, S. J. 2013. Multilocus sequence typing of *Cronobacter* spp. from powdered infant formula and milk powder production factories. *International Dairy Journal*, **30**, 1–7.

Stenfors Arnesen, L. P., Fagerlund, A. & Granum, P. E. 2008. From soil to gut: *Bacillus cereus* and its food poisoning toxins. *FEMS Microbiology Reviews*, **32**, 579–606.

Stewart, J. C. & Seiberling, D. A. 1996. Clean in place. *Chemical Engineering*, **102**, 72–79.

Storgårds, E., Tapani, K., Hartwall, P., Saleva, R. & Suihko, M. L. 2006. Microbial attachment and biofilm formation in brewery bottling plants. *Journal of the American Society of Brewing Chemists*, **64**, 8–15.

Tang, X., Flint, S. H., Bennett, R. J., Brooks, J. D. & Morton, R. H. 2009a. Biofilm growth of individual and dual strains of *Klebsiella oxytoca* from the dairy industry on ultrafiltration membranes. *Journal of Industrial Microbiology and Biotechnology*, **36**, 1491–7.

Tang, X., Flint, S. H., Brooks, J. D. & Bennett, R. J. 2009b. Factors affecting the attachment of micro-organisms isolated from ultrafiltration and reverse osmosis membranes in dairy processing plants. *Journal of Applied Microbiology*, **107**, 443–51.

Tang, X., Flint, S. H., Bennett, R. J. & Brooks, J. D. 2010. The efficacy of different cleaners and sanitisers in cleaning biofilms on UF membranes used in the dairy industry. *Journal of Membrane Science*, **352**, 71–5.

Tauveron, G., Slomianny, C., Henry, C. & Faille, C. 2006. Variability among *Bacillus cereus* strains in spore surface properties and influence on their ability to contaminate food surface equipment. *International Journal of Food Microbiology*, **110**, 254–62.

Teh, K. H., Flint, S., Palmer, J., Andrewes, P., Bremer, P. & Lindsay, D. 2012. Proteolysis produced within biofilms of bacterial isolates from raw milk tankers. *International Journal of Food Microbiology*, **157**, 28–34.

Teh, K. H., Lindsay, D., Palmer, J., Andrewes, P., Bremer, P. & Flint, S. 2013. Lipolysis within single culture and co-culture biofilms of dairy origin. *International Journal of Food Microbiology*, **163**, 129–35.

Teh, K. H., Flint, S., Palmer, J., Andrewes, P., Bremer, P. & Lindsay, D. 2014a. Biofilm – an unrecognised source of spoilage enzymes in dairy products? *International Dairy Journal*, **34**, 32–40.

Teh, K. H., Lindsay, D., Palmer, J., Andrewes, P., Bremer, P. J. & Flint, S. H. 2014b. Proteolysis in ultra-heat-treated skim milk after exposure to multispecies biofilms under conditions modelling a milk tanker. *International Journal of Dairy Technology*, **67**, 176–81.

Thomason, M. K., Fontaine, F., De Lay, N. & Storz, G. 2012. A small RNA that regulates motility and biofilm formation in response to changes in nutrient availability in *Escherichia coli*. *Molecular Microbiology*, **84**, 17–35.

Van Acker, J., De Smet, F., Muyldermans, G., Bougatef, A., Naessens, A. & Lauwers, S. 2001. Outbreak of necrotizing enterocolitis associated with *Enterobacter sakazakii* in powdered milk formula. *Journal of Clinical Microbiology*, **39**, 293–7.

Vilain, S., Pretorius, J. M., Theron, J. & Brözel, V. S. 2009. DNA as an adhesin: *Bacillus cereus* requires extracellular DNA to form biofilms. *Applied and Environmental Microbiology*, **75**, 2861–8.

Wijman, J. G. E., De Leeuw, P. P. L. A., Moezelaar, R., Zwietering, M. H. & Abee, T. 2007. Air-liquid interface biofilms of Bacillus cereus: formation, sporulation, and dispersion. *Applied and Environmental Microbiology*, **73**, 1481–8.

Yegorenkova, I. V., Tregubova, K. V., Matora, L. Y., Burygin, G. L. & Ignatov, V. V. 2011. Biofilm formation by *Paenibacillus polymyxa* strains differing in the production and rheological properties of their exopolysaccharides. *Current Microbiology*, **62**, 1554–9.

Yoon, J. & Lund, D. B. 1989. Effect of operating conditions, surface coatings and pretreatment on milk fouling in a plate heat exchanger. In: Kessler, H. G. (ed.) Fouling and Cleaning in Food Processing. University of Munich, Germany, pp. 59–80.

Zuo, R. 2007. Biofilms: strategies for metal corrosion inhibition employing microorganisms. *Applied Microbiology and Biotechnology*, **76**, 1245–53.

5 Raw Milk Quality Influenced by Biofilms and the Effect of Biofilm Growth on Dairy Product Quality

Koon Hoong Teh[1], Steve Flint[1] and Phil Bremer[2]

[1]Institute of Food, Nutrition and Human Health, Massey University, Palmerston North, New Zealand

[2]Department of Food Science, University of Otago, Dunedin, New Zealand

5.1 Introduction

The international growth in the consumption of dairy products has led to the development of a large, state-of-the art dairy industry that can handle and process large volumes of milk and milk products. The quality of dairy products, such as pasteurised milk, cheese and butter, and clinical products, such as phospholipid concentrates, gangliosides and colostrum, is influenced by the quality of raw milk, which in turn is determined by what occurs on dairy farms and during the transportation and storage of raw milk.

The quality of raw milk can have a major impact on the manufacturing of dairy products, by influencing product yields and functionality and the occurrence of sensory defects, such as bitterness and rancidity (Shah, 1994; Guinotthomas *et al.*, 1995; Celestino *et al.*, 1997b; Chen *et al.*, 2003). Raw milk is a perishable product that can easily be compromised by both operational factors (handling, transportation, temperature abuse) and natural factors (microbial contaminants, naturally occurring enzymes).

An important measure of raw milk quality is the number or count of microorganisms present. The microorganisms in raw milk can originate from multiple and diverse sources, including the cow itself, the dairy farm environment, milking equipment, raw milk storage tanks and milk transport vehicles, as well as raw milk storage silos and processing equipment (e.g. separators) at the dairy manufacturing plant (Figure 5.1) (LeJeune *et al.*, 2001; Coorevits *et al.*, 2008; Vacheyrou *et al.*, 2011). Biofilms have been shown to play an important role in microbial contamination from each of these sources, and can have an impact on the final quality of dairy products.

Biofilms in the Dairy Industry, First Edition.
Edited by Koon Hoong Teh, Steve Flint, John Brooks and Geoff Knight.

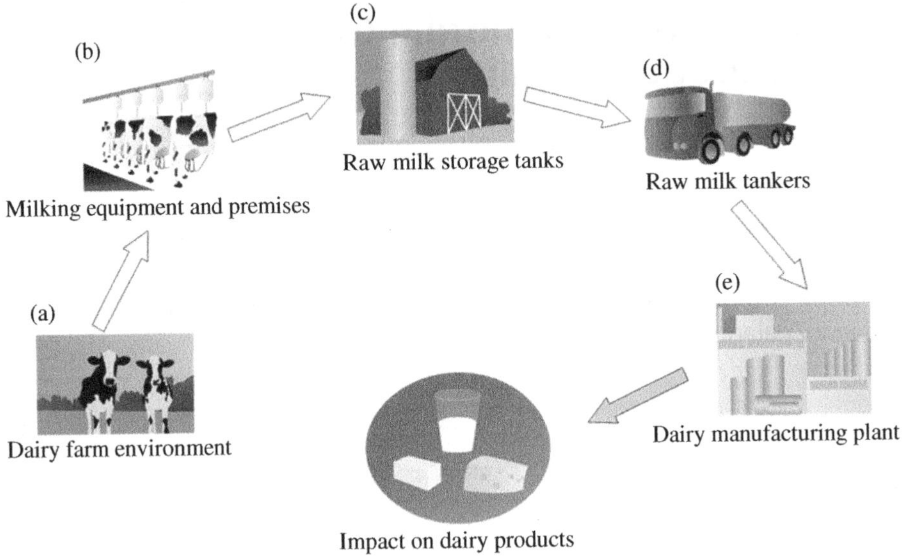

Figure 5.1 Microbial contamination of raw milk can occur at every stage of the supply chain, from its production on the dairy farm up to its processing into a dairy product. Contamination sources include (a) the dairy farm environment, (b) milking equipment and premises, (c) raw milk storage tanks, (d) raw milk transport tankers, and (e) the raw milk silos and handling equipment at the dairy manufacturing plant.

5.2 Composition of raw milk

Milk is composed of water, protein, fat, carbohydrate and minerals. The fat is present as fat globules, which are suspended in the water phase and are surrounded by a membrane consisting of phospholipids and proteins. Caseins, which are categorised into four groups – αS1-, αS2-, β- and κ- casein – make up approximately 80% of the total protein in milk. The remaining proteins are the whey proteins. Lactose, a disaccharide containing a molecule of glucose and a molecule of galactose, is the main carbohydrate in milk. The components in milk are subjected to enzymatic reactions that can either be beneficial or detrimental to the final products.

The composition of raw milk can vary with the breed of cow, type of feed, condition of the animal and the season. The ratio of the components is important in the manufacture of a number of dairy products, as it can affect the manufacturing efficiency, yield, composition and quality of the dairy products (Bruhn & Franke, 1991; Guinee *et al.*, 2007). To maintain consistency and meet compositional specifications, the protein to fat ratio (PFR) of raw milk is often standardised to a narrow range for the manufacture of a number of dairy products.

5.3 Measurement of raw milk quality

The quality of milk is measured by the standard plate count (SPC) and somatic cell count (SCC). The lower the SPC and SCC, the better the quality of the raw milk. The use of incentive programs for dairy farmers has proven to be an effective way of reducing the SCC in raw milk without disrupting milk production (Nightingale *et al.*, 2008). Lower SCC and SPC can be achieved in raw milk when the milk yield on the farm increases (Berry *et al.*, 2006), due to the dilution of cell counts in larger volumes of raw milk yield per cow.

The SPC is generally influenced by specific groups of bacteria, which may include mesophilic, pathogenic and spoilage microorganisms, as well as thermoduric, psychrotrophic and mastitis-causing bacteria from the environment (Holm *et al.*, 2004; Jayarao *et al.*, 2004). Microbial contamination of raw milk initially occurs through low-level contamination on the farm. On-farm sources can include teats (skin flora and bacteria associated with soil, plant and faecal material), milking equipment and raw milk storage tanks. Poor hygiene of the udder will lead to increased numbers of bacteria in the raw milk, as dirty udders can harbour a wide variety of bacteria (Jayarao *et al.*, 2004). Persistent milk residues in the milking equipment may harbour thermoduric bacteria (Holm *et al.*, 2004). During milking, thermoduric bacteria associated with these residues can detach and enter the raw milk, leading to an increase in the SPC (Hayes *et al.*, 2001). Finally, poor handling of raw milk during transportation and storage may lead to an increase in the microbial population, especially of microorganisms responsible for spoilage (Hayes *et al.*, 2001; Holm *et al.*, 2004; Jayarao *et al.*, 2004). Recontamination of processed milk with psychrotrophic bacteria can occur after pasteurisation in the dairy plant, especially during filling (Eneroth *et al.*, 1998).

The SCC can be used to gauge the health of the dairy herd, as a high count of SCC is a good indicator of mastitis. Somatic cells are released when the parenchyma of the mammary gland is infected. The inflammation of the udder is caused by a wide range of bacteria; however, the most common bacteria infecting the udder are *Staphylococcus aureus* and *Streptococcus* species. The most common species of *Streptococcus* isolated from raw milk is *Streptococcus uberis*, followed by *Streptococcus agalactiae* (Hayes *et al.*, 2001; Zadoks *et al.*, 2004; Howard, 2006). Increased levels of mastitis-causing pathogens in the udder can lead to a direct increase in the SPC of the raw milk, as well as contamination of milking cups by pathogens (Zadoks *et al.*, 2004). Therefore, cleaning of teats pre- and post-milking is important to reducing the bacterial load of raw milk, as well as inflammatory infections in dairy herds (Jayarao *et al.*, 2004).

5.4 Regulations and guidelines for the production of raw milk

In order to produce quality finished products, the bacterial counts of raw milk should be low. Regulations and guidelines for maintaining raw milk quality vary for different countries. Most of these regulations and guidelines require raw milk to be kept at low temperatures to prevent microbial growth.

5.4.1 *In Europe*

A consolidated version of Regulation (EC) No 853/2004 as at 1 July 2013 (EU, 2013) lays down provisions on hygiene rules for foods of animal origin; Annex III Section XI thereof contains specific rules for the production and placing on the market of raw milk, colostrum, dairy products and colostrum-based products.

Milking equipment and raw milk storage tanks must be constructed of materials with surfaces that can be easily cleaned, so as to limit the chance of contamination. Before milking, the teats, udder and adjacent parts of the cow need to be cleaned; however, teat dips or spray can only be used after authorisation or registration in accordance with the procedures laid down in Directive 98/9/EC of the European Parliament and of the Council of 16 February 1998 (EU, 2013).

Raw milk and colostrum must be cooled immediately after collection to a temperature of 8 °C or lower, in the case of daily collection, or to 6 °C if not collected daily (EU, 2013). Colostrum may be frozen until collection. During transportation of raw milk and colostrum to the treatment and/or processing establishment, the cold chain must be maintained and the temperature of the cooled milk must not exceed 10 °C, unless the milk is used for the manufacture of dairy products that require warmer temperatures or has been collected within 2 hours of milking prior to processing.

5.4.2 *In the United States*

The USDA standard (USDA, 2011), as of July 2011, requires milking premises to be of adequate size and to be arranged to permit normal sanitary milking operations. The milking premises are also required to be kept clean and to have procedures in place to prevent access by other animals. In addition, equipment used for milking and for handling of raw milk should be maintained in good, clean, working condition. The milking equipment must be properly cleaned and sanitised after each milking session, while the udders and teats of cows need to be cleaned before milking, by washing, wiping or any other sanitary method. In the event that a cow produces abnormal milk, the milking equipment must be cleaned and sanitised immediately after milking.

The milk stored in a raw milk storage tank needs to be cooled to 4 °C or lower within 2 hours after milking and maintained at 10 °C or lower prior to transfer to the dairy manufacturing plant. The tank should be easily accessible for cleaning and servicing, should not be located over a floor drain or under a ventilator and should not be accessible by animals and fowl.

5.4.3 *In New Zealand*

Dairy farm operators in New Zealand are required to follow regulations under the Animal Products Act 1999 and Animal Products (Dairy) Regulations 2005 (MAF, 2005; MPI, 2014). This is to ensure the quality and safety of the milk. The Animal Products Act requires dairy operators to implement an approved risk management programme that identifies, controls, manages, eliminates or minimises hazards and other risks during milking. As of July 2011, Animal Products (Dairy Processing Specifications) Notice 2011 specified the requirements for processing of dairy material and dairy products (MAF, 2011). The general requirements state that the milking premises need to be kept clean and tidy, and free from birds, rodents, insects and other pests. The milking premises must only be used to milk animals with clean teats, which must be cleaned with approved detergents and sanitisers in accordance with regulation 24(1) (d) of the Animal Products (Dairy) Regulation 2005 (MAF, 2005). The milking premises need to be cleaned in a way that minimises the risk of contaminating the milk with detergents or sanitisers.

The raw milk must be filtered and cooled to 7 °C or below within 3 hours of the completion of milking. The temperature of raw milk must be maintained at 7 °C or below until the collection of additional milk from the next milking session.

The approved process is currently under review, with the following requirements to be in place by 2018 (Ministry of Primary Industries):

1. Raw milk must be cooled to 10 °C or below within 4 hours of the commencement of milking.
2. Raw milk must be cooled to 6 °C or below within 6 hours of the commencement of milking and within 2 hours of the completion of milking.
3. Raw milk must be held at or below 6 °C until collection or the next milking.
4. Raw milk must not exceed 10 °C during subsequent milkings.

In situations where there is continuous milking, such as in an automated milking system, the milk must enter the bulk milk tank at 6 °C or below. Continuous milking is defined as milking for 6 hours or longer from the time that milk first enters any bulk milk tank.

Farm dairy operators must have an auditable system that confirms milk cooling requirements are met. As a minimum, milk cooling performance must be monitored and recorded on at least three occasions per dairy season, including:

1. Within the first two months of lactation, once the full herd has calved.
2. About the time of peak milk production.
3. February each year.

Where electronic data-capture and recording systems are installed, it is recommended that such systems should be capable of holding delivery line and bulk milk tank temperature data for the previous 30 days for both milk and cleaning regimes.

5.5 Microbial profile of raw milk and its effect on the dairy industry

The indigenous microbial community in raw milk plays an important role in the dairy industry and can influence the value of the final dairy product. The microbial community of raw milk is diverse and can include pathogenic and spoilage microorganisms, as well as beneficial or technological bacteria, which include starter and probiotic bacteria. A large number of other bacterial species have been detected in raw milk that do not appear to have any impact on the quality of dairy products.

The microbial community in bovine raw milk is highly diverse (Table 5.1) and is often dominated by lactic acid-producing bacteria (*Lactococcus*, *Streptococcus*, *Lactobacillus*, *Leuconsostoc* and *Enterococcus* species) and psychrotrophic bacteria (*Pseudomonas*, *Acinetobacter* and *Aeromonoas* species) (Quigley *et al.*, 2013). The diversity of the microbial community in raw milk can be assessed by either the isolation of bacteria using agar-based

Table 5.1 The diversity of microbial community detect in bovine raw milk. Adapted from Quigley *et al.* (2013).

Genera of bacteria detected in bovine raw milk		
Acidobacteria	*Enterobacter*	*Pantoea*
Acinetobacter	*Enterococcus*	*Paracoccus*
Adhaeribacter	*Escherichica*	*Phyllobacterium*
Aerococcus	*Facklamia*	*Propionibacterium*
Aeromonas	*Frigoribacterium*	*Proteobacteria*
Achrombacter	*Hafnia*	*Providencia*
Arthrobacter	*Halomonas*	*Pseudoclavibacter*
Bacillus	*Janibacter*	*Pseudomonas*
Bacteriodetes	*Janthinobacterium*	*Psychrobacter*
Bosea	*Klebsiella*	*Rahnella*
Brachybacterium	*Kocuria*	*Ralstonia*
Bradyrhizobium	*Lactobacillus*	*Renibacterium*
Brevibacterium	*Lactococcus*	*Rhodoccus*
Campylobacter	*Leconostoc*	*Rothia*
Caryophanon	*Leucobacter*	*Salmonella*
Chryseobacterium	*Leuconostoc*	*Serratia*
Clavibacter	*Listeria*	*Sphingomonas*
Clostridium	*Microbacterium*	*Stapylococcus*
Comamonas	*Micrococcus*	*Stenotrophomonas*
Corynebacterium	*Nocadioides*	*Streptococcus*
Deinococcus	*Ochrobacterum*	*Thauera*
Delftia	*Orinthinicoccus*	*Trichococcus*
Dermacoccus	*Paenibacillus*	*Yania*
Empedobacter	*Pandoraea*	*Yersina*

methods followed by characterisation using phenotypic or genotypic methods, or the isolation of bacterial genomic DNA from raw milk followed by application of molecular methods such as Denaturing Gradient Gel Electrophoresis (DGGE), Single-strand conformation polymorphism (SSCP), generation of a clone library or pyro sequencing (Quigley *et al.*, 2013).

5.5.1 *Spoilage microorganisms in raw milk*

Spoilage microorganisms have an economic impact on the dairy industry because of their ability to decrease the sensory properties and yield of final milk products. In the most severe cases, the growth of spoilage microorganisms and the production of metabolic byproducts and/or extracellular enzymes can result in dairy products becoming unfit for sale. In cheese making, for example, the yield and quality of cheese is reduced when the number of spoilage bacteria in the milk becomes greater than 10^6 CFU/ml (Ledenbach & Marshal, 2010).

Psychrotrophic bacteria and heat-stable spoilage enzymes

Most of the bacteria responsible for the spoilage of dairy products are psychrotrophic, and many of these have the ability to grow at low temperatures (<4 °C), either as planktonic cells or within biofilms; they also have the ability to produce extracellular enzymes (Nörnberg *et al.*, 2011; Teh *et al.*, 2011). The secretion of bacterial enzymes is a complex process that is influenced by a variety of environmental factors, such as oxygen concentration, temperature and iron content, as well as bacterial population factors, such as quorum sensing and phase variation (Jaspe *et al.*, 2000; Woods *et al.*, 2001; Haddadi *et al.*, 2005; Nicodème *et al.*, 2005; van den Broek *et al.*, 2005; Liu *et al.*, 2007). The secretion of bacterial enzymes usually peaks during the mid to late exponential phase or the early stationary phase of bacterial population growth. Members of the genus *Pseudomonas* are very common examples of psychrotrophic spoilage bacteria associated with milk and other dairy products. Pseudomonads isolated from raw milk have been reported to produce proteases with molecular weights ranging from 39 to 45 kDa (Marchand *et al.*, 2009a). The amount of protease secreted varies among different species of *Pseudomonas*. For example, the amount produced by *P. chlororaphis* is greater than the amount produced by *P. fluorescens* (Nicodème *et al.*, 2005). Bacterial species belonging to genera other than *Pseudomonas*, such as *Bacillus*, *Micrococcus*, *Aerococcus*, *Serratia* and *Lactococcus*, also have the potential to spoil dairy products through the production of spoilage enzymes.

Heat treatment processes used in the dairy industry, such as pasteurisation, will inactivate vegetative cells of psychrotrophic bacteria but will not inactivate heat-stable enzymes. The heat-stability of the bacterial enzymes increases when multiple heat-stable enzymes, such as proteases, are present (Chopra & Mathur, 1985). Heat-stable bacterial enzymes can remain active throughout the storage of dairy products, and only trace amounts of bacterial enzymes may be required to cause spoilage (Shah, 1994).

Bacterial enzymes secreted by psychrotrophic bacteria have been found to reduce the shelf-life of ultra-high-temperature (UHT) milk during storage at room temperature (Celestino *et al.*, 1997a; Sørhaug & Stepaniak, 1997; Barbano *et al.*, 2006). The reduction in shelf-life of dairy products is likely the result of the degradation of milk casein by the different types of bacterial protease that remain active after heat treatment (Fairbairn & Law, 1986; Grufferty & Fox, 1988; Åkerstedt *et al.*, 2012). This degradation of the casein micelle structure causes the coagulation of the milk (Fairbairn & Law, 1986). In another study, whole milk powder manufactured from fresh raw milk was shown to have a lower concentration of free fatty acids (FFAs) than milk powder manufactured from raw milk that had been stored at 4 °C for 2 days (Celestino *et al.*, 1997a). The increased concentration of FFAs in milk powder made from the stored milk was believed to be caused by lipases, secreted by psychrotrophic bacteria during storage of the raw milk, breaking down the milk fat. Besides causing lipolysis of milk fat, heat-stable lipases are also known to reduce the stability of milk foam in beverages such as cappuccino (Huppertz, 2010).

Biofilms as a source of spoilage enzymes

Some bacteria that produce heat-stable enzymes also have the ability to form biofilms (Teh *et al.*, 2011). Enzymes produced by microorganisms growing in biofilms can remain attached to or be trapped within the biofilm matrix, or they can be released from the biofilm (Bagge *et al.*, 2004; Khajanchi *et al.*, 2009; Rajendran *et al.*, 2010). The pores within the biofilms can provide microenvironments for enzymatic activities, as well as protection for both the bacterial cells and the enzymes against hazardous conditions (Li *et al.*, 2006; Licitra *et al.*, 2007; Rosche *et al.*, 2009; Wang & Chen, 2009). Furthermore, enzymes within biofilms have been reported to occupy particular niches (Iwashita *et al.*, 2001). Within *Aspergillus* biofilms, for example, β-glucosidases tended to be bound within the surfaces of cells when *A. kawachii* is grown in biofilms, whereas in solid-state fermentation, the β-glucosidases tend to be extracellular (Iwashita *et al.*, 1998). The authors suggested that the extracellular soluble polysaccharide from *A. kawachii* influenced the stability and the localisation of the β-glucosidases within the biofilm (Iwashita *et al.*, 2001). In another study, proteases were found to be tightly bound to cell walls within sludge biofilms, while α-amylase and α-glucosidase were immobilised in the matrix as cell-free enzymes (Yu *et al.*, 2007).

The dispersal of cells and cell clumps (containing cells and enzymes) from mature biofilms into milk during storage and processing may further increase the risk of spoilage (Teh *et al.*, 2014). Bacterial cells embedded within alginate beads have been used as a model of a dispersed biofilm matrix (Xu *et al.*, 1996) and immobilised cells within alginate beads have been shown to secrete enzymes (Zakaria *et al.*, 1992). In addition, it is possible that the biofilm matrix that binds cells together in dispersed clumps may protect both the enzymes and the bacterial cells from inactivation during processing.

The realisation that biofilms may be a potential source of spoilage enzymes is a relatively recent concept, derived from studies showing that the proteolytic and lipolytic activity of bacteria within dairy biofilms was greater than that of bacteria in a planktonic state (Teh *et al.*, 2012, 2013). It has similarly been reported for fungal and wastewater biofilms that enzymatic activities were greater than in their planktonic counterparts (Frølund *et al.*, 1995; Gamarra *et al.*, 2010). In fungal biofilms, Gamarra *et al.* (2010) reported that even though the biomass of *A. niger* from a biofilm fermentation was lower than that of the biomass in a submerged or solid-state fermentation, the yield of cellulase was significantly higher in the biofilm. This increase in activity can be explained by the differences in metabolic activities and physiologies of biofilms and planktonic cells (Oosthuizen *et al.*, 2001; Wang & Chen, 2009; Gamarra *et al.*, 2010). In addition, co-culture biofilms had higher enzymatic activity than their corresponding single-culture biofilms (Abate *et al.*, 1999; Teh *et al.*, 2012).

Enzyme production by bacteria within biofilms has been shown to depend upon several factors, including temperature, growth mode and nutrient availability (Teh *et al.*, 2012, 2013). It has been shown that *P. fluorescens*, which is psychrotrophic, is able to grow and produce proteases at 37 °C only when it is in a biofilm, and not in a planktonic state (Teh *et al.*, 2012). This can be explained by the ability of cells in biofilms to grow at higher temperatures than their planktonic counterparts (Rogers *et al.*, 1994; Nilsson *et al.*, 2011). Furthermore, stress response may facilitate the production of enzymes, where the accumulation of enzymes in biofilms increases the ability of cells to survive in biofilms (Budhani & Struthers, 1998; Spector & Kenyon, 2012; Thomason *et al.*, 2012). For example, lipolytic activity was found to be higher in *S. aureus* biofilm when it was grown in a nutrient-limited environment rather than a nutrient-rich environment (Teh *et al.*, 2013).

Enzyme production in biofilms may also be influenced by quorum sensing (Khiyami *et al.*, 2006). Quorum-sensing signal molecules have been shown to be responsible for the production of proteases and biofilms (Swift *et al.*, 1999; Liu *et al.*, 2007; Khajanchi *et al.*, 2009). Interestingly, while quorum sensing has been reported to influence the production of enzymes in biofilms, correlations between the number of quorum-sensing signal molecules (such as *N*-acylhomoserine lactone (AHL)) present, biofilm formation and enzyme production have not been reported (Khajanchi *et al.*, 2009; Marchand *et al.*, 2009b).

Detection of bacterial proteases and lipases

Early detection of bacterial enzymes in raw milk is crucial in preventing the escalation of spoilage defects during storage, when the commercial investment (e.g. distribution) and potential losses (product recall) are at their greatest. For example, the recall of 11 Blue Slim Line Brick Packs of Pura (a brand name of National Foods) was required in Australia in 2000 when UHT dairy products started to curdle and produce off-odours and gas, which resulted in swelling of the cartons (http://www.recalls.gov.au/content/index.phtml/itemId/955901, last accessed 12 March 2015).

Several methods are available for the detection of bacterial enzymes, including general detection using electrophoresis, high-performance liquid chromatography (HPLC) and spectrophotometric, fluorimetric and immunological methods (Teh *et al.*, 2014). The quantification of proteolysis can be determined by gel electrophoresis, in which the intensity of the clearing of milk protein and/or large polypeptides is measured (Chove *et al.*, 2011). Electrophoresis has also been used to identify the molecular weight of bacterial protease, and the proteolysis of the substrates (Recio *et al.*, 1997; Marchand *et al.*, 2009b). For example, the byproducts of the proteolysis of milk proteins such as γ1-, γ2- and γ3- caseins can be detected using electrophoresis (Recio *et al.*, 1997). However, the limitations of this method are the requirement for the use of hazardous chemicals, the poor resolution of low–molecular-weight peptides and the long time it takes to produce a result (Chove *et al.*, 2011).

HPLC is regarded as a simple, reproducible, accurate and sensitive method by which to detect proteolysis. It has been used to differentiate proteolysis by plasmin and bacterial protease in UHT milk (Chen *et al.*, 2003; Chove *et al.*, 2011). HPLC can also be used to quantify the hydrolytic products of lipolysis, such as FFAs and mono- and diglycerides, during the incubation of lipase with an ester substrate; however, gas chromatography (GC) is generally preferred over HPLC as GC is more sensitive. GC may be used to separate and quantify the hydrolytic products of lipase (Louwrier *et al.*, 1996; Patel *et al.*, 1996). The advantage of HPLC over GC analysis on lipolysis is that GC analysis requires fatty acids (FAs) to be derivaitised before chromatographic separation (Thomason *et al.*, 1999). Sample preparation for HPLC is simple, requiring incubation of the lipase with a substrate emulsion. A chloroform–methanol mixture is then used to stop the reaction and to extract the reaction products. Normal phase separation using silica columns enables FFAs from mixed triacylglycerols to be eluted in one peak, which adds to the sensitivity of the method. By modifying the analytical column conditions, the substrates and products of the lipase reaction can be monitored. However, the cost of the equipment and the difficulty of finding suitable standards for quantification limit its use for routine testing (Chen *et al.*, 2003; Chove *et al.*, 2011).

Spectrophotometric and fluorimetric methods have been used to measure proteolysis using modified substrates such as synthetic chromogenic (azocaseins) and fluorogenic substrates (fluorescein-thiocarbamoyl-β-casein) (Recio *et al.*, 1997; Chen *et al.*, 2003,). Fluorescamine has been used to quantify the number of free peptides produced by proteolysis: the fluorogenic compound reacts with the free peptides to form a highly fluorescent product (Le *et al.*, 2006). The advantage of fluorescamine is that it is simple, rapid and sensitive to low levels of protease (Chove *et al.*, 2011), while the acyl esters of the fluorescent compound 4-methylumbelliferone (4-MU) can be used as a substrate in detecting lipolysis in skim milk, skim milk powder, whey powder and whey protein concentrate (Fitzgerald & Deeth, 1983). With this assay, the activity is expressed as the amount of 4-MU released per unit time, where the increased fluorescence indicates lipolysis. Immunological methods such as the enzyme-linked immunosorbent assay (ELISA) are very sensitive; however, ELISA may overestimate the amount of active enzyme, as it cannot differentiate between active and inactive enzymes. It has been suggested that a combination of ELISA and spectrophotometric assays may be suitable for use in quality control during processing (Chen *et al.*, 2003) and in detecting the early stages of spoilage in dairy products.

Spore-forming bacteria

Historically, the spore-forming bacteria responsible for causing spoilage of dairy products belonged to the genus *Bacillus*, which comprised Gram-positive and Gram-variable, aerobic, rod-shaped bacteria that produce heat-resistant spores. The vegetative cells of *Bacillus* species were 0.5×1.2 to $2.5 \times 10\,\mu m$, and occured singly or in chains (Schoeni & Wong, 2005). The genus was also very diverse and included psychrotrophic, mesophilic and thermophilic species. Developments in microbial taxonomy led to a revision of the genus *Bacillus* and the creation of a large number of new genera, which include *Geobacillus, Anoxybacillus* and *Paenibacillus* (Ash *et al.*, 1991, 1993; Pikuta *et al.*, 2000; Nazina *et al.*, 2001). Many of the spore-forming bacteria of concern to the dairy industry have remained within the genus *Bacillus*, including *B. cereus* and *B. licheniformis*. Others, such as *G. stearothermophilus* and *A. flavithermus*, have been transferred to new genera.

Spore-forming bacteria are ubiquitous in the farm environment and can be isolated from a wide variety of materials, including feed, bedding materials, manure, silage, soils and milking shed wash water, all of which come into contact with the teats of cows (te Giffel *et al.*, 2002; Howard, 2006; Magnusson *et al.*, 2006, 2007; Huck *et al.*, 2008). In fact, the teats appear to be one of the primary routes by which bacteria, and in particular spores, enter raw milk. There is a large diversity of bacterial spores, which can survive heating at 100 °C for 30 minutes and germinate when conditions are ideal (Scheldeman *et al.*, 2005). These spoilage bacteria can contaminate milk in both vegetative and spore state.

Teats that have been in contact with soil are likely to be contaminated with more bacterial spores than are teats that have come into contact with faecal or bedding materials (Vissers *et al.*, 2007b). The transmission of relatively small amounts of contaminated soils (1–13 mg/l) into raw milk can increase the concentration of bacterial spores to more than $3\log_{10}$ spores/l

(Vissers *et al.*, 2007a). High concentrations of bacterial spores in bedding materials during housing can also lead to increased concentrations of spores in milk. Feeding of cows with silage, which can contain greater than $5 \log_{10}$ spores/g, can lead to concentrations of greater than $4 \log_{10}$ spores/g in faecal materials and increases the risk of milk being contaminated with bacterial spores (Magnusson *et al.*, 2007).

Contamination of raw milk by spore-forming bacteria usually demonstrates a seasonal influence. In the northern hemisphere, where cows are often housed over the winter months, there can be large seasonal variations in the types and levels of spore-forming bacteria in raw milk. For example, cows that are housed indoors have a lower prevalence of contamination by spore-forming bacteria than cows on pasture (Slaghuis *et al.*, 1997). In addition, *B. cereus*, which is associated with soil, is usually found in raw milk during summer, while *B. licheniformis*, which is associated with bedding material, is usually found in winter (Crielly *et al.*, 1994; Sutherland & Murdoch, 1994; Svensson *et al.*, 1999, 2004).

A very important property of bacterial spores is their ability to survive many of the heat treatments, such as pasteurisation, that are employed during manufacturing of dairy products. One example of spoilage of dairy products caused by *Bacillus* and related genera is flat sour spoilage of evaporated milk, caused by acid production during fermentation of carbohydrates (Kalogridou-vassiliadou, 1992). Spores can germinate during the storage of pasteurised and UHT-treated products and cause off-flavours or curdling of milk (Ranieri *et al.*, 2009; De Jonghe *et al.*, 2010). Spore-forming bacteria that contaminate dairy products may originate from raw milk, but it is believed they also originate from post-pasteurisation contamination and from growth within the manufacturing process (Svensson *et al.*, 1999, 2004, 2006; Banyko & Vyletelova, 2009).

Psychrotrophic spore-forming bacilli belonging to the genus *Paenibacillus* have recently been identified as a cause of spoilage of pasteurised milk (Figure 5.2). The spores of these bacteria survive pasteurisation and, in the absence of Gram-negative post-pasteurisation contaminants, such as pseudomonads, can germinate and grow in pasteurised milk at refrigeration temperatures (Ranieri & Boor, 2010). A recent study showed that the spore-forming population in pasteurised milk shifts from one dominated by *Bacillus* to one dominated by *Paenibacillus* after 10 days of refrigerated storage (Martin *et al.*, 2011). Strains of *Paenibacillus* can be differentiated from *Bacillus* by phenotypic characteristics, such as the ability to grow at 6 °C and produce β-galactosidase, and using molecular methods, such as real-time polymerase chain reaction (RT-PCR) (Ivy *et al.*, 2012; Ranieri *et al.*, 2012). Paenibacilli are present in very low numbers in raw milk, but their numbers increase during prolonged storage of pasteurised milk. This may be the result of germination of spores and the ability of spores to grow at low temperatures. They have been isolated from faecal materials, raw milk and pasteurised dairy products (Scheldeman *et al.*, 2004; Velaquez *et al.*, 2004; Coorevits *et al.*, 2008).

It has been suggested that the source of paenibacilli in dairy products is the raw milk and that contamination can occur at any of a number of points along the production chain, from milking on the dairy farm to transportation of raw milk to the processing plant (Huck *et al.*, 2007a). Furthermore, paenibacilli can also be found in processing plants, which can result in pre- or post-pasteurisation of dairy products (Huck *et al.*, 2007b). Certain species of *Paenibacillus* have the ability to produce exopolysaccharides and to form biofilms (Timmusk *et al.*, 2005; Aguilera *et al.*, 2008), which may increase their persistence in the dairy farm and milking premises, as some strains of *Paenibacillus* are known to persist for extended periods

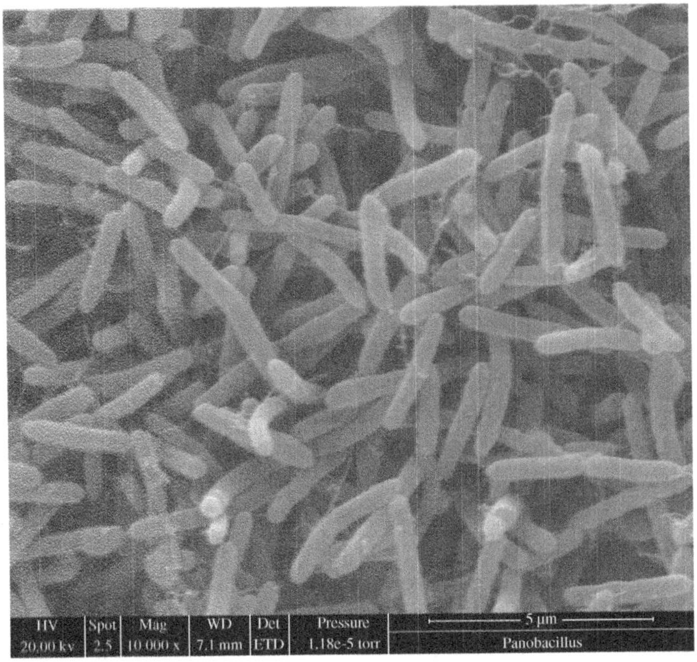

Figure 5.2 Scanning electron microscopy (SEM) of *Paenibacillus* species.

of time in dairy manufacturing plants (Durak *et al.*, 2006). Understanding the effect of *Paenibacillus* species on raw milk quality and their ability to form biofilms on dairy farms may potentially reduce the source of contamination, which might contribute to the overall quality of dairy products.

5.5.2 *Foodborne pathogens*

Pathogenic bacteria found in raw milk that have been responsible for food poisoning outbreaks include *Campylobacter jejuni*, *Salmonella* species, *Listeria monocytogenes*, *Escherichia coli* and *S. aureus* (Oliver *et al.*, 2009; Claeys *et al.*, 2013).

Campylobacter

Campylobacteriosis is caused by the consumption of food products contaminated by either *C. jejuni* or *C. coli.* It is usually associated with the consumption of contaminated poultry, but *C. jejuni/coli* have been found in raw milk, inadequately pasteurised milk and cheese (Hussain *et al.*, 2007). At least 39 campylobacteriosis outbreaks associated with the consumption of raw milk were reported worldwide between 1970 and 2010 (Claeys *et al.*, 2013). In the United States, most reported cases of campylobacteriosis are due to the consumption of raw milk (Taylor *et al.*, 2013).

The number of *Campylobacter* in dairy cattle faeces has been reported to increase during spring and autumn, which suggests that dairy-related campylobacteriosis is more prevalent during these seasons (Stanley & Jones, 2003; Taylor *et al.*, 2013). Even though only a small proportion of a dairy herd may be shedding high numbers of *C. jejuni* (>10^5 CFU/g), cross-contamination can occur through transmission via their hides, water troughs or grazing pasture. This may result in the contamination of raw milk during milking, as *Campylobacter* species have been isolated from in-line milk filters (Stanley & Jones, 2003; Serraino *et al.*, 2013).

Although *C. jejuni* is a microaerophilic microorganism, it has been found in the dairy environment, which suggests that *C. jejuni* can survive in an aerobic environment, possibly within biofilms. *C. jejuni* has been reported to form and survive in a mixed-species biofilm (Teh *et al.*, 2010), and its survival is enhanced in pre-established biofilms (Hanning *et al.*, 2008). Studies have shown that isolates of *C. jejuni* obtained from poultry processing plants can form biofilms (Hanning *et al.*, 2008; Kudirkiene *et al.*, 2012). However, no studies have been carried out to determine whether *C. jejuni* strains isolated from dairy environments have such an ability. Furthermore, the effects of the biofilm-forming capability of *C. jejuni* on campylobacteriosis cases associated with the consumption of raw milk have yet to be investigated.

Salmonella

Nontyphoidal salmonellosis is one of the leading foodborne illnesses in England, Wales, Australia and the United States (Scallan *et al.*, 2011). There have been several *Salmonella* outbreaks associated with the consumption of raw milk and cheese made from inadequately pasteurised milk or from raw milk, such as Cotija (a Mexican-style aged cheese), Morbier (a French semi-soft cheese) and Cheddar (De Valk *et al.*, 2000; Mazurek *et al.*, 2004; Austin *et al.*, 2008; Duynhoven *et al.*, 2009).

Salmonella has been found on conventional and organic dairy farms (Fossler *et al.*, 2004; Van Kessel *et al.*, 2011; Jackson *et al.*, 2012). Even with good hygienic practice during milking, it is difficult to eliminate *Salmonella* contamination of raw milk, because of the potential presence of faecal material (Van Kessel *et al.*, 2011). For example, in a 16-herd study from four states in the United States, at least 10% of the faecal material was found to be positive for *Salmonella*, accounting for 56% of the overall dairy herd (Callaway *et al.*, 2005). In another study, the prevalence rate of *Salmonella* detected in faecal material over a 2-year period ranged from 8.4 to 88% of the dairy herd (Van Kessel *et al.*, 2007). Although the cattle frequently shed *Salmonella*, it is difficult to isolate infected cows for treatment as they are asymptomatic. The prevalence of *Salmonella* species in the dairy herds was found to be associated with increased herd size, historical clinical salmonellosis, poor farm management and transmission between farms (Kabagambe *et al.*, 2000; Adhikari *et al.*, 2009). An initial *Salmonella*-free dairy herd may be contaminated with *Salmonella* species by the introduction of new cows brought from infected herds (Nielsen *et al.*, 2007). However, proper farm management practices, such as an initial quarantine of new cows and nutrient management, may reduce the risk of infection (Losinger *et al.*, 1995).

The prevalence of *Salmonella* species in the dairy farm environment may also result from the ability of *Salmonella* strains to attach and form biofilms on stainless steel surfaces and rubber surfaces, which may increase the risk of contamination of raw milk (Steenackers *et al.*, 2012). Most of the salmonellosis associated with the consumption of raw milk is believed to originate from faecal material. However, biofilms associated with the dairy farm environment/

equipment, and which contain strains of *Salmonella*, may also be one of the contributing factors to salmonellosis outbreaks. *Salmonella* species have been found in milk from raw milk storage tanks (Van Kessel *et al.*, 2011), which suggests that *Salmonella* species may proliferate and form a biofilm if a raw milk storage tank is not properly cleaned.

Listeria monocytogenes

Listeria monocytogenes is an important foodborne pathogen that is known to tolerate harsh environmental conditions, such as high salt concentrations (up to 14 %) and low water activity (0.92 a_w), and to grow over a wide range of temperatures (−1.5 to 45 °C) and pH levels (4 to 9) (Lundén *et al.*, 2004). It is responsible for causing listeriosis, which has an average case-fatality rate of 20–30% (Swaminathan & Gerner-Smidt, 2007). Pregnant women, newborn babies, elderly people and immunocompromised people are the most susceptible to listeriosis (McLauchlin *et al.*, 2004). *L. monocytogenes* can be found in raw milk and has been responsible for a number of outbreaks associated with the consumption of soft cheeses, such as Brie (Lundén *et al.*, 2004; Swaminathan & Gerner-Smidt, 2007). The detection rate of *L. monocytogenes* in milk from raw milk storage tanks ranges from 2.8 to 16.0%, with the highest detection rate found for in-line milk filters (Oliver *et al.*, 2009; Santorum *et al.*, 2012). In a dairy study by the National Animal Health Monitoring System, the most common serotypes of *L. monocytogenes* isolated from raw milk storage tanks and in-line milk filters were 1/2a, 1/2b and 4b, and 89% of the *L. monocytogenes* strains were considered to be potential human pathogens (Van Kessel *et al.*, 2011).

Factors that may contribute to contamination of raw milk by *L. monocytogenes* are farm management, feed, herd sizes, geographical locations, seasons, animal housing and milking premises (Husu *et al.*, 1990; Antognoli *et al.*, 2009; Scallan *et al.*, 2011). Feed was found to be a major source of both pathogenic and nonpathogenic species of *Listeria* on four dairy farms of different sizes, with a higher risk of contamination of raw milk associated with a larger herd size (Husu *et al.*, 1990; Antognoli *et al.*, 2009). In addition, poor cow hygiene and dirty milking equipment might increase the contamination of raw milk by *L. monocytogenes*, as surface runoff and yard dust or debris were found to have the highest *L. monocytogenes* content (Husu *et al.*, 1990; Sanaa *et al.*, 1993; Fox *et al.*, 2009). *L. monocytogenes* has also been isolated from milking equipment, which suggests that biofilms on milking equipment may be an important source of *L. monocytogenes* (Latorre *et al.*, 2010). Milk may become contaminated with *L. monocytogenes* during milking, resulting in post-processing contamination of the final dairy product (Waak *et al.*, 2002; Weiler *et al.*, 2013). It is important to detect the source of *L. monocytogenes* in the milking premises in order to minimise the potential route of *L. monocytogenes* contamination in raw milk and the dairy manufacturing plant.

Escherichia coli

Pathogenic strains of *E. coli*, including Shiga toxin-producing *E. coli* (STEC) strains belonging to serotype 0157:H7, have caused sporadic cases and outbreaks of foodborne disease associated with the consumption of raw milk, cheese and yoghurts (Farrokh *et al.*, 2013). STECs produce Shiga toxins, a family of bacteriophage-encoded cytotoxins known to cause diarrhoea, haemorrhagic colitis and haemolytic uremic syndrome (HUS) (Reilly & WHO Consultation Prevention Control, 1998).

Raw milk may be contaminated with pathogenic and nonpathogenic *Escherichia coli*. Various regional and national surveys performed in the United States indicate that from 0.8 to 3.8% of raw milk from raw milk storage tanks is contaminated with STEC (Van Kessel *et al.*, 2011). There are several possible routes for the transmission of *E. coli*, including excretion in the faecal material of cows, bedding material, the dairy farm environment, other animals, feeds and drinking water (LeJeune *et al.*, 2001; Eriksson *et al.*, 2005; Williams *et al.*, 2005; Chase-Topping *et al.*, 2008; Fremaux *et al.*, 2008). Raw milk can also be contaminated with STEC as a result of subclinical mastitis (Lira *et al.*, 2004; Hussein & Sakuma, 2005). For example, at least 20 strains isolated from 2000 milk samples from cows with clinical and subclinical mastitis have been found to be positive for the Shiga toxin-producing gene (Lira *et al.*, 2004). STEC can also contaminate raw milk during milking from dirty teats (Hussein & Sakuma, 2005). STEC is known to form biofilms on stainless steel; this is influenced by exopolysaccharide production, nutrient availability and temperature (Ryu *et al.*, 2004). Therefore, STEC may persist in the milking environment as biofilms.

STEC may also survive in raw milk and raw milk products. STEC can survive at refrigeration temperatures, as it has been shown to have higher resistance to cold stress than nonpathogenic *E. coli* due to the activity of the rpoS gene, which regulates the expression of proteins involved in homeoviscous adaption during cold shock (Vidovic *et al.*, 2011). Furthermore, STEC can survive in cultured or fermented dairy products made from raw milk because of its ability to survive in stressful environments, where it is able to upregulate its stress-response genes (Farrokh *et al.*, 2013). In addition, STEC biofilms can release toxins, depending on environmental conditions (Villegas *et al.*, 2013). Shiga toxin is heat-stable and pasteurisation of milk may not be sufficient to inactive it (Rasooly & Do, 2010). The presence of STEC biofilms on dairy farm or milking premises may be a potential source of toxin.

Staphylococcus aureus

Outbreaks of foodborne disease caused by *S. aureus* have been associated with the consumption of milk and dairy products (Altekruse *et al.*, 1998; De Buyser *et al.*, 2001). The source of these bacteria in milk is likely to be strains of *S. aureus* causing mastitis in cows (Kerouanton *et al.*, 2007; Guimaraes *et al.*, 2013). *S. aureus* can be shed into milk from infected cows in high numbers.

Staphylococcal food poisoning is caused by enterotoxins that are produced during the growth of *S. aureus*. The growth of *S. aureus* and the production of enterotoxin are influenced by several factors, including incubation temperature, pH, water activity, salt concentration and redox potential (Hennekinne *et al.*, 2012). The presence of enterotoxin in milk is normally caused by high counts of *S. aureus* in raw milk, arising from temperature abuse (Le Loir *et al.*, 2003; Guimaraes *et al.*, 2013). Even after milk has been subjected to a proper heat treatment that inactivates vegetative cells of *S. aureus*, the heat-stable enterotoxin may still persist.

At least 21 types of staphylococcal enterotoxin have been identified, with molecular weights ranging from 22 to 29 kDa (Schelin *et al.*, 2011). Staphylococcal enterotoxins all have similar structural and biological properties, and they belong to a group of pyrogenic toxin superantigens

(PTSAgs) encoded on phage, pathogenicity islands, bacterial chromosomes and plasmids (Schelin *et al.*, 2011). Staphylococcal enterotoxins are heat-resistant. Only a small amount (10–20 ng) of enterotoxin is required to cause staphylococcal food poisoning (Asao *et al.*, 2003; Le Loir *et al.*, 2003).

Strains of *S. aureus* have been shown to form biofilms, which increases the ability of these bacteria to survive and persist on surfaces (Götz, 2002; Gutierrez *et al.*, 2012). Enterotoxin production may be induced during the dispersal of cells from staphylococcal biofilms. For example, staphylococcal enterotoxin D (SED) expression was found to increase during the activation of the *agr* system – a quorum-sensing system that is associated with the dispersal of cells from staphylococcal biofilms (Boles & Horswill, 2008; Wallin-Carlquist *et al.*, 2010; Márta *et al.*, 2011). The formation of biofilms by *S aureus* and the associated production of enterotoxins in dairy farm milking systems have not been investigated.

5.5.3 *Beneficial bacteria*

Lactic acid-producing bacteria as starter cultures

The lactic acid-producing bacteria (LAB) are a group of bacteria of benefit to the dairy industry (Quigley *et al.*, 2011). LAB have been widely studied and many species and strains are used as starter cultures for the manufacture of cheese and yoghurt. For example, the development of cheese flavour is influenced by the types of starter culture employed, as well as by the type and composition of milk and the cheese-making conditions (Steele *et al.*, 2013). LAB possess a wide range of hydrolytic enzymes, which hydrolyse milk proteins and peptides to short peptides and amino acids in cheese, and thus contribute to the development of the cheese flavour during ripening (Williams & Banks, 1997; Bouton *et al.*, 1998; Sousa *et al.*, 2001).

Lactococcus lactis subsp. *lactis* and *Lactococcus lactis* subsp. *cremoris*, LAB commonly used as commercial starter cultures, are also commonly found in raw milk (Lafarge *et al.*, 2004; Quigley *et al.*, 2011). A large number of other LAB can also be found in dairy farm environments and in raw milk. *L. lactis* subsp. *lactis* is usually isolated from the general environment, while *L. lactis* subsp. *cremoris* is isolated from the dairy manufacturing environment (Salama *et al.*, 1995; Corrole *et al.*, 1998).

The naturally occurring LAB in raw milk may contribute to desirable flavours found in artisanal cheeses such as Grana Padano and Parmigiano Reggiano (Yu *et al.*, 2007; Neviani *et al.*, 2009). For example, in Grana Padano, an Italian cheese made from raw cow's milk, the natural whey starter cultures have at least 10^7 cells/ml of LAB, which are composed predominately of thermophilic lactobacilli, followed by thermophilic heterofermentative lactobacilli and occasionally *Streptococcus thermophilus* (Yu *et al.*, 2007; Rossetti *et al.*, 2008; Santarelli *et al.*, 2008; Neviani *et al.*, 2009).

Some LAB can form biofilms on artisanal cheese-making equipment, such as aging boards and milk vats. The cheese-making equipment is also known to provide a good source of LAB, which is important for the development of specific characteristics of the cheese (Mariani *et al.*, 2007; Lortal *et al.*, 2009; Didienne *et al.*, 2012; Feligini *et al.*, 2012). For example, artisanal cheese such as Ragusano (a Sicilian cheese) and Salers (a French cheese) is made using traditional methods in which raw milk is curdled in wooden vats known as 'tina' and 'gerle', respectively (Licitra *et al.*, 2007; Lortal *et al.*, 2009; Didienne *et al.*, 2012). These cheeses are manufactured without the use of commercial starter cultures; the naturally occurring bacteria present in the raw milk and on the surfaces of the wooden vats provide natural starter cultures (Licitra *et al.*, 2007). The types of LAB present on these wooden vats vary, with 'gerle' having more diverse strains than 'tina'. The types of LAB present on 'gerle' include *Lactobacillus casei, L. lactis, Leuconostoc pseudomesenteroides, Lactococcus garvieae, Leuconostoc mesenteroides* and *Lactobacillus plantarum*, whereas the dominant species on 'tina' is *S. thermophilus*, followed by *L. lactis*, *Lactobacillus delbrueckii* subsp. *lactis* and *Lactobacillus acidophilus* (Licitra *et al.*, 2007; Lortal *et al.*, 2009; Didienne *et al.*, 2012). Only *L. lactis* is found in both the 'tina' and 'gerle' wooden vats (Lortal *et al.*, 2009; Didienne *et al.*, 2012).

Bacteriocins of LABs

Aside from their use as starter cultures, many strains of LAB are known for the production of bacteriocins (Servin, 2004). Bacteriocins produced by LAB are generally active against a wide range of pathogenic bacteria, and can be used to improve the safety and quality of dairy products (Servin, 2004). Bacteriocins are classified into four main classes and a total of eight subclasses: IA, IB and IC; IIa, IIb and IIc; IIIa and IIIb; and IV (Snyder & Worobo, 2014). Class I bacteriocins are also known as lantibiotics (lanthionine-containing antibiotics). They are small peptides of 19–38 amino acid residues in length, and are further divided into three subclasses: Class IA are flexible, linear peptides; Class IB are rigid, globular peptides; and Class IC are multicomponent lantibiotics (Altena *et al.*, 2000; Cleveland *et al.*, 2001; Cotter *et al.*, 2005; Nes *et al.*, 2007). Class II bacteriocins are known as non-lantibotics or unmodified peptides. They include the pediocin-like, antilisterial bacteriocins (Class IIa), two-peptide bacteriocins (Class IIb) and thiol-containing bacteriocins (Class IIc) (Snyder & Worobo, 2014). Class III bacteriocins are heat-stable and are divided into Class IIIa, bacteriolytic and Class IIIb, non-lytic bacteriocins. Class IV bacteriocins are cyclic post-translationally modified bacteriocins (Heng & Tagg, 2006; Snyder & Worobo, 2014).

Bacteriocins can be added to food directly, indirectly as biopreservatives (purified or semipurified bacteriocins) or indirectly as byproducts of fermentation (Messaoudi *et al.*, 2013). There are six factors that need to be considered before a bacteriocin can be applied to food: (i) 'Generally Recognised as Safe' (GRAS) status; (ii) a broad spectrum or specific inhibition against pathogens; (iii) heat-stability; (iv) absence of health risks; (v) benefits to the food product, such as improved quality, safety and flavour; and (vi) high specific activity (Holzapfel *et al.*, 1995). Bacteriocins have been commercially applied as a biopreservative in ricotta-type cheese to control foodborne pathogens such as *L. monocytogenes* (Davies *et al.*, 1997).

Bacteriocins are capable of reducing the number of attached bacteria during the early stages of attachment and biofilm formation (Minei *et al.*, 2008; Winkelstroter *et al.*, 2011). For example, the number of attached *L. monocytogenes* on stainless steel was reduced during the first 6 hours of incubation with either *Lactobacillus sakei* 1 or its bacteriocin (Winkelstroter *et al.*, 2011). Similarly, attachment of *L. monocytogenes* was reduced when it was grown with *Enterococcus faecium* in dual-species biofilms (Minei *et al.*, 2008). However, in the latter case, attachment of *L. monocytogenes* was restored during prolonged incubation, possibly due to the reduced susceptibility of strains or unspecific mechanisms such as nutrient competition and acid production (Alves *et al.*, 2006; Hammami *et al.*, 2009; Winkelstroter *et al.*, 2011).

5.6 Biofilms at dairy farms

5.6.1 *General characteristics of biofilms*

The occurrence of biofilms in the dairy industry may result in economic loss due to low quality/yields, food spoilage or food safety problems and difficulties in cleaning and maintaining hygiene. These biofilms found in dairy farm environments and on equipment used for milking, storage and transportation of raw milk are composed of a variety of bacteria, including pathogenic and spoilage bacteria, all interacting in a microbial community (Sutherland & Murdoch, 1994). The ability of bacteria to survive exposure to hazardous conditions, such as during cleaning and sanitation, is increased when they are present within a biofilm. For example, *S. thermophilus* and *L. monocytogenes* demonstrated a greater resistance to heat and sanitisers in the presence of organic material when grown in a biofilm than when grown in suspension (Frank & Koffi, 1990; Flint *et al.*, 2002). In addition, co-culture biofilms of *P. fluorescens* and *B. cereus* were more resistant than planktonic cells to chlorine dioxide-based sanitisers (Lindsay *et al.*, 2002). The increased resistance was associated with growing on a surface and possibly with a change in the physiology of the cell (Frank & Koffi, 1990; Dhir & Dodd, 1995; Steward *et al.*, 2006). Once biofilms are established, they are very difficult to remove, due to their physicohemical properties (Hood & Zottola, 1997).

Bacterial attachment and biofilm formation can occur at any stage in the production chain, from the dairy farm through to the raw milk silo at the dairy manufacturing plant. Bacteria originating from the farm have been found on surfaces in dairy manufacturing plants (Huck *et al.*, 2008). This may be due to the transfer of bacteria from mature biofilms in the dairy environment into raw milk during milking, and later attachment to the processing line downstream (Flint *et al.*, 1997; Wijman *et al.*, 2007; Latorre *et al.*, 2010).

5.6.2 *Cows*

Cows, and in particular their teats, are a potential source of microbial contamination of raw milk (Bell, 1997). The teats of cows can contain high numbers of bacteria and a highly diverse bacterial population, which varies between farms (Braem *et al.*, 2012; Monsallier *et al.*, 2012; Verdier-Metz *et al.*, 2012). This may be influenced by farm management, including indoor/outdoor feeding, bedding material and hygiene practices (Hagi *et al.*, 2010; Vacheyrou *et al.*, 2011). For example, bacteria that are associated with bedding material can contaminate the teats, which may result in the contamination of raw milk (Vacheyrou *et al.*, 2011). However, not all bacteria found attached to teats contaminate the raw milk. For example, bacteria belonging to genera such as *Solobacterium*, *Clavibacter* and *Arcanobacterium* are found on the teats of cows but do not compete well against other microflora in milk (Verdier-Metz *et al.*, 2012). The bacterial load found on the hides of cows ranges from 4 to 13 $\log_{10}$ CFU/cm^2 (Bell, 1997; Small *et al.*, 2005), while *E. coli* present on the hide can range from 2 to 8 $\log_{10}$ CFU/cm^2 (Bacon *et al.*, 2000). The high bacterial load on cows may contaminate milking premises during milking and subsequently form biofilms in the premises and on equipment.

5.6.3 *Milking equipment and raw milk storage tanks*

Poor hygiene on the farm can lead to a proliferation of bacteria during later stages of the milking process (Villar *et al.*, 1996). The bacteria in raw milk may attach to and grow on the dairy equipment, from where they can be released into the milk (Sharma & Anand, 2002; Shi & Zhu, 2009). Dirty milking equipment and raw milk storage tanks can also facilitate the formation of bacterial biofilms, due to the influence that milk residues have on cell and spore attachment and on bacterial growth (Speers & Gilmour, 1985; Al-Makhlafi *et al.*, 1994; Murphy & Boor, 2000). Materials commonly used in the construction of milking premises and milking equipment, including stainless steel, glass, rubber, polystyrene and glass, have been shown to support cell and spore attachment and biofilm formation (Czechowski, 1990; Mafu *et al.*, 1990; Suárez *et al.*, 1992). Biofilms that develop in milking equipment and raw milk storage tanks can act as a chronic source of microbial contamination for raw milk (Barnes *et al.*, 1999; Latorre *et al.*, 2010). In addition, enzymes produced by bacteria in growing biofilms can contaminate raw milk and can have an impact on the quality of dairy products.

Conditioning of a surface can affect the rate and the extent of bacterial attachment to that surface (Denyer *et al.*, 1993). Gram-negative bacteria attach more readily on to surfaces conditioned with milk residue than do Gram-positive bacteria (Suárez *et al.*, 1992). Even though surface conditioning may reduce the initial microbial attachment, with prolonged incubation, the number of bacterial cells attached can increase through growth. Scratches on the surfaces of materials have been shown to be associated with bacterial attachment (Wirtanen *et al.*, 1995).

Milking equipment may act as a reservoir and entry point for potentially heat-resistant spores entering raw milk after heat treatment, possibly due to the formation of biofilms in areas that are difficult to access for cleaning (Scheldeman *et al.*, 2005). Bacterial spores may have different surface characteristics, which can influence their attachment to the substrate. However, there is no simple relationship between individual physiochemical interactions and adhesion of spores to a surface (Seale *et al.*, 2008). Given sufficient time, both the attached bacterial cells and bacterial spores on milking equipment may proliferate and form biofilms, which can subsequently act as a source of contamination to the milk.

During milking, both the biofilm and metabolites produced within it, such as enzymes, can disperse from the milking equipment into the fluid raw milk and subsequently reduce the quality of the raw milk through enzymatic degradation.

The in-line milk filters attached to milking equipment, which are used to remove particles from raw milk (e.g. soil particles and vegetation), may be another important source of microbial contamination. Foodborne pathogens have been isolated from in-line milk filters, which suggests filters may contribute to contamination of raw milk by pathogenic bacteria (Stanley & Jones, 2003; Van Kessel *et al.*, 2011; Serraino *et al.*, 2013). In-line milk filters must be changed frequently, to minimise clogging and rupture, and should not be reused, as biofilms may form on them. Furthermore, in a recent study, about 50% of the bacterial isolates found on raw milk storage tanks were able to produce spoilage enzymes and form biofilms, which suggests inadequate cleaning of the tanks (Flach *et al.*, 2014). This finding highlights the importance of an effective cleaning regime in raw milk storage tanks in minimising biofilm formation and consequent enzyme production.

Equipment used for milking and storage of raw milk on the farm should be designed to minimise the accumulation of milk deposits and the formation of biofilms when in use. Equipment should also be designed so that it can be effectively cleaned and sanitised, and an effective cleaning and sanitation regime must be established. The efficiency of cleaning and sanitation is influenced by a number of factors, including the type and concentration of cleaning chemicals, water hardness, cleaning temperature and the duration of cleaning applications. For example, a high level of water hardness can hinder the effectiveness of cleaning and sanitation chemicals, as the high concentration of ions in the water can react with the caustic cleaning agents to produce precipitation (Cords *et al.*, 2001). Softening agents are often added to water used for the preparation of cleaning and sanitising solutions (Elmoslemany *et al.*, 2009). Cleaning at a suboptimal temperature (e.g. 47–53 °C) can also reduce the overall efficiency of cleaning processes (Latorre *et al.*, 2010). There are several reasons why cleaning temperatures might be below optimal, such as an incorrect temperature setting or the use of an inefficient heating system (Bava *et al.*, 2011). Finally, bacterial spores have been detected on processing equipment surfaces after cleaning (Flint *et al.*, 1997; Elmoslemany *et al.*, 2009), so removal and inactivation of spores should be considered when designing the cleaning and sanitation regime.

5.6.4 *Raw milk tanker*

Starting with good quality raw milk is very important, particularly when raw milk must be transported over long distances. Biofilms can form on the internal surfaces of raw milk tankers, with bacteria originating in the raw milk collected from farms. If surfaces in milk tankers are not adequately cleaned, bacteria in biofilms that survive cleaning and sanitation can grow, produce enzymes and contaminate subsequent batches of raw milk.

A number of factors will influence biofilm formation on the internal surfaces of milk tankers, including whether the storage tank is single- or double-skinned, whether it has a refrigeration system and how well it has been cleaned and sanitised. It has been shown that the internal surface temperatures of a single-skinned milk tanker during the transportation of raw milk from dairy farm to dairy manufacturing plant are within the ideal range for the proliferation of psychrotrophic bacteria (Teh, 2013). The internal surface temperatures of milk tankers fluctuate during raw milk transportation and are season-dependent. The upper surfaces of milk tankers tend to be the hottest. This suggests that bacteria that come into contact with the upper surfaces of milk tankers (Figure 5.3a), through the splashing of raw milk, are more likely to encounter temperatures suitable for growth and biofilm formation.

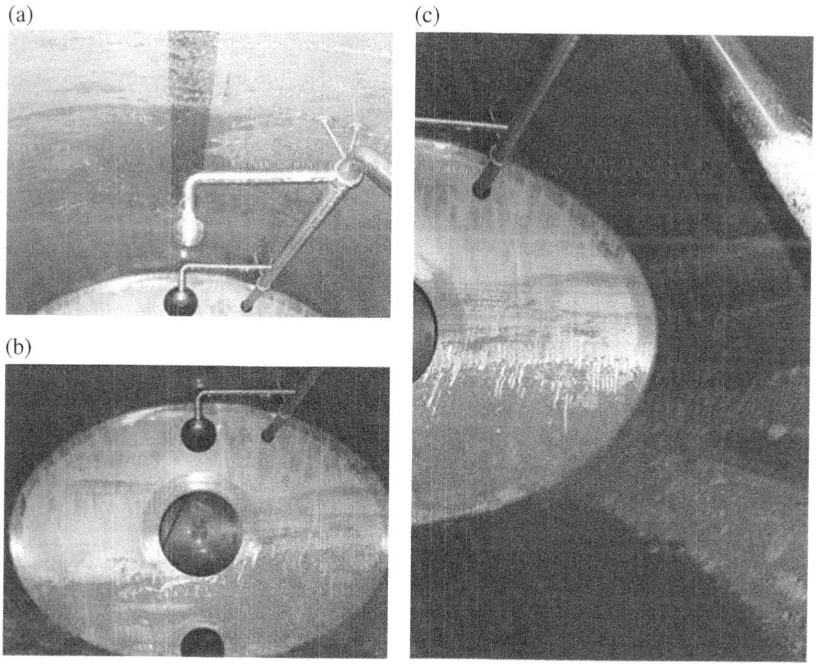

Figure 5.3 Milk residues in a raw milk tanker after raw milk collection: (a) upper part; (b) partition; (c) side.

Furthermore, the interface between air and raw milk along the inside walls of milk tankers (Figure 5.3b,c) appears to be prone to the accumulation of milk deposits and is a potentially important site for biofilm formation.

Bacteria isolated from milk tankers have the ability to form biofilms and can produce either heat-stable proteases, lipases or both (Teh *et al.*, 2011), which can be released into the raw milk. Bacteria that detach from biofilms in milk tankers will contribute to the SPC and may colonise the surfaces of processing equipment at dairy manufacturing plants (Rollet *et al.*, 2009). Most of the bacteria that grow in biofilms in milk tankers will be inactivated by heat treatments applied in dairy processes; however, the heat-stable enzymes may retain their activity and contaminate the end-products. The predominant microbial populations in biofilms that develop in processing equipment will differ to those found in biofilms that develop in raw milk tankers, owing to the change in environmental conditions and microbial interactions (Martiny *et al.*, 2003; Elias & Banin, 2012). While psychrotrophic bacteria will be dominant in biofilms that develop during the cold storage of raw milk, the dominant microorganisms in processing lines will shift to bacteria favoured by the warmer local conditions.

5.7 Conclusion

The safety and quality of raw milk have important economic consequences for dairy manufacturers. The prevalence of biofilms at the dairy farm and in the transport chain may be an unrecognised source of pathogens and spoilage enzymes. Controlling biofilms on the dairy farm and during transportation of raw milk may improve the overall safety and quality of dairy products.

References

Abate, C. M., Castro, G. R., Sineriz, F. & Callieri, D. A. S. 1999. Production of amylolytic enzymes by *Bacillus amyloliquefaciens* in pure culture and in co-culture with *Zymomonas mobilis*. *Biotechnology Letters*, **21**, 249–52.

Adhikari, B., Besser, T. E., Gay, J. M., Fox, L. K., Davis, M. A., Cobbold, R. N., Berge, A. C. B. & Hancock, D. D. 2009. The role of animal movement, including off-farm rearing of heifers, in the interherd transmission of multidrug-resistant *Salmonella*. *Journal of Dairy Science*, **92**, 4229–38.

Aguilera, M., Quesada, M. T., Del Águila, V. G., Morillo, J. A., Rivadeneyra, M. A., Ramos-Cormenzana, A. & Monteoliva-Sánchez, M. 2008. Characterisation of *Paenibacillus jamilae* strains that produce exopolysaccharide during growth on and detoxification of olive mill wastewaters. *Bioresource Technology*, **99**, 5640–4.

Åkerstedt, M., Wredle, E., Lam, V. & Johansson, M. 2012. Protein degradation in bovine milk caused by *Streptococcus agalactiae*. *Journal of Dairy Research*, **79**, 297–303.

Al-Makhlafi, H., Mcguire, J. & Daeschel, M. 1994. Influence of preadsorbed milk proteins on adhesion of *Listeria monocytogenes* to hydrophobic and hydrophilic silica surfaces. *Applied and Environmental Microbiology*, **60**, 3560–5.

Altekruse, S. F., Timbo, B. B., Mowbray, J. C., Bean, N. H. & Potter, M. E. 1998. Cheese-associated outbreaks of human illness in the United States, 1973 to 1992: sanitary manufacturing practices protect consumers. *Journal of Food Protection*, **61**, 1405–7.

Altena, K., Guder, A., Cramer, C. & Bierbaum, G. 2000. Biosynthesis of the lantibiotic mersacidin: Organization of a type B lantibiotic gene cluster. *Applied and Environmental Microbiology*, **66**, 2565–71.

Alves, V. F., Martinez, R. C. R., Lavrador, M. A. S. & De Martinis, E. C. P. 2006. Antilisterial activity of lactic acid bacteria inoculated on cooked ham. *Meat Science*, **74**, 623–7.

Antognoli, M. C., Lombard, J. E., Wagner, B. A., Mccluskey, B. J., Van Kessel, J. S. & Karns, J. S. 2009. Risk factors associated with the presence of viable *Listeria monocytogenes* in bulk tank milk from US Dairies. *Zoonoses and Public Health*, **56**, 77–83.

Asao, T., Kumeda, Y., Kawai, T., Shibata, T., Oda, H., Haruki, K., Nakazawa, H. & Kozaki, S. 2003. An extensive outbreak of staphylococcal food poisoning due to low-fat milk in Japan: estimation of enterotoxin A in the incriminated milk and powdered skim milk. *Epidemiology and Infection*, **130**, 33–40.

Ash, C., Farrow, J. A. E., Wallbanks, S. & Collins, M. D. 1991. Phylogenetic heterogeneity of the genus *Bacillus* revealed by comparative analysis of small-subunit-ribosomal RNA sequences. *Letters in Applied Microbiology*, **13**, 202–6.

Ash, C., Priest, F. G. & Collins, M. D. 1993. Molecular identification of rRNA group 3 bacilli (Ash, Farrow, Wallbanks and Collins) using a PCR probe test. Proposal for the creation of a new genus *Paenibacillus*. *Antonie Van Leeuwenhoek*, **64**, 253–60.

Austin, C., Saathoff-Huber, L., Bordson, M., Dobbins, C., Gross, C., Marishta, K., Carlson, F., Maurice, G. & Trevino, I. C. 2008. Outbreak of multidrug-resistant *Salmonella enterica* serotype Newport infections associated with consumption of unpasteurised Mexican-style aged cheese – Illinois, March 2006-April 2007 (Reprinted from *MMWR*, vol 57, pg 432–435, 2008). *Jama-Journal of the American Medical Association*, **299**, 2850–1.

Bacon, R. T., Belk, K. E., Sofos, J. N., Clayton, R. P., Reagan, J. O. & Smith, G. C. 2000. Microbial populations on animal hides and beef carcasses at different stages of slaughter in plants employing multiple-sequential interventions for decontamination. *Journal of Food Protection*, **63**, 1080–6.

Bagge, N., Hentzer, M., Andersen, J. B., Ciofu, O., Givskov, M. & Hoiby, N. 2004. Dynamics and spatial distribution of beta-lactamase expression in *Pseudomonas aeruginosa* biofilms. *Antimicrobial Agents and Chemotherapy*, **48**, 1168–74.

Banyko, J. & Vyletelova, M. 2009. Determining the source of *Bacillus cereus* and *Bacillus licheniformis* isolated from raw milk, pasteurised milk and yoghurt. *Letters in Applied Microbiology*, **48**, 318–23.

Barbano, D. M., Ma, Y. & Santos, M. V. 2006. Influence of raw milk quality on fluid milk shelf life. *Journal of Dairy Science*, **89**, E15–19.

Barnes, L. M., Lo, M. F., Adams, M. R. & Chamberlain, A. H. L. 1999. Effect of milk proteins on adhesion of bacteria to stainless steel surfaces. *Applied and Environmental Microbiology*, **65**, 4543–8.

Bava, L., Zucali, M., Sandrucci, A., Brasca, M., Vanoni, L., Zanini, L. & Tamburini, A. 2011. Effect of cleaning procedure and hygienic condition of milking equipment on bacterial count of bulk tank milk. *Journal of Dairy Research*, **78**, 211–19.

Bell, R. G. 1997. Distribution and sources of microbial contamination on beef carcasses. *Journal of Applied Microbiology*, **82**, 292–300.

Berry, D. P., O'Brien, B., O'Callaghan, E. J., Sullivan, K. O. & Meaney, W. J. 2006. Temporal trends in bulk tank somatic cell count and total bacterial count in Irish dairy herds during the past decade. *Journal of Dairy Science*, **89**, 4083–93.

Boles, B. R. & Horswill, A. R. 2008. AGR-mediated dispersal of *Staphylococcus aureus* biofilms. *Plos Pathogens*, **4**.

Bouton, Y., Guyot, P. & Grappin, R. 1998. Preliminary characterization of microflora of Comte cheese. *Journal of Applied Microbiology*, **85**, 123–31.

Braem, G., De Vliegher, S., Verbist, B., Heyndrickx, M., Leroy, F. & De Vuyst, L. 2012. Culture-independent exploration of the teat apex microbiota of dairy cows reveals a wide bacterial species diversity. *Veterinary Microbiology*, **157**, 383–90.

Bruhn, J. C. & Franke, A. A. 1991. Raw milk composition and cheese yields in California: 1987 and 1988. *Journal of Dairy Science*, **74**, 1108–14.

Budhani, R. K. & Struthers, J. K. 1998. Interaction of *Streptococcus pneumoniae* and *Moraxella catarrhalis*: Investigation of the indirect pathogenic role of beta-lactamase-producing moraxellae by use of a continuous-culture biofilm system. *Antimicrobial Agents and Chemotherapy*, **42**, 2521–6.

Callaway, T. R., Keen, J. E., Edrington, T. S., Baumgard, L. H., Spicer, L., Fonda, E. S., Griswold, K. E., Overton, T. R., Vanamburgh, M. E., Anderson, R. C., Genovese, K. J., Poole, T. L., Harvey, R. B. & Nisbet, D. J. 2005. Fecal prevalence and diversity of *Salmonella* species in lactating dairy cattle in four states. *Journal of Dairy Science*, **88**, 3603–8.

Celestino, E. L., Iyer, M. & Roginski, H. 1997a. The effects of refrigerated storage of raw milk on the quality of whole milk powder stored for different periods. *International Dairy Journal*, **7**, 119–27.

Celestino, E. L., Iyer, M. & Roginski, H. 1997b. Reconstituted UHT-treated milk: effects of raw milk, powder quality and storage conditions of UHT milk on its physico-chemical attributes and flavour. *International Dairy Journal*, **7**, 129–40.

Chase-Topping, M., Gally, D., Low, C., Matthews, L. & Woolhouse, M. 2008. Super-shedding and the link between human infection and livestock carriage of *Escherichia coli* O157. *Nature Reviews Microbiology*, **6**, 904–12.

Chen, L., Daniel, R. M. & Coolbear, T. 2003. Detection and impact of protease and lipase activities in milk and milk powders. *International Dairy Journal*, **13**, 255–75.

Chopra, A. K. & Mathur, D. K. 1985. Purification and characterization of heat-stable proteases from *Bacillus stearothermophilus* RM–67. *Journal of Dairy Science*, **68**, 3202–11.

Chove, L. M., Grandison, A. S. & Lewis, M. J. 2011. Comparison of methods for analysis of proteolysis by plasmin in milk. *Journal of Dairy Research*, **78**, 184–90.

Claeys, W. L., Cardoen, S., Daube, G., De Block, J., Dewettinck, K., Dierick, K., De Zutter, L., Huyghebaert, A., Imberechts, H., Thiange, P., Vandenplas, Y. & Herman, L. 2013. Raw or heated cow milk consumption: Review of risks and benefits. *Food Control*, **31**, 251–62.

Cleveland, J., Montville, T. J., Nes, I. F. & Chikindas, M. L. 2001. Bacteriocins: safe, natural antimicrobials for food preservation. *International Journal of Food Microbiology*, **71**, 1–20.

Coorevits, A., De Jonghe, V., Vandroemme, J., Reekmans, R., Heyrman, J., Messens, W., De Vos, P. & Heyndrickx, M. 2008. Comparative analysis of the diversity of aerobic spore-forming bacteria in raw milk from organic and conventional dairy farms. *Systematic and Applied Microbiology*, **31**, 126–40.

Cords, B. R., Dychdala, G. R. & Richter, F. 2001. Cleaning and Sanizting in Milk Production and Processing. Marcel Dekker Inc, New York, NY.

Corroler, D., Mangin. I., Desmasures, N. & Gueguen, M. 1998. An ecological study of Lactococci isolated from raw milk in the Camembert cheese registered Designation of Origin area. *Applied and Environmental Microbiology*, **64**, 4729–35.

Cotter, P. D., Hill, C. & Ross, R. P. 2005. Bacteriocins: developing innate immunity for food. *Nature Reviews Microbiology*, **3**, 777–88.

Crielly, E. M., Logan, N. A. & Anderton, A. 1994. Studies on the *Bacillus* flora of milk and milk products. *Journal of Applied Bacteriology*, **77**, 256–63.

Czechowski, M. H. 1990. Bacterial attachment to Buna-N gaskets in milk processing equipment. *Australian Journal of Dairy Technology*, **45**, 113–14.

Davies, E. A., Bevis, H. E. & Delvesbroughton, J. 1997. The use of the bacteriocin, nisin, as a preservative in ricotta-type cheeses to control the food-borne pathogen *Listeria monocytogenes*. *Letters in Applied Microbiology*, **24**, 343–6.

De Buyser, M. L., Dufour, B., Maire, M. & Lafarge, V. 2001. Implication of milk and milk products in food-borne diseases in France and in different industrialised countries. *International Journal of Food Microbiology*, **67**, 1–17.

De Jonghe, V., Coorevits, A., De Block, J., Van Coillie, E., Grijspeerdt, K., Herman, L., De Vos, P. & Heyndrickx, M. 2010. Toxinogenic and spoilage potential of aerobic spore-formers isolated from raw milk. *International Journal of Food Microbiology*, **136**, 318–25.

De Valk, H., Delarocque-Astagneau, E., Colomb, G., Ple, S., Godard, E., Vaillant, V., Haeghebaert, S., Bouvet, P., Grimont, F., Grimont, P. & Desenclos, J. C. 2000. A community-wide outbreak of *Salmonella enterica* serotype *Typhimurium* infection associated with eating a raw milk soft cheese in France. *Epidemiology and Infection*, **124**, 1–7.

Denyer, S. P., Hanlon, H. W. & Davies, M. C. 1993. Mechanisms of microbial adherence. In: Denyer, S. P., Sussman, M. & Gorman, S. P. (eds) Microbial Biofilms: Formation and Control. John Wiley & Sons, Chichester.

Dhir, V. K. & Dodd, C. E. R. 1995. Susceptibility of suspended and surface attached *Salmonella enteritidis* to biocides and elevated temperatures. *Applied and Environmental Microbiology*, **61**, 1731–8.

Didienne, R., Defargues, C., Callon, C., Meylheuc, T., Hulin, S. & Montel, M.-C. 2012. Characteristics of microbial biofilm on wooden vats ('gerles') in PDO Salers cheese. *International Journal of Food Microbiology*, **156**, 91–101.

Durak, M. Z., Fromm, H. I., Huck, J. R., Zadoks, R. N. & Boor, K. J. 2006. Development of molecular typing methods for *Bacillus* spp. and *Paenibacillus* spp. isolated from fluid milk products. *Journal of Food Science*, **71**, M50–6.

Duynhoven, Y. T. H. P. V., Isken, L. D., Borgen, K., Besselse, M., Soethoudt, K., Haitsma, O., Mulder, B., Notermans, D. W., Jonge, R. D., Kock, P., Pelt, W. V., Stenvers, O. & Steenbergen, J. V. 2009. A prolonged outbreak of *Salmonella* typhimurium infection related to an uncommon vehicle: hard cheese made from raw milk. *Epidemiology and Infection*, **137**, 1548–57.

Elias, S. & Banin, E. 2012. Multi-species biofilms: living with friendly neighbors. *Fems Microbiology Reviews*, **36**, 990–1004.

Elmoslemany, A. M., Keefe, G. P., Dohoo, I. R. & Jayarao, B. M. 2009. Risk factors for bacteriological quality of bulk tank milk in Prince Edward Island dairy herds. Part 1: Overall risk factors. *Journal of Dairy Science*, **92**, 2634–43.

Eneroth, A., Christiansson, A., Brendehaug, J. & Molin, G. 1998. Critical contamination sites in the production line of pasteurised milk, with reference to the psychrotrophic spoilage flora. *International Dairy Journal*, **8**, 829–34.

Eriksson, E., Aspan, A., Gunnarsson, A. & Vagsholm, I. 2005. Prevalence of verotoxin-producing *Escherichia coli* (VTEC) O157 in Swedish dairy herds. *Epidemiology and Infection*, **133**, 349–58.

EU. 2013. Consolidated version of Regulation (EC) No 853/2004.

Fairbairn, D. J. & Law, B. A. 1986. Proteinases of psychrotrophic bacteria: their production, properties, effects and control. *Journal of Dairy Research*, **53**, 139–77.

Farrokh, C., Jordan, K., Auvray, F., Glass, K., Oppegaard, H., Raynaud, S., Thevenot, D., Condron, R., De Reu, K., Govaris, A., Heggum, K., Heyndrickx, M., Hummerjohann, J., Lindsay, D., Miszczycha, S., Moussiegt, S., Verstraete, K. & Cerf, O. 2013. Review of shiga-toxin-producing *Escherichia coli* (STEC) and their significance in dairy production. *International Journal of Food Microbiology*, **162**, 190–212.

Feligini, M., Panelli, S., Buffoni, J. N., Bonacina, C., Andrighetto, C. & Lombardi, A. 2012. Identification of microbiota present on the surface of Taleggio cheese using PCR-DGGE and RAPD-PCR. *Journal of Food Science*, **77**, M609–15.

Fitzgerald, C. H. & Deeth, H. C. 1983. Factors influencing lipolysis by skim milk cultures of some psychrotrophic microogranisms. *Australian Journal of Dairy Technology*, **38**, 97–103.

Flach, J., Grzybowski, V., Toniazzo, G. & Corçã. G. 2014. Adhesion and production of degrading enzymes by bacteria isolated from biofilms in raw milk cooling tanks. *Food Science and Technology (Campinas)*, **34**, 571–6.

Flint, S. H., Bremer, P. J. & Brooks, J. D. 1997. Biofilms in dairy manufacturing plant – description, current concerns and methods of control. *Biofouling*, **11**, 81–97.

Flint, S., Brooks, J., Bremer, P., Walker, K. & Hausman, E. 2002. The resistance to heat of thermo-resistant *streptococci* attached to stainless steel in the presence of milk. *Journal of Industrial Microbiology & Biotechnology*, **28**, 134–6.

Fossler, C. P., Wells, S. J., Kaneene, J. B., Ruegg, P. L., Warnick, L. D., Bender, J. B., Godden, S. M., Halbert, L. W., Campbell, A. M. & Zwald, A. M. G. 2004. Prevalence of *Salmonella* spp on conventional and organic dairy farms. *Javma-Journal of the American Veterinary Medical Association*, **225**, 567–73.

Fox, E., O'Mahony, T., Clancy, M., Dempsey, R., O'Brien, M. & Jordan, K. 2009. *Listeria monocytogenes* in the Irish dairy farm environment. *Journal of Food Protection*, **72**, 1450–6.

Frank, J. F. & Koffi, R. A. 1990. Surface – adherent growth of *Listeria monocytogenes* is associated with increased resistance to surfactant sanititzers and heat. *Journal of Food Protection*, **53**, 550–4.

Fremaux, B., Prigent-Combaret, C. & Vernozy-Rozand, C. 2008. Long-term survival of shiga toxin-producing *Escherichia coli* in cattle effluents and environment: an updated review. *Veterinary Microbiology*, **132**, 1–18.

Frølund, B., Griebe, T. & Nielsen, P. H. 1995. Enzymatic activity in the activated sludge floc matrix. *Applied Microbiology and Biotechnology*, **43**, 755–61.

Gamarra, N. N., Villena, G. K. & Gutierrez-Correa, M. 2010. Cellulase production by *Aspergillus niger* in biofilm, solid-state, and submerged fermentations. *Applied Microbiology and Biotechnology*, **87**, 545–51.

Götz, F. 2002. Staphylococcus and biofilms. *Molecular Microbiology*, **43**, 1367–78.

Grufferty, M. B. & Fox, P. F. 1988. Milk alkaline proteinase. *Journal of Dairy Research*, **55**, 609–30.

Guimaraes, F. D. F., Nobrega, D. B., Richini-Pereira, V. B., Marson, P. M., De Figueiredo Pantoja, J. C. & Langoni, H. 2013. Enterotoxin genes in coagulase-negative and coagulase-positive staphylococci isolated from bovine milk. *Journal of Dairy Science*, **96**, 2866–72.

Guinee, T. P., Mulholland, E. O., Kelly, J. & Callaghan, D. J. O. 2007. Effect of protein-to-fat ratio of milk on the composition, manufacturing efficiency, and yield of cheddar cheese. *Journal of Dairy Science*, **90**, 110–23.

Guinotthomas, P., Alammoury, M. & Laurent, F. 1995. Effects of storage conditions on the composition of raw milk. *International Dairy Journal*, **5**, 211–23.

Gutierrez, D., Delgado, S., Vazquez-Sanchez, D., Martinez, B., Lopez Cabo, M., Rodriguez, A., Herrera, J. J. & Garcia, P. 2012. Incidence of *Staphylococcus aureus* and analysis of associated bacterial communities on food industry surfaces. *Applied and Environmental Microbiology*, **78**, 8547–54.

Haddadi, K., Moussaoui, F., Hebia, I., Laurent, F. & Le Roux, Y. 2005. *Escherichia coli* proteolytic activity in milk and casein breakdown. *Reproduction Nutrition Development*, **45**, 485–96.

Hagi, T., Kobayashi, M. & Nomura, M. 2010. Molecular-based analysis of changes in indigenous milk microflora during the grazing period. *Bioscience Biotechnology and Biochemistry*, **74**, 484–7.

Hammami, I., Rhouma, A., Jaouadi, B., Rebai, A. & Nesme, X. 2009. Optimization and biochemical characterization of a bacteriocin from a newly isolated *Bacillus subtilis* strain 14B for biocontrol of *Agrobacterium* spp. strains. *Letters in Applied Microbiology*, **48**, 253–60.

Hanning, I., Jarquin, R. & Slavik, M. 2008. *Campylobacter jejuni* as a secondary colonizer of poultry biofilms. *Journal of Applied Microbiology*, **105**, 1199–208.

Hayes, M. C., Ralyea, R. D., Murphy, S. C., Carey, N. R., Scarlett, J. M. & Boor, K. J. 2001. Identification and characterization of elevated microbial counts in bulk tank raw milk. *Journal of Dairy Science*, **84**, 292–8.

Heng, N. C. K. & Tagg, J. R. 2006. What's in a name? Class distinction for bacteriocins. *Nature Reviews Microbiology*, **4**.

Hennekinne, J.-A., De Buyser, M.-L. & Dragacci, S. 2012. *Staphylococcus aureus* and its food poisoning toxins: characterization and outbreak investigation. *Fems Microbiology Reviews*, **36**, 815–36.

Holm, C., Jepsen, J., Larsen, M. & Jespersen, L. 2004. Predominant microflora of downgraded Danish bulk tank milk. *Journal of Dairy Science*, **87**, 1151–7.

Holzapfel, W. H., Geisen, R. & Schillinger, U. 1995. Biological preservation of foods with reference to protective cultures, bacteriocins and food-grade enzymes. *International Journal of Food Microbiology*, **24**, 343–62.

Hood, S. K. & Zottola, E. A. 1997. Isolation and identification of adherent gram-negative microorganisms from four meat-processing facilities. *Journal of Food Protection*, **60**, 1135–8.

Howard, P. 2006. Mastitis pathogens present in bulk tank milk from seven dairy herds in the Waikato region, New Zealand. *New Zealand Veterinary Journal*, **54**, 41–3.

Huck, J. R., Hammond, B. H., Murphy, S. C., Woodcock, N. H. & Boor, K. J. 2007a. Tracking spore-forming bacterial contaminants in fluid milk-processing systems. *Journal of Dairy Science*, **90**, 4872–83.

Huck, J. R., Woodcock, N. H., Ralyea, R. D. & Boor, K. J. 2007b. Molecular subtyping and characterization of psychrotolerant endospore-forming bacteria in two New York state fluid milk processing systems. *Journal of Food Protection*, **70**, 2354–64.

Huck, J. R., Sonnen, M. & Boor, K. J. 2008. Tracking heat-resistant, cold-thriving fluid milk spoilage bacteria from farm to packaged product. *Journal of Dairy Science*, **91**, 1218–28.

Huppertz, T. 2010. Foaming properties of milk: a review of the influence of composition and processing. *International Journal of Dairy Technology*, **63**, 477–88.

Hussain, I., Mahmood, M. S., Akhtar, M. & Khan, A. 2007. Prevalence of *Campylobacter* species in meat, milk and other food commodities in Pakistan. *Food Microbiology*, **24**, 219–22.

Hussein, H. S. & Sakuma, T. 2005. Shiga toxin-producing *Escherichia coli*: pre- and postharvest control measures to ensure safety of dairy cattle products. *Journal of Food Protection*, **68**, 199–207.

Husu, J. R., Seppanen, J. T., Sivela, S. K. & Rauramaa, A. L. 1990. Contamination of raw milk by *Listeria monocytogenes* on dairy farms. *Zentralblatt fur Veterinarmedizin. Reihe B. Journal of Veterinary Medicine Series* B, **37**, 268–75.

Ivy, R. A., Ranieri, M. L., Martin, N. H., Bakker, H. C. D., Xavier, B. M., Wiedmann, M. & Boor, K. J. 2012. Identification and characterization of psychrotolerant sporeformers associated with fluid milk production and processing. *Applied and Environmental Microbiology*, **78**, 1853–64.

Iwashita, K., Todoroki, K., Kimura, H., Shimoi, H. & Ito, K. 1998. Purification and characterization of extracellular and cell wall bound beta-glucosidases from *Aspergillus kawachii*. *Bioscience Biotechnology and Biochemistry*, **62**, 1938–46.

Iwashita, K., Shimoi, H. & Ito, K. 2001. Extracellular soluble polysaccharide (ESP) from *Aspergillus kawachii* improves the stability of extracellular beta-glucosidases (EX-1 and EX-2) and is involved in their localization. *Journal of Bioscience and Bioengineering*, **91**, 134–40.

Jackson, E. E., Erten, E. S., Neeraj, M., Graham, T. E., Larkin, J. W., Blodgett, R. J., Schlesser, J. E. & Reddy, R. M. 2012. Detection and enumeration of four foodborne pathogens in raw commingled silo milk in the United States. *Journal of Food Protection*, **75**, 1382–93.

Jaspe, A., Palacios, P., Fernandez, L. & Sanjose, C. 2000. Effect of extra aeration on extracellular enzyme activities and ATP concentration of dairy *Pseudomonas fluorescens*. *Letters in Applied Microbiology*, **30**, 244–8.

Jayarao, B. M., Pillai, S. R., Sawant, A. A., Wolfgang, D. R. & Hegde, N. V. 2004. Guidelines for monitoring bulk tank milk somatic cell and bacterial counts. *Journal of Dairy Science*, **87**, 3561–73.

Kabagambe, E. K., Wells, S. J., Garber, L. P., Salman, M. D., Wagner, B. & Fedorka-Cray, P. J. 2000. Risk factors for fecal shedding of *Salmonella* in 91 US dairy herds in 1996. *Preventive Veterinary Medicine*, **43**, 177–94.

Kalogridou-vassiliadou, D. 1992. Biochemical activities of *Bacillus* species isolated from flat sour evaporated milk. *Journal of Dairy Science*, **75**, 2681–6.

Kerouanton, A., Hennekinne, J. A., Letertre, C., Petit, L., Chesneau, O., Brisabois, A. & De Buyser, M. L. 2007. Characterization of *Staphylococcus aureus* strains associated with food poisoning outbreaks in France. *International Journal of Food Microbiology*, **115**, 369–75.

Khajanchi, B. K., Sha, J., Kozlova, E. V., Erova, T. E., Suarez, G., Sierra, J. C., Popov, V. L., Horneman, A. J. & Chopra, A. K. 2009. N-Acylhomoserine lactones involved in quorum sensing control the type VI secretion system, biofilm formation, protease production, and *in vivo* virulence in a clinical isolate of *Aeromonas hydrophila*. *Microbiology-SGM*, **155**, 3518–31.

Khiyami, M. A., Pometto, A. L. & Kennedy, W. J. 2006. Ligninolytic enzyme production by *Phanerochaete chrysosporium* in plastic composite support biofilm stirred tank bioreactors. *Journal of Agricultural and Food Chemistry*, **54**, 1693–8.

Kudirkiene, E., Cohn, M. T., Stabler, R. A., Strong, P. C. R., Serniene, L., Wren, B. W., Nielsen, E. M., Malakauskas, M. & Brondsted, L. 2012. Phenotypic and genotypic characterizations of *Campylobacter jejuni* isolated from the broiler meat production process. *Current Microbiology*, **65**, 398–406.

Lafarge, V., Ogier, J. C., Girard, V., Maladen, V., Leveau, J. Y., Gruss, A. & Delacroix-Buchet, A. 2004. Raw cow milk bacterial population shifts attributable to refrigeration. *Applied and Environmental Microbiology*, **70**, 5644–50.

Latorre, A. A., Van Kessel, J. S., Karns, J. S., Zurakowski, M. J., Pradhan, A. K., Boor, K. J., Jayarao, B. M., Houser, B. A., Daugherty, C. S. & Schukken, Y. H. 2010. Biofilm in milking equipment on a dairy farm as a potential source of bulk tank milk contamination with *Listeria monocytogenes*. *Journal of Dairy Science*, **93**, 2792–802.

Le Loir, Y., Baron, F. & Gautier, M. 2003. *Staphylococcus aureus* and food poisoning. *Genetics and Molecular Research: GMR*, **2**, 63–76.

Le, T. X., Datta, N. & Deeth, H. C. 2006. Sensitive HPLC method for measuring bacterial proteolysis and proteinase activity in UHT milk. *Food Research International*, **39**, 823–30.

Ledenbach, L. H. & Marshal, R. T. 2010. Microbiological Spoilage of Dairy Products. Springer, New York, NY.

LeJeune, J. T., Besser, T. E. & Hancock, D. D. 2001. Cattle water troughs as reservoirs of *Escherichia coli* O157. *Applied and Environmental Microbiology*, **67**, 3053–7.

Li, X. Z., Webb, J. S., Kjelleberg, S. & Rosche, B. 2006. Enhanced benzaldehyde tolerance in *Zymomonas mobilis* biofilms and the potential of biofilm applications in fine-chemical production. *Applied and Environmental Microbiology*, **72**, 1639–44.

Licitra, G., Ogier, J. C., Parayre, S., Pediliggieri, C., Carnemolla, T. M., Falentin, H., Madec, M. N., Carpino, S. & Lortal, S. 2007. Variability of bacterial biofilms of the 'tina' wood vats used in the ragusano cheese-making process. *Applied and Environmental Microbiology*, **73**, 6980–7.

Lindsay, D., Brozel, V. S., Mostert, J. F. & Von Holy, A. 2002. Differential efficacy of a chlorine dioxide-containing sanitizer against single species and binary biofilms of a dairy-associated *Bacillus cereus* and a *Pseudomonas fluorescens* isolate. *Journal of Applied Microbiology*, **92**, 352–61.

Lira, W. M., Macedo, C. & Marin, J. M. 2004. The incidence of shiga toxin-producing *Escherichia coli* in cattle with mastitis in Brazil. *Journal of Applied Microbiology*, **97**, 861–6.

Liu, M., Wang, H. & Griffiths, M. W. 2007. Regulation of alkaline metalloprotease promoter by N-acyl homoserine lactone quorum sensing in *Pseudomonas fluorescens*. *Journal of Applied Microbiology*, **103**, 2174–84.

Lortal, S., Di Blasi, A., Madec, M.-N., Pediliggieri, C., Tuminello, L., Tanguy, G., Fauquant, J., Lecuona, Y., Campo, P., Carpino, S. & Licitra, G. 2009. Tina wooden vat biofilm: a safe and highly efficient lactic acid bacteria delivering system in PDO Ragusano cheese making. *International Journal of Food Microbiology*, **132**, 1–8.

Losinger, W. C., Wells, S. J., Garber, L. P., Hurd, H. S. & Thomas, L. A. 1995. Management factors related to *Salmonella* shedding by dairy heifers. *Journal of Dairy Science*, **78**, 2464–72.

Louwrier, A., Drtina, G. J. & Klibanov, A. M. 1996. On the issue of interfacial activation of lipase in nonaqueous media. *Biotechnology and Bioengineering*, **50**, 1–5.

Lundén, J., Tolvanen, R. & Korkeala, H. 2004. Human listeriosis outbreaks linked to dairy products in Europe. *Journal of dairy science*, **87**, E6–12.

MAF. 2005. Animal Products (Dairy) Regulations 2005. Ministry of Agriculture and Forestry, New Zealand.

MAF. 2011. Animal Products (Dairy Processing Specifications) Notice 2011. Ministry of Agriculture and Forestry, New Zealand.

Mafu, A. A., Roy, D., Goulet, J. & Magny, P. 1990. Attachment of *Listeria monocytogenes* to stainless steel, glass, polypropylene, and rubber surfaces after short contact times. *Journal of Food Protection*, **53**, 742–6.

Magnusson, M., Christiansson, A., Svensson, B. & Kolstrup, C. 2006. Effect of different premilking manual teat-cleaning methods on bacterial spores in milk. *Journal of Dairy Science*, **89**, 3866–75.

Magnusson, M., Christiansson, A. & Svensson, B. 2007. *Bacillus cereus* spores during housing of dairy cows: factors affecting contamination of raw milk. *Journal of Dairy Science*, **90**, 2745–54.

Marchand, S., Heylen, K., Messens, W., Coudijzer, K., De Vos, P., Dewettinck, K., Herman, L., De Block, J. & Heyndrickx, M. 2009a. Seasonal influence on heat-resistant proteolytic capacity of *Pseudomonas lundensis* and *Pseudomonas fragi*, predominant milk spoilers isolated from Belgian raw milk samples. *Environmental Microbiology*, **11**, 467–82.

Marchand, S., Vandriesche, G., Coorevits, A., Coudijzer, K., De Jonghe, V., Dewettinck, K., De Vos, P., Devreese, B., Heyndrickx, M. & De Block, J. 2009b. Heterogeneity of heat-resistant proteases from milk *Pseudomonas* species. *International Journal of Food Microbiology*, **133**, 68–77.

Mariani, C., Briandet, R., Chamba, J. F., Notz, E., Carnet-Pantiez, A., Eyoug, R. N. & Oulahal, N. 2007. Biofilm ecology of wooden shelves used in ripening the French raw milk smear cheese Reblochon de Savoie. *Journal of Dairy Science*, **90**, 1653–61.

Márta, D., Wallin-Carlquist, N., Schelin, J., Borch, E. & Radstrom, P. 2011. Extended staphylococcal enterotoxin D expression in ham products. *Food Microbiology*, **28**, 617–20.

Martin, N. H., Ranieri, M. L., Murphy, S. C., Ralyea, R. D., Wiedmann, M. & Boor, K. J. 2011. Results from raw milk microbiological tests do not predict the shelf-life performance of commercially pasteurised fluid milk. *Journal of Dairy Science*, **94**, 1211–22.

Martiny, A. C., Jorgensen, T. M., Albrechtsen, H. J., Arvin, E. & Molin, S. 2003. Long-term succession of structure and diversity of a biofilm formed in a model drinking water distribution system. *Applied and Environmental Microbiology*, **69**, 6899–907.

Mazurek, J., Salehi, E., Propes, D., Holt, J., Bannerman, T., Nicholson, L. M., Bundesen, M., Duffy, R. & Moolenaar, R. L. 2004. A multistate outbreak of *Salmonella enterica* serotype *typhimurium* infection linked to raw milk consumption – Ohio, 2003. *Journal of Food Protection*, **67**, 2165–70.

McLauchlin, J., Mitchell, R. T., Smerdon, W. J. & Jewell, K. 2004. *Listeria monocytogenes* and listeriosis: a review of hazard characterisation for use in microbiological risk assessment of foods. *International Journal of Food Microbiology*, **92**, 15–33.

Messaoudi, S., Manai, M., Kergourlay, G., Prevost, H., Connil, N., Chobert, J. M. & Dousset, X. 2013. *Lactobacillus salivarius*: bacteriocin and probiotic activity. *Food Microbiology*, **36**, 296–304.

Minei, C. C., Gomes, B. C., Ratti, R. P., D'Angelis, C. E. M. & De Martinis, E. C. P. 2008. Influence of peroxyacetic acid and nisin and coculture with *Enterococcus faecium* on *Listeria monocytogenes* biofilm formation. *Journal of Food Protection*, **71**, 634–8.

MPI. 2014. Animal Products Act 1999. Ministry for Primary Industries, New Zealand.

Monsallier, F., Verdier-Metz, I., Agabriel, C., Martin, B. & Montel, M.-C. 2012. Variability of microbial teat skin flora in relation to farming practices and individual dairy cow characteristics. *Dairy Science & Technology*, **92**, 265–78.

Murphy, S. C. & Boor, K. J. 2000. Trouble-shooting sources and causes of high bacteria counts in raw milk. *Dairy, Food and Environmental Sanitation*, **20**, 606–11.

Nazina, T. N., Tourova, T. P., Poltaraus, A. B., Novikova, E. V., Grigoryan, A. A., Ivanova, A. E., Lysenko, A. M., Petrunyaka, V. V., Osipov, G. A., Belyaev, S. S. & Ivanov, M. V. 2001. Taxonomic study of aerobic thermophilic bacilli: descriptions of *Geobacillus subterraneus* gen. nov., sp. nov. and *Geobacillus uzenensis* sp. nov. from petroleum reservoirs and transfer of *Bacillus stearothermophilus Bacillus thermocatenulatus, Bacillus thermoleovorans, Bacillus kaustophilus, Bacillus thermoglucosidasius and Bacillus thermodenitrificans to Geobacillus* as the new combinations *G. stearothermophilus, G. thermocatenulatus, G. thermoleovorans, G. kaustophilus, G. thermoglucosidasius* and *G. thermodenitrificans. International Journal of Systematic and Evolutionary Microbiology*, **51**, 433–46.

Nes, I. F., Diep, D. B. & Holo, H. 2007. Bacteriocin diversity in *Streptococcus* and *Enterococcus*. *Journal of Bacteriology*, **189**, 1189–98.

Neviani, E., Lindner, J. D. D., Bernini, V. & Gatti, M. 2009. Recovery and differentiation of long ripened cheese microflora through a new cheese-based cultural medium. *Food Microbiology*, **26**, 240–5.

Nicodème, M., Grill, J. P., Humbert, G. & Gaillard, J. L. 2005. Extracellular protease activity of different *Pseudomonas* strains: dependence of proteolytic activity on culture conditions. *Journal of Applied Microbiology*, **99**, 641–8.

Nielsen, L. R., Warnick, L. D. & Greiner, M. 2007. Risk factors for changing test classification in the Danish surveillance program for *Salmonella* in dairy herds. *Journal of Dairy Science*, **90**, 2815–25.

Nightingale, C., Dhuyvetter, K., Mitchell, R. & Schukken, Y. 2008. Influence of variable milk quality premiums on observed milk quality. *Journal of Dairy Science*, **91**, 1236–44.

Nilsson, R. E., Ross, T. & Bowman, J. P. 2011. Variability in biofilm production by *Listeria monocytogenes* correlated to strain origin and growth conditions. *International Journal of Food Microbiology*, **150**, 14–24.

Nörnberg, M., Mentges, M. L., Silveira, S. T., Tondo, E. C. & Brandelli, A. 2011. A psychrotrophic *Burkholderia cepacia* strain isolated from refrigerated raw milk showing proteolytic activity and adhesion to stainless steel. *Journal of Dairy Research*, **78**, 257–62.

Oliver, S. P., Boor, K. J., Murphy, S. C. & Murinda, S. E. 2009. Food safety hazards associated with consumption of raw milk. *Foodborne Pathogens and Disease*, **6**, 793–806.

Oosthuizen, M. C., Steyn, B., Lindsay, D., Brozel, V. S. & Von Holy, A. 2001. Novel method for the proteomic investigation of a dairy-associated *Bacillus cereus* biofilm. *FEMS Microbiology Letters*, **194**, 47–51.

Patel, M. T., Nagarajan, R. & Kilara, A. 1996. Hydrolysis of milk fat by lipase in solvent-free phospholipid reverse micellar media. *Journal of Food Science*, **61**, 33–8.

Pikuta, E., Lysenko, A., Chuvilskaya, N., Mendrock, U., Hippe, H., Suzina, N., Nikitin, D., Osipov, G. & Laurinavichius, K. 2000. *Anoxybacillus pushchinensis* gen. nov., sp nov., a novel anaerobic, alkaliphilic, moderately thermophilic bacterium from manure, and description of *Anoxybacillus falvithermus* comb. nov. *International Journal of Systematic and Evolutionary Microbiology*, **50**, 2109–17.

Quigley, L., O'Sullivan, O., Beresford, T. P., Ross, R. P., Fitzgerald, G. F. & Cotter, P. D. 2011. Molecular approaches to analysing the microbial composition of raw milk and raw milk cheese. *International Journal of Food Microbiology*, **150**, 81–94.

Quigley, L., O'Sullivan, O., Stanton, C., Beresford, T. P., Ross, R. P., Fitzgerald, G. F. & Cotter, P. D. 2013. The complex microbiota of raw milk. *FEMS Microbiology Reviews*, **37**, 664–98.

Rajendran, R., Robertson, D. P., Hodge, P. J., Lappin, D. F. & Ramage, G. 2010. Hydrolytic enzyme production is associated with *Candida Albicans* biofilm formation from patients with Type 1 diabetes. *Mycopathologia*, **170**, 229–35.

Ranieri, M. L. & Boor, K. J. 2010. Tracking and eliminating sporeformers in dairy systems. *Australian Journal of Dairy Technology*, **65**, 74–80.

Ranieri, M. L., Huck, J. R., Sonnen, M., Barbano, D. M. & Boor, K. J. 2009. High temperature, short time pasteurisation temperatures inversely affect bacterial numbers during refrigerated storage of pasteurised fluid milk. *Journal of Dairy Science*, **92**, 4823–32.

Ranieri, M. L., Ivy, R. A., Mitchell, W. R., Call, E., Masiello, S. N., Wiedmann, M. & Boor, K. J. 2012. Real-time PCR detection of *Paenibacillus* spp. in raw milk to predict shelf life performance of pasteurised fluid milk products. *Applied and Environmental Microbiology*, **78**, 5855–63.

Rasooly, R. & Do, P. M. 2010. Shiga toxin Stx2 is heat-stable and not inactivated by pasteurisation. *International Journal of Food Microbiology*, **136**, 290–4.

Recio, I., Amigo, L. & Lopezfandino, R. 1997. Assessment of the quality of dairy products by capillary electrophoresis of milk proteins. *Journal of Chromatography B*, **697**, 231–42.

Reilly, A. & WHO Consultation Prevention Control. 1998. Prevention and control of enterohaemorrhagic *Escherichia coli* (EHEC) infections: memorandum from a WHO meeting. *Bulletin of the World Health Organization*, **76**, 245–55.

Rogers, J., Dowsett, A. B., Dennis, P. J., Lee, J. V. & Keevil, C. W. 1994. Influence of temperature and plumbing material selection on biofilm formation and growth of *Legionella pneumophila* in a model potable water-system containing complex microbial flora. *Applied and Environmental Microbiology*, **60**, 1585–92.

Rollet, C., Gal, L. & Guzzo, J. 2009. Biofilm-detached cells, a transition from a sessile to a planktonic phenotype: a comparative study of adhesion and physiological characteristics in *Pseudomonas aeruginosa*. *FEMS Microbiology Letters*, **290**, 135–42.

Rosche, B., Li, X. Z., Hauer, B., Schmid, A. & Buehler, K. 2009. Microbial biofilms: a concept for industrial catalysis? *Trends in Biotechnology*, **27**, 636–43.

Rossetti, L., Fornasari, M. E., Gatti, M., Lazzi, C., Neviani, E. & Giraffa, G. 2008. Grana Padano cheese whey starters: microbial composition and strain distribution. *International Journal of Food Microbiology*, **127**, 168–71.

Ryu, J. H., Kim, H. & Beuchat, L. R. 2004. Attachment and biofilm formation by *Escherichia coli* O157:H7 on stainless steel as influenced by exopolysaccharide production, nutrient availability, and temperature. *Journal of Food Protection*, **67**, 2123–31.

Salama, M. S., Musafija-Jeknic, T., Sandine, W. E. & Giovannoni, S. J. 1995. An ecological study of lactic acid bacteria: isolation of new strains of *Lactotoccus* including *lactococcus lactis* subspecies *cremoris*. *Journal of Dairy Science*, **78**, 1004–17.

Sanaa, M., Poutrel, B., Menard, J. L. & Serieys, F. 1993. Risk-factors associated with contamination of raw milk by *Listeria monocytogenes* in dairy farms. *Journal of Dairy Science*, **76**, 2891–8.

Santarelli, M., Gatti, M., Lazzi, C., Bernini, V., Zapparoli, G. A. & Neviani, E. 2008. Whey starter for Grana Padano cheese: effect of technological parameters on viability and composition of the microbial community. *Journal of Dairy Science*, **91**, 883–91.

Santorum, P., Garcia, R., Lopez, V. & Martinez-Suarez, J. V. 2012. Review. Dairy farm management and production practices associated with the presence of *Listeria monocytogenes* in raw milk and beef. *Spanish Journal of Agricultural Research*, **10**, 360–71.

Scallan, E., Hoekstra, R. M., Angulo, F. J., Tauxe, R. V., Widdowson, M.-A., Roy, S. L., Jones, J. L. & Griffin, P. M. 2011. Foodborne illness acquired in the United States – major pathogens. *Emerging Infectious Diseases*, **17**, 7–15.

Scheldeman, P., Goossens, K., Rodriguez-Diaz, M., Pil, A., Goris, J., Herman, L., De Vos, P., Logan, N. A. & Heyndrickx, M. 2004. *Paenibacillus lactis* sp. nov., isolated from raw and heat-treated milk. *International Journal of Systematic and Evolutionary Microbiology*, **54**, 885–91.

Scheldeman, P., Pil, A., Herman, L., De Vos, P. & Heyndrickx, M. 2005. Incidence and diversity of potentially highly heat-resistant spores isolated at dairy farms. *Applied and Environmental Microbiology*, **71**, 1480–94.

Schelin, J., Wallin-Carlquist, N., Cohn, M. T., Lindqvist, R., Barker, G. C. & Radstrom, P. 2011. The formation of *Staphylococcus aureus* enterotoxin in food environments and advances in risk assessment. *Virulence*, **2**, 580–92.

Schoeni, J. L. & Wong, A. C. L. 2005. *Bacillus cereus* food poisoning and its toxins. *Journal of Food Protection*, **68**, 636–48.

Seale, R. B., Flint, S. H., Mcquillan, A. J. & Bremer, P. J. 2008. Recovery of spores from thermophilic dairy bacilli and effects of their surface characteristics on attachment to different surfaces. *Applied and Environmental Microbiology*, **74**, 731–7.

Serraino, A., Florio, D., Giacometti, F., Piva, S., Mion, D. & Zanoni, R. G. 2013. Presence of *Campylobacter* and *Arcobacter* species in in-line milk filters of farms authorized to produce and sell raw milk and of a water buffalo dairy farm in Italy. *Journal of Dairy Science*, **96**, 2801–7.

Servin, A. L. 2004. Antagonistic activities of *lactobacilli* and *bifidobacteria* against microbial pathogens. *FEMS Microbiology Reviews*, **28**, 405–40.

Shah, N. P. 1994. Psychrotrophs in milk – a review. *Milchwissenschaft – Milk Science International*, **49**, 432–7.

Sharma, M. & Anand, S. K. 2002. Biofilms evaluation as an essential component of HACCP for food/dairy processing industry – a case. *Food Control*, **13**, 469–77.

Shi, X. & Zhu, X. 2009. Biofilm formation and food safety in food industries. *Trends in Food Science & Technology*, **20**, 407–13.

Slaghuis, B. A., Giffel, M. C. T., Beumer, R. R. & Andre, G. 1997. Effect of pasturing on the incidence of *Bacillus cereus* spores in raw milk. *International Dairy Journal*, **7**, 201–5.

Small, A., Wells-Burr, B. & Buncic, S. 2005. An evaluation of selected methods for the decontamination of cattle hides prior to skinning. *Meat Science*, **69**, 263–8.

Snyder, A. B. & Worobo, R. W. 2014. Chemical and genetic characterization of bacteriocins: antimicrobial peptides for food safety. *Journal of the Science of Food and Agriculture*, **94**, 28–44.

Sørhaug, T. & Stepaniak, L. 1997. Psychrotrophs and their enzymes in milk and dairy products: quality aspects. *Trends in Food Science and Technology*, **8**, 35–41.

Sousa, M. J., Ardo, Y. & Mcsweeney, P. L. H. 2001. Advances in the study of proteolysis during cheese ripening. *International Dairy Journal*, **11**, 327–45.

Spector, M. P. & Kenyon, W. J. 2012. Resistance and survival strategies of *Salmonella enterica* to environmental stresses. *Food Research International*, **45**, 455–81.

Speers, J. G. S. & Gilmour, A. 1985. The influence of milk and milk components on the attachment of bacteria to farm dairy equipment surfaces. *Journal of Applied Bacteriology*, **59**, 325–32.

Stanley, K. & Jones, K. 2003. Cattle and sheep farms as reservoirs of *Campylobacter*. *Journal of Applied Microbiology*, **94**, 104S–13S.

Steele, J., Broadbent, J. & Kok, J. 2013. Perspectives on the contribution of lactic acid bacteria to cheese flavor development. *Current Opinion in Biotechnology*, **24**, 135–41.

Steenackers, H., Hermans, K., Vanderleyden, J. & De Keersmaecker, S. C. J. 2012. *Salmonella* biofilms: an overview on occurrence, structure, regulation and eradication. *Food Research International*, **45**, 502–31.

Steward, P. S., Folsom, J. P., Williamson, K. S., Franklin, M. J., Boegli, L. & James, G. A. 2006. Gentetic basis of *Pseudomonas aeruginosa* biofilm antibiotic tolerance. 6th Biofilm Conference on Biofilms, 2006, Miami, USA, p. 52.

Suárez, B., Ferreiros, C. M. & Criado, M. T. 1992. Adherence of psychrotrophic bacteria to dairy equipment surfaces. *Journal of Dairy Research*, **59**, 381–8.

Sutherland, A. D. & Murdoch, R. 1994. Seasonal occurrence of psychrotrophic *Bacillus* species in raw milk, and studies on the interactions with mesphilic *Bacillus* sp. *International Journal of Food Microbiology*, **21**, 279–92.

Svensson, B., Eneroth, A., Brendehaug, J. & Christiansson, A. 1999. Investigation of *Bacillus cereus* contamination sites in a dairy plant with RAPD-PCR. *International Dairy Journal*, **9**, 903–12.

Svensson, B., Ekelund, K., Ogura, H. & Christiansson, A. 2004. Characterisation of *Bacillus cereus* isolated from milk silo tanks at eight different dairy plants. *International Dairy Journal*, **14**, 17–27.

Svensson, B., Monthan, A., Shaheen, R., Andersson, M. A., Salkinoja-Salonen, M. & Christiansson, A. 2006. Occurrence of emetic toxin producing *Bacillus cereus* in the dairy production chain. *International Dairy Journal*, **16**, 740–9.

Swaminathan, B. & Gerner-Smidt, P. 2007. The epidemiology of human listeriosis. *Microbes and Infection*, **9**, 1236–43.

Swift, S., Lynch, M. J., Fish, L., Kirke, D. F., Tomas, J. M., Stewart, G. & Williams, P. 1999. Quorum sensing-dependent regulation and blockade of exoprotease production in *Aeromonas hydrophila*. *Infection and Immunity*, **67**, 5192–9.

Taylor, E. V., Herman, K. M., Ailes, E. C., Fitzgerald, C., Yoder, J. S., Mahon, B. E. & Tauxe, R. V. 2013. Common source outbreaks of *Campylobacter* infection in the USA, 1997–2008. *Epidemiology and Infection*, **141**, 987–96.

Te Giffel, M. C., Wagendorp, A., Herrewegh, A. & Driehuis, F. 2002. Bacterial spores in silage and raw milk. *Antonie van Leeuwenhoek*, **81**, 625–30.

Teh, K. H. 2013. Enzymes produced by bacteria within biofilms of dairy origin and their effect on dairy products. PhD thesis, Massey University.

Teh, K. H., Flint, S. & French, N. 2010. Biofilm formation by *Campylobacter jejuni* in controlled mixed-microbial populations. *International Journal of Food Microbiology*, **143**, 118–24.

Teh, K. H., Flint, S., Palmer, J., Lindsay, D., Andrewes, P. L. & Bremer, P. 2011. Thermo-resistant enzyme-producing bacteria isolated from the internal surfaces of raw milk tankers. *International Dairy Journal*, **21**, 742–7.

Teh, K. H., Flint, S., Palmer, J., Andrewes, P., Bremer, P. & Lindsay, D. 2012. Proteolysis produced within biofilms of bacterial isolates from raw milk tankers. *International Journal of Food Microbiology*, **157**, 28–34.

Teh, K. H., Lindsay, D., Palmer, J., Andrewes, P. L., Bremer, P. & Flint, S. 2013. Lipolysis within single culture and co-culture biofilms of dairy origin. *International Journal of Food Microbiology*, **163**, 129–35.

Teh, K. H., Flint, S., Palmer, J., Andrewes, P., Bremer, P. & Lindsay, D. 2014. Biofilm – an unrecognised source of spoilage enzymes in dairy products? *International Dairy Journal*, **34**, 32–40.

Thomason, C. A., Delaguis, P. J. & Mazza, G. 1999. Detection and measurement of microbial lipase activity: a review. *Critical Reviews in Food Science and Nutrition*, **39**, 165–87.

Thomason, M. K., Fontaine, F., De Lay, N. & Storz, G. 2012. A small RNA that regulates motility and biofilm formation in response to changes in nutrient availability in *Escherichia coli*. *Molecular Microbiology*, **84**, 17–35.

Timmusk, S., Grantcharova, N. & Wagner, E. G. H. 2005. *Paenibacillus polymyxa* invades plant roots and forms biofilms. *Applied and Environmental Microbiology*, **71**, 7292–300.

USDA. 2011. Milk for Manufacturing Purposes and its Production and Processing. US Department of Agriculture.

Vacheyrou, M., Normand, A.-C., Guyot, P., Cassagne, C., Piarroux, R. & Bouton, Y. 2011. Cultivable microbial communities in raw cow milk and potential transfers from stables of sixteen French farms. *International Journal of Food Microbiology*, **146**, 253–62.

Van den Broek, D., Bloemberg, G. V. & Lugtenberg, B. 2005. The role of phenotypic variation in rhizosphere *Pseudomonas* bacteria. *Environmental Microbiology*, **7**, 1686–97.

Van Kessel, J. S., Karns, J. S., Wolfgang, D. R., Hovingh, E. & Schukken, Y. H. 2007. Longitudinal study of a clonal, subclinical outbreak of *Salmonella enterica* subsp. *enterica* serovar *Cerro* in a US dairy herd. *Foodborne Pathogens and Disease*, **4**, 449–61.

Van Kessel, J. S., Karns, J. S., Lombard, J. E. & Kopral, C. A. 2011. Prevalence of *Salmonella enterica*, *Listeria monocytogenes*, and *Escherichia coli* virulence factors in bulk tank milk and in-line filters from US dairies. *Journal of Food Protection*, **74**, 759–68.

Velaquez, E., De Miguel, T., Poza, M., Rivas, R., Rossello-Mora, R. & Villa, T. G. 2004. *Paenibacillus favisporus* sp nov., a xylanolytic bacterium isolated from cow faeces. *International Journal of Systematic and Evolutionary Microbiology*, **54**, 59–64.

Verdier-Metz, I., Gagne, G., Bornes, S., Monsallier, F., Veisseire, P., Delbes-Paus, C. & Montel, M.-C. 2012. Cow teat skin, a potential source of diverse microbial populations for cheese production. *Applied and Environmental Microbiology*, **78**, 326–33.

Vidovic, S., Mangalappalli-Illathu, A. K. & Korber, D. R. 2011. Prolonged cold stress response of *Escherichia coli* O157 and the role of rpoS. *International Journal of Food Microbiology*, **146**, 163–9.

Villar, A., Garcia, J. A., Iglesias, L., Garcia, M. L. & Otero, A. 1996. Application of principal component analysis to the study of microbial populations in refrigerated raw milk from farms. *International Dairy Journal*, **6**, 937–45.

Villegas, N. A., Baronetti, J., Albesa, I., Polifroni, R., Parma, A., Etcheverria, A., Becerra, M., Padola, N. & Paraje, M. 2013. Relevance of biofilms in the pathogenesis of shiga-toxin-producing *Escherichia coli* infection. *The Scientific World Journal*, Article ID 607258.

Vissers, M. M. M., Giffel, M. C. T., Driehuis, F., De Jong, P. & Lankveld, J. M. G. 2007a. Minimizing the level of *Bacillus cereus* spores in farm tank milk. *Journal of Dairy Science*, **90**, 3286–93.

Vissers, M. M. M., Giffel, M. C. T., Driehuis, F., De Jong, P. & Lankveld, J. M. G. 2007b. Predictive modeling of *Bacillus cereus* spores in farm tank milk during grazing and housing periods. *Journal of Dairy Science*, **90**, 281–92.

Waak, E., Tham, W. & Danielsson-Tham, M. L. 2002. Prevalence and fingerprinting of *Listeria monocytogenes* strains isolated from raw whole milk in farm bulk tanks and in dairy plant receiving tanks. *Applied and Environmental Microbiology*, **68**, 3366–70.

Wallin-Carlquist, N., Marta, D., Borch, E. & Radstrom, P. 2010. Prolonged expression and production of *Staphylococcus aureus* enterotoxin A in processed pork meat. *International Journal of Food Microbiology*, **141**, S69–74.

Wang, Z. W. & Chen, S. 2009. Potential of biofilm-based biofuel production. *Applied Microbiology and Biotechnology*, **83**, 1–18.

Weiler, C., Ifland, A., Naumann, A., Kleta, S. & Noll, M. 2013. Incorporation of *Listeria monocytogenes* strains in raw milk biofilms. *International Journal of Food Microbiology*, **161**, 61–8.

Wijman, J. G. E., De Leeuw, P. P. L. A., Moezelaar, R., Zwietering, M. H. & Abee, T. 2007. Air–liquid interface biofilms of *Bacillus cereus*: formation, sporulation, and dispersion. *Applied and Environmental Microbiology*, **73**, 1481–8.

Williams, A. G. & Banks, J. M. 1997. Proteolytic and other hydrolytic enzyme activities in non-starter lactic acid bacteria (NSLAB) isolated from Cheddar cheese manufactured in the United Kingdom. *International Dairy Journal*, **7**, 763–74.

Williams, A. P., Avery, L. M., Killham, K. & Jones, D. L. 2005. Persistence of *Escherichia coli* O157 on farm surfaces under different environmental conditions. *Journal of Applied Microbiology*, **98**, 1075–83.

Winkelstroter, L. K., Gomes, B. C., Thomaz, M. R. S., Souza, V. M. & Martinis, E. C. P. D. 2011. *Lactobacillus sakei* 1 and its bacteriocin influence adhesion of *Listeria monocytogenes* on stainless steel surface. *Food Control*, **22**, 1404–7.

Wirtanen, G., Ahola, H. & Mattila-Sandholm, T. 1995. Evaluation of cleaning procedures in elimination of biofilm from stainless steel surfaces in open process equipment. *Food and Bioproducts Processing*, **73**, 9–16.

Woods, R. G., Burger, M., Beven, C. A. & Beacham, I. R. 2001. The aprX-lipA operon of *Pseudomonas fluorescens* B52: a molecular analysis of metalloprotease and lipase production. *Microbiology UK*, **147**, 345–54.

Xu, X. M., Stewart, P. S. & Chen, X. 1996. Transport limitation of chlorine disinfection of *Pseudomonas aeruginosa* entrapped in alginate beads. *Biotechnology and Bioengineering*, **49**, 93–100.

Yu, G.-H., He, P.-J., Shao, L.-M. & Lee, D.-J. 2007. Enzyme activities in activated sludge flocs. *Applied Microbiology and Biotechnology*, **77**, 605–12.

Zadoks, R. N., Gonzalez, R. N., Boor, K. J. & Schukken, Y. H. 2004. Mastitis-causing *streptococci* are important contributors to bacterial counts in raw bulk tank milk. *Journal of Food Protection*, **67**, 2644–50.

Zakaria, Z., Razak, C. N. A., Ampon, K., Basri, M., Yunus, W. M. Z., Shirai, Y., Salleh, A. B. & Hashimoto, K. 1992. Optimum conditions for the production of lipase by alginate-immobilized bacteria. *Journal of General and Applied Microbiology*, **38**, 429–38.

6 Thermoresistant Streptococci

Steve Flint[1], John Brooks[2] and Phil Bremer[3]

[1]Institute of Food, Nutrition and Human Health, Massey University, Palmerston North, New Zealand
[2]School of Applied Sciences, Faculty of Health and Environmental Sciences, Auckland University of Technology, Auckland, New Zealand
[3]Department of Food Science, University of Otago, Dunedin, New Zealand

6.1 Characteristics of *Streptococcus thermophilus* and *S. macedonicus*

Streptococcus thermophilus is a thermoresistant bacterium that belongs to a group referred to within the dairy industry as the thermophilic lactic acid bacteria (LAB). It is not a true thermophile as it has a growth range of 20–50 °C and an optimum temperature of approximately 42 °C. *Streptococcus macedonicus* was first described by Tsakalidou *et al.* (1998) as encompassing strains isolated from Greek cheese. Soon thereafter, Flint *et al.* (1999a) characterised some isolates of streptococci obtained from milk biofilms, for which the species name *Streptococcus waius* was proposed. Isolates of both species were later examined by Mora *et al.* (2002), who found the two were synonyms and reclassified all strains of *S. waius* as *S. macedonicus*. *S. macedonicus* has a growth range of 24–52 °C and an optimum growth temperature of 39 °C. In addition, *S. macedonicus* can grow in up to 7% NaCl (Pearce & Flint, 2002).

Thermoresistant streptococci ferment a limited number of sugars and do not utilise arginine. They may therefore survive well in a mixed biofilm environment, where other microorganisms break down more complex sugars or proteins to provide arginine. However, most biofilm studies have been conducted on single species.

6.2 Biofilms of thermoresistant streptococci in dairy manufacturing equipment

The growth of thermoresistant streptococci in industrial and pilot-scale cheese-milk pasteurisation equipment has been demonstrated by a number of studies (Hup *et al.*, 1979; Bouman *et al.*, 1982; De Jong *et al.*, 2002; Knight *et al.*, 2004). The number of thermoresistant streptococci in pasteurised cheese-milk increases over the duration of a production run and, for long production runs (>8 hours), may reach levels where it causes problems with cheese

Biofilms in the Dairy Industry, First Edition.
Edited by Koon Hoong Teh, Steve Flint, John Brooks and Geoff Knight.

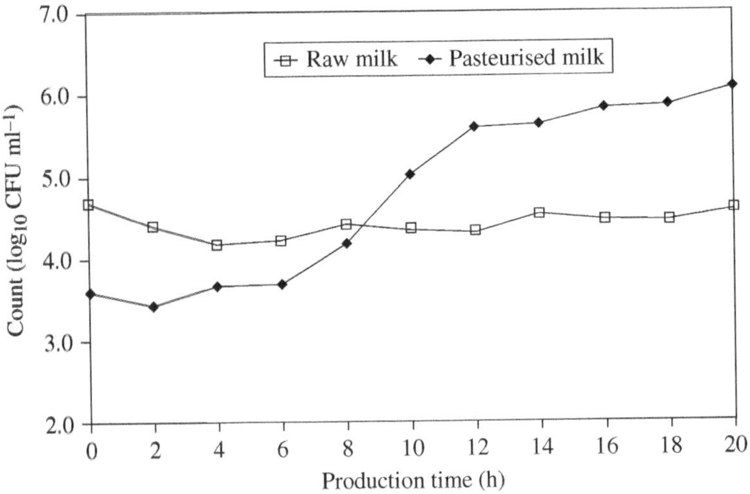

Figure 6.1 Bacterial counts for raw and pasteurised milk measured on M17 agar (37 °C, 48 hours) throughout a 20 hour-long production run of a pilot-scale pasteurisation plant. The increased counts observed after 8 hours were due to growth of thermoresistant streptococci on heat-exchange surfaces on the downstream side of the regenerative section of the pasteuriser (G. Knight, unpublished).

quality. An example of the changes in levels of thermoresistant streptococci in pasteurised milk during a production run of a pilot-scale pasteurisation plant (3000 l/h) is shown in Figure 6.1.

Thermoresistant streptococci are able to grow during cheese making and can cause quality issues. In one manufacturing plant, the levels of thermoresistant streptococci in pasteurised cheese-milk were greater than 6.0 $\log_{10}$ cells/ml after 7–8 hours of operation. The Gouda cheese produced from this milk had an unsuitable texture and an unclean, yeasty flavour (Hup *et al.*, 1979).

The length of time it takes for milk to travel through pasteurisation equipment (2–3 minutes) is too short for the increase in counts to be explained by bacterial growth in milk. The increase in counts has been attributed to biofilm formation on the surfaces of the pasteurisation equipment. Specifically, thermoresistant streptococci grow on surfaces on the downstream (pasteurised) side of the regenerative section, in the temperature range 30–50 °C (Hup *et al.*, 1979; Lehmann *et al.*, 1990; Knight *et al.*, 2004). Evidence for this localisation of growth comes from line studies of pasteurisation equipment, where increases in counts for thermoresistant streptococci occur after milk has cooled to below 50 °C (Knight *et al.*, 2004), and from swabbing of surfaces (Figure 6.2). Bouman *et al.* (1982) found thermoresistant streptococci attached to stainless steel surfaces at levels of 7.0 $\log_{10}$ cells/cm^2 in a model pasteuriser.

The growth of thermoresistant streptococci in a biofilm has been reported to be slower in raw milk than in pasteurised milk, although attachment was the same in both (Driessen *et al.*, 1984). This suggests there may be growth inhibitory factors in raw milk that are destroyed by heat treatment. Knight *et al.* (2004) also detected growth of thermoresistant streptococci on the upstream (raw milk) side of the regenerative section, although at a slower rate than was found on the downstream side. They attributed the slower growth rate of thermoresistant streptococci in this region to competition from the microflora present in the raw milk. Further

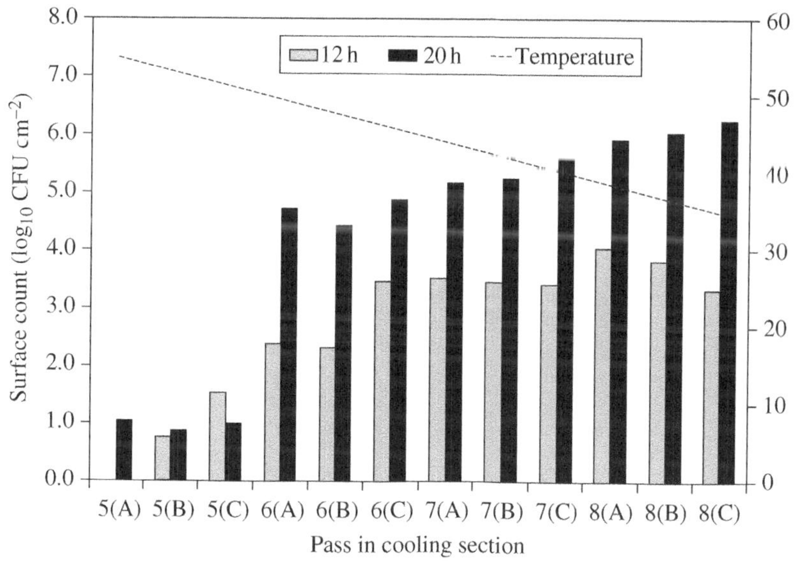

Figure 6.2 Counts obtained on M17 Agar (37 °C for 48 hours) for surface swabs of heat-exchange plates on the downstream side of the regenerative section. Results are for passes 5, 6, 7 and 8 of the plate heat exchanger, in which milk was cooled from 55 to 35 °C. Bulk milk temperatures are indicated on the graph. Surface temperatures were approximately 2.5 °C lower. Three areas (10 × 10 cm) were swabbed on each plate (A, B and C), with A at the warmer end, B in the middle and C at the cooler end of the plate (G. Knight, unpublished).

studies are required to determine the importance of different components of milk and other dairy fluids in the formation of biofilms by thermoresistant streptococci.

In a pilot-scale plant, biofilms of thermoresistant streptococci reached a steady state after approximately 12 hours at 42.5 °C when growth and detachment rates were equivalent (4.9 log_{10} cells/cm^2/s) (Lee *et al.*, 1997).

S. thermophilus appears to be a persistent and prolific natural contaminant of cheese vats used in the manufacture of traditional cheeses from raw milk, where it provides some of the natural starter microflora (Settanni *et al.*, 2012). Problems with *S. thermophilus* in cheese manufacture depend on the type of cheese being manufactured. In natural, raw-milk cheeses manufactured in wooden vats, *S. thermophilus* dominates the microflora in the vats (Licitra *et al.*, 2007; Settanni *et al.*, 2012). This is believed to prevent pathogen contamination.

In summary, thermoresistant streptococci may be beneficial or detrimental in the manufacture of dairy products. In both situations, the colonisation of the manufacturing plant is an important prerequisite to the influence of these bacteria on the dairy industry (Figure 6.3).

6.3 Attachment of thermoresistant streptococci to surfaces

Attachment of cells to surfaces is a very important step in the development of biofilms by thermoresistant streptococci in dairy manufacturing equipment. Cell surface charge, hydrophobicity and cell surface materials (e.g. polysaccharide and proteins) are thought to be involved in the attachment of bacteria to surfaces (Marshall *et al.*, 1971; Doyle *et al.*, 1990; Neu, 1992; Hood & Zottola, 1995), with the importance of each factor varying between species. Flint *et al.* (1997)

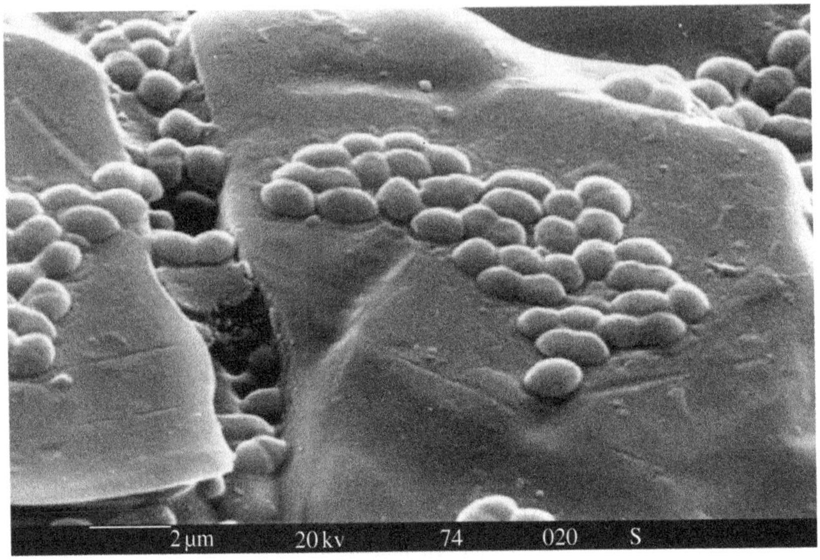

Figure 6.3 Thermoresistant streptococci colonising stainless steel.

were unable to demonstrate a relationship between the attachment within 30 minutes of *S. thermophilus* strains to stainless steel surfaces and any of these factors, with the exception of cell surface proteins.

Although it was not possible to demonstrate a relationship between the magnitude of the cell surface charge and attachment over a 30-minute period, surface charge may still influence the attachment process (Flint *et al.*, 1997). The predominantly negative charge on the bacterial surface is likely to cause repulsion of bacteria from surfaces that are negatively charged, such as stainless steel. The studies of Van der Mei *et al.* (1993), using microelectrophoresis to measure zeta-potentials, concluded that the thermoresistant dairy streptococci were only slightly negatively charged; therefore, the repulsion between cell and substrate surfaces may be low.

The hydrophobicity of individual strains of thermoresistant streptococci differ, but there is no obvious relationship between hydrophobicity and cell attachment to a stainless steel substrate (Flint *et al.*, 1997). Jameson *et al.* (1995) found a similar result, demonstrating that hydrophobicity and surface proteins did not affect the attachment of *Streptococcus oralis* to a salivary pellicle.

Extracellular polymeric substances (EPS) are believed to be important in the attachment of some bacteria to surfaces and have been termed 'adhesive polymers' (Neu, 1992). Marshall *et al.* (1971) suggested that EPS play a role in both initial and irreversible attachment. Some authors believe that EPS can promote a preconditioning of the surface, making attachment more favourable (Oliveira *et al.*, 1994). Herald and Zottola (1989) showed that compounds that bind to or disrupt carbohydrates, such as sodium metaperiodate, Cetavlon and concanavalin A, all decrease the attachment of *Pseudomonas fragi* to stainless steel. However, Allison and Sutherland (1987) found that the presence of polysaccharide did not affect attachment, with a polysaccharide-producing wild-type strain and a non-polysaccharide-producing mutant attaching equally well to glass.

EPS play little part in the initial attachment of cells of thermoresistant streptococci, according to trials quantifying attachment following treatment of the cells with chemicals that disrupt cell surface polysaccharides (Flint *et al.*, 1997). The treatments used were lysosyme, which dissolves bacterial cell wall mucopolysaccharides by hydrolysing the β (1–4) linkages between N-acetyl-amino-2-deoxy-D glucose residues (Windholz, 1983); sodium metaperiodate, which oxidises vicinal hydroxyl groups of component monosaccharides (Gopal & Reilly, 1995); and TCA, which extracts peptidoglycan-associated cell wall polymers (Heckels & Virji, 1988). Disruption of the polysaccharides was examined by measuring the EPS before and after treatment. None of the treatments resulted in a consistent difference in the attachment of thermoresistant streptococci isolates to stainless steel, even after the disrupted polysaccharides were removed by sonication.

One important observation was the heterogeneity of the results obtained with each treatment (Flint *et al.*, 1997). For each, some strains demonstrated increased cell attachment to surfaces and others showed decreased attachment. There was also no consistency between the treatments. For example, while one treatment resulted in an increase in cell attachment for an individual strain, another resulted in a decrease. Both *S. thermophilus* and *S. macedonicus* strains were affected. This heterogeneity in responses means that it is important to screen strains using more than one method in order to produce meaningful results.

6.4 The role of cell surface proteins in attachment of thermoresistant streptococci

In biological systems, adhesion interactions between host tissues and bacterial cells are mediated by proteins on the bacterial cell surface (Jenkinson, 1994). For example, site-specific colonisation by oral streptococci results from interplay between the host cell receptors and expression of bacterial adhesins. The importance of proteins in the attachment of cells to abiotic surfaces is less well documented. One study demonstrated that the attachment of *Azospirillum brasilense* to glass and polystyrene surfaces was correlated with protein concentration at the cell surface (Dufrêne *et al.*, 1996).

The removal of cell surface proteins from *S. thermophilus* isolates, using SDS or trypsin treatments, led to reductions in cells attachment, suggesting that cell surface proteins played an important role in the initial attachment phase (Flint *et al.*, 1997). The reductions in attachment were similar for both treatments with each of the 11 isolates included in the study. Total cell counts for treated and untreated cells used in attachment experiments were similar, indicating that reductions in attachment were not simply a result of the treatments reducing total cell numbers.

Treatments to remove cell surface proteins (and treatments to disrupt polysaccharides) can inactivate bacterial cells. An investigation was performed to determine whether cell inactivation caused by the SDS and trypsin treatments was responsible for the reduced levels of cell attachment. The attachment of viable cells of *S. thermophilus* to stainless steel surfaces was compared with attachment of cells inactivated by heat, ultraviolet (UV) light and formaldehyde. In all cases, inactivated cells attached to stainless steel surfaces at levels similar to those of viable cells, indicating that cell viability does not influence attachment of *S. thermophilus* to stainless steel surfaces (Flint *et al.*, 1997). This observation is in contrast to that from the study of Czechowski (1990), which found attachment of *Pseudomonas fluorescens* was reduced by up to 99% following the inactivation of cells by heat.

The interaction of cell surface proteins with the substrate may involve electrostatic and/or hydrophobic interactions. Paul and Jeffrey (1985) found that treatment with proteolytic enzymes decreased the hydrophobicity of *Vibrio proteolytica,* as determined by the attachment to polystyrene and the microbial adherence to hydrocarbons (MATH) test. Similarly, in the study of Flint *et al.* (1997), treatment with trypsin decreased the hydrophobicity of *S. thermophilus,* probably as a result of the removal of hydrophobic protein groups from the surfaces of cells. In this case, changes in hydrophobicity could not be related to cell attachment. In fact, with the exception of specific forms of bacteria, such as *Bacillus* spores (Wiencek *et al.*, 1991), there is no clear evidence in the literature that hydrophobicity is a strong predictor of attachment to solid surfaces.

Changes in cell surface structures following treatment with trypsin were visualised by examining cells with transmission electron microscopy (TEM). Cell surface structures such as fibril tufts, which have been observed on oral streptococci and have been implicated in attachment (Weerkamp *et al.*, 1986), were not observed on the surface of *S. thermophilus* (H) (Flint *et al.*, 1997).

Changes to cellular proteins following treatment with SDS or trypsin were demonstrated using SDS-PAGE (Flint *et al.*, 1997). The most obvious change was the loss of a polypeptide of approximately 55 kDa following treatment of cells with SDS. The N-terminal sequence of this protein matched that of β-lactoglobulin, although the molecular weight of the protein was three times that of β-lactoglobulin. Extracts of proteins from the cell wall of *S. thermophilus* (H) produced two bands via SDS-PAGE, one of which was confirmed as being part of the 'attachment protein' by a positive reaction with antisera to the 55 kDa polypeptide. Unfortunately, the N-terminal sequence of this polypeptide could not be determined. Evidence for the role of the 'attachment protein' was provided by attachment assays, which demonstrated that attachment of *S. thermophilus* cells to stainless steel was inhibited in the presence of antibodies to the 'attachment protein' and by TEM imaging of immunolabelled cells, showing localisation of the immunolabel at the cell surface (S. H. Flint, unpublished). An investigation into the source of the 'attachment protein' was also performed; it concluded that the protein could not have originated from the growth medium and was likely produced by the cells.

The 'attachment protein' appears to be similar to β-lactoglobulin, a protein that appears to have an affinity for surfaces; for example, it is known to be involved in fouling in dairy processing plants (Jeurnink *et al.*, 1996). Exposure of stainless steel to β-lactoglobulin inhibited subsequent attachment of *S. thermophilus* (H) (Flint *et al.*, 1997). Here, it appears that this protein blocked attachment sites on the surface, essentially competing with the 'attachment protein' for these sites. Similar observations were made by Bourassa *et al.* (1996), who found that unidentified whey proteins reduced the attachment of *Lactococcus lactis* subsp. *diacetylactis.* This conflicts with the hypothesis that coating or conditioning of surfaces with proteins assists in the attachment of microbial cells (Kirtley & Mcguire, 1989; Marshall, 1996). The influence of conditioning on bacterial attachment to surfaces may depend on the type of protein(s) and the bacteria involved.

6.5 Biofilm growth

There is a lack of information on the actual growth kinetics of the thermoresistant streptococci in a biofilm. Evidence from the levels of thermoresistant strepcococci being released from a pasteuriser suggests that growth is rapid. In one manufacturing plant, the levels of thermoresistant streptococci in milk from the pasteuriser reached over 6.0 $\log_{10}$ cells/ml after 7–8 hours of operation (Hup *et al.*, 1979). Knight *et al.* (2004) found that numbers of *S. thermophilus* released in the milk during pasteurisation through a pilot-scale pasteuriser reached 10^6 CFU/ml within 10 hours.

The development of biofilms consisting of thermoresistant streptococci in pasteurisers and thermalisers has caused contamination of cheese-milk, resulting in associated problems with cheese quality. Gouda cheese produced from this milk with levels of thermoresistant streptococci >10^6 cells/ml had an 'unsuitable' texture and an 'unclean yeasty flavour' (Hup *et al.*, 1979). The amount of time the milk was in the pasteuriser was too short for the increase in numbers to be due to bacterial growth, and therefore the presence of a contaminating biofilm was postulated (Driessen & Bouman, 1979). The attachment of bacteria and protein to the plates of pasteurisers occurred in the temperature range 30–50 °C (Hup *et al.*, 1979), with colonisation reported to be localised in the regeneration section of pasteurisers (Lehmann *et al.*, 1990). Bacteria associated with the cooling section of a model pasteuriser were found to be attaching directly to the stainless steel at levels of 7.0 $\log_{10}$ cells/cm^2 (Bouman *et al.*, 1982). In a pilot-scale plant, biofilms of thermoresistant streptococci reached a steady state after approximately 12 hours at 42.5 °C when the growth and detachment rates were equivalent (4.9 $\log_{10}$ cells/cm^2/s^1) (Lee *et al.*, 1997). Rademacher *et al.* (1995) reported that the attachment and growth of thermoresistant bacteria on the plate surfaces of a pasteuriser depended on the number of bacteria in the milk before pasteurisation, and colonisation of a pasteuriser was often associated with thermal treatment (thermisation) of the milk before pasteurisation.

6.6 Strategies to control thermoresistant streptococci

Regular cleaning using cleaning-in-place (CIP) systems is the most accepted method for control of biofilms of thermoresistant streptococci among dairy manufacturers. During the cleaning of dairy processing equipment, cleaning solutions are circulated at temperatures in the range 65–80 °C. The total cleaning time can vary from 2 to 3 hours, so equipment will be exposed to elevated temperatures for a significant period. For biofilms that develop in locations where cleaning solutions do not reach (e.g. in contact points and gaskets), exposure to elevated temperatures experienced during cleaning may be the only control mechanism.

However, standard cleaning systems are not always effective in controlling biofilms of thermoresistant streptococci, with both viable and dead cells seen on stainless steel surfaces after cleaning (Flint *et al.*, 1999b). A suitable approach may be to look for cleaning and sanitation treatments that effect cell removal.

6.6.1 *Influence of heat*

The sensitivity of thermoresistant streptococci is affected by environmental factors, including the presence of milk solids and attachment to surfaces. The sensitivity to heat of thermoresistant streptococci is not affected by attachment to stainless steel. However, when cells adhered in the

Table 6.1 Summary of results for the heat treatment of *S. thermophilus* (H).

Cell status	D-value (minutes) at 60 °C	Z-value
Planktonic in water	2.0	9.9
Planktonic in milk	14	7.6
Attached in water	2.2	9.4
Attached in milk	8.1	10.7
Biofilm (18 hours)	1.7	Not done

presence of skim milk, the sensitivity of *S. thermophilus* (H) decreased, with a fivefold increase in the D-value at 60 °C (Table 6.1) (Flint *et al.*, 2002). Although test samples with adhered cells were rinsed, residual milk protein associated with the cell and the substrate may have protected the cells from heat. A sevenfold increase in the D-value at 60 °C was also observed for planktonic cells in milk (Table 6.1). This is consistent with the effect of organic material on microbial resistance to heat (Joslyn, 1983). The potential for thermoresistant streptococci to survive heat treatment in a dairy manufacturing plant is therefore greater than that expected in other environments.

An increase was observed in the resistance to heat of 12 and 24 hour-biofilms of thermoresistant streptococci (grown in skim milk in a continuous-flow laboratory reactor) compared with planktonic cells in water. However, the D-values for 12 and 24 hour-biofilm cells at 60 °C were less than those observed for cells adhered recently in the presence of skim milk. Therefore, there is no evidence that the formation of a biofilm of these organisms produces any materials that would protect these bacteria from heat, as the increased resistance to heat may be explained by the effect of milk protein. Increased resistance to heat of biofilm cells has been reported for other organisms. For *Listeria monocytogenes* (Frank & Koffi, 1990), increased resistance to heat was associated with the amount of growth on the substrate; for *Salmonella enteritidis* (Dhir & Dodd, 1995), increased resistance to heat was believed to result from a change in the physiology of the cell induced by attachment.

The presence of milk also affects the correlation values for the thermal death curves for the thermoresistant streptococci (Flint *et al.*, 2002). The correlation values were less for planktonic cells in milk and cells adhered in the presence of milk than for planktonic and adhered cells in water. This may reflect variations in the distribution of bacteria in the milk, with clumping around milk proteins resulting in the protection of cells and associated variation in the susceptibility to heat. This hypothesis needs further investigation.

The sensitivity to heat of *S. thermophilus* suggests that the 70 °C for 30 minutes used in the routine cleaning of dairy manufacturing plants should be adequate to inactivate the cells in a biofilm. In practice, it may be difficult to maintain this temperature in a large plant, and a reduction in the temperature may enable survival. Accumulated organic material, including biofilm that has not been removed over a succession of manufacturing runs, may provide additional protection to the cells, beyond the large increase in the D-value seen in the presence of milk proteins.

A novel approach that uses temperature to control the growth of thermoresistant streptococci was proposed by Knight *et al.* (2004). This involves the periodic application of a 'temperature step change' to the location in the regenerative section of a pilot-scale cheese-milk pasteuriser where biofilms of thermoresistant streptococci develop. During a 'temperature step change', the growth region (35–50 °C) is subjected to a temperature of 55 °C for 10 minutes. The period between 'temperature step changes' is 60 minutes. Operating under step-change conditions, the time in which an increase in numbers of thermoresistant streptococci in the pasteurised milk was detected increased from around 8–10 up to 20 hours.

6.6.2 *Influence of cleaning and sanitation*

Although cleaning and sanitation of the dairy manufacturing plant are the main methods by which to control biofilms of thermophilic streptococci, the nature of biofilms is that they have a greater resistance to cleaning systems than planktonic cells. The increased resistance to sanitisers of biofilm cells compared with the planktonic cells of many species is widely reported (Le Chevallier *et al.*, 1988; Yu & McFeters, 1994).

In order to determine the effect of current industrial cleaning programmes, a manufacturing plant was monitored using epifluorescence and conductance detection methods (Flint *et al.*, 1999a). Biofilms of thermoresistant streptococci in a dairy manufacturing plant persisted after routine cleaning procedures. Although this plant was cleaned frequently (every 6 hours), the total numbers of bacteria, detected by epifluorescence microscopy, before and after cleaning were similar, demonstrating that the routine cleaning procedure was removing few of the bacteria from the substrate. The total numbers of bacteria for most sampling periods were (3–4 $\log_{10}$ cells/cm^2), suggesting that the operating conditions, in particular the short operating time, prevented bacteria reaching the high levels (10^7 cells/cm^2) recorded by others (Bouman *et al.*, 1982).

The increased resistance to chemical sanitisers of thermoresistant streptococci cells attached to a surface or grown in a biofilm was demonstrated by Flint *et al.* (1999b). Attached cells of two strains of *S. thermophilus* (H and 48) had similar levels of susceptibility to sanitisers to that of planktonic cells, with the exception that attached cells of strain H were more resistant to cetyl trimethyl ammonium bromide (CTAB) than sodium hypochlorite. Biofilm cells of both strains were more resistant to both sanitisers, with strain 48 more resistant to the chlorine sanitiser than strain H. The increased resistance of biofilm cells to sanitisers may be caused by organic material (i.e. milk protein) associated with the biofilm matrix (e.g. polysaccharides) or protective mechanisms from other aspects associated with colonisation of a surface (Bridier *et al.*, 2011; Wirtanen & Mattila-Sandholm, 1992).

With either sanitiser, biofilms of both strains of *S. thermophilus* survived the normal concentrations (200 ppm sodium hypochlorite, 25 ppm CTAB) used in dairy manufacturing plants. The survival of any cells following treatment with sanitisers will allow the rapid regeneration of a biofilm, increasing the risk of contamination of the manufacturing plant and products.

The effects of a range of chemical treatments against thermoresistant streptococci, attached to or grown as biofilms on stainless steel, were evaluated by Flint *et al.* (1999b). These experiments confirmed that acid and caustic treatments failed to remove thermoresistant streptococci from the surface of stainless steel. Treatments that affected proteins, particularly a treatment with proteolytic enzymes, were most effective in

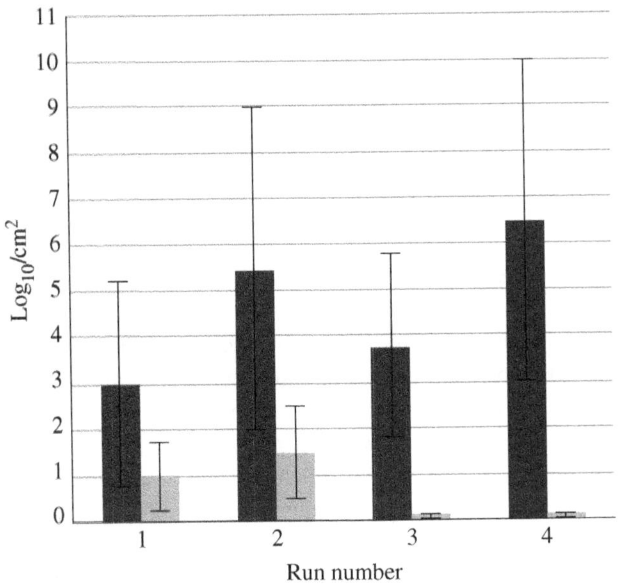

Figure 6.4 Comparison between the effectiveness of standard cleaning techniques and a proteolytic enzyme-based cleaner (Paradigm) in removing a biofilm of *S. thermophlus* (H) from stainless steel. Surface counts were viable cell counts, determined using a conductance technique (Malthus Microbiological Growth Analyser). ■, before cleaning; ■, after cleaning. (1) 1.8% NaOH 75 °C 30 minutes; 1% HNO_3 75 °C 30 minutes (industry-standard CIP procedure); (2) 1.8% NaOH 75 °C 30 minutes (frequently used short industry cleaning procedure); (3) 0.08% Paradigm 60 °C 30 minutes; 1.8% NaOH 75 °C 30 minutes; (4) 0.08% Paradigm 60 °C 30 minutes; 1% HNO_3 75 °C 30 minutes.

reducing the total number of cells (attached or in a biofilm) on stainless steel (Figure 6.4). This corresponds with previous work demonstrating the importance of proteins associated with the cell surface in attachment (Flint *et al.*, 1997). Further improvements in reducing bacteria may be obtained by fine tuning these proteolytic treatments, by turbulent flow and by optimisation of the concentration of the enzyme and the time of exposure to the enzyme. The concept of using enzymes for the removal of biofilms is not new; polysaccharide-hydrolysing enzymes have also been found to be effective in removing biofilms of strains of staphylococci and pseudomonads from steel and polypropylene (Johansen *et al.*, 1997).

The laboratory trials were followed by tests on biofilms in a pilot-scale plant (Flint *et al.*, 1999). The enhanced removal and successful inactivation of biofilms of thermoresistant streptococci using a commercial proteolytic enzyme cleaner in the pilot-scale trial suggest that this may be a realistic alternative procedure for routine use in a milk pasteurising plant. No viable cells were detected (detection limit approximately 1 cell/cm^2) on the stainless steel following enzyme cleaning, although some cells were still detected by epifluorescence microscopy. This suggests that the enzyme cleaner inactivates the cells. The pilot-scale work could be extended to demonstrate the effect of proteolytic enzyme cleaners in removing naturally occurring (rather than seeded) biofilms in dairy manufacturing plants during sequential manufacturing runs.

6.7 Conclusion

The formation of biofilms of thermoresistant streptococci in dairy manufacturing plant has been shown to involve two species: *S. thermophiles*, and *S. macedonicus*. Attachment of representatives of this group of bacteria to the stainless steel substrate has been shown to be mediated by a protein associated with the cell surface. Trials have demonstrated that proteolytic enzyme cleaners have the potential to improve the cleaning methods presently used in dairy manufacturing plants.

References

Allison, D. A. & Sutherland, I. W. 1987. The role of exopolysaccharides in attachment of freshwater bacteria. *Journal of General Microbiology*, **133**, 1319–27.

Bouman, S., Lund, D. B., Driessen, F. M. & Schmidt, D. G. 1982. Growth of thermoresistant streptococci and deposition of milk constituents on plates of heat exchangers during long operating times. *Journal of Food Protection*, **45**, 806–12.

Bourassa, S., Vuillemard, J.-C. & Rouxhet, P. 1996. Attachment of *Lactococcus lactis diacetylactis* to surfaces. In: Wijffels, R. H., Buitelaar, R. M., Bucke, C. & Tramper, J. (eds) *Immobilized Cells: Basics and Applications*. Elsevier Science, Amsterdam, pp. 84–9.

Bridier, A., Briandet, R., Thomas, V. & Dubois-Brissonnet, F. 2011. Resistance of bacterial biofilms to disinfectants: a review. *Biofouling*, **27**, 1017–32.

Czechowski, M. H. 1990. Cleaning and sanitation. *Australian Journal of Dairy Technology*, **45**, 38–9.

De Jong, P., Te Giffel, M. C. & Kiezebrink, E. A. 2002. Prediction of the adherence, growth and release of microorganisms in production chains. *International Journal of Food Microbiology*, **17**, 13–25.

Dhir, V. K. & Dodd, E. R. 1995. Susceptibility of suspended and surface-attached *Salmonella enteritidis* to biocides and elevated temperatures. *Applied and Environmental Microbiology*, **61**, 1731–8.

Doyle, R. J., Rosenberg, M. & Drake, D. 1990. Hydrophobicity of oral bacteria. In: Doyle, R. J. & Rosenberg, M. (eds) Microbial Cell Surface Hydrophobicity. American Society of Microbiology, Washington, DC, pp. 387–419.

Driessen, F. M. & Bouman, S. 1979. Growth of thermoresistant streptococci in cheese milk pasteurizers II – experiments with a model pasteurizer. *Zuivelzicht*, **71**, 1062–4.

Driessen, F. M., De Vries, J. & Kingma, F. 1984. Attachment and growth of thermoresistant streptococci on stainless steel during heat treatment of milk. *Journal of Food Protection*, **47**, 848–52.

Dufrêne, Y. F., Vermeiren, H., Vanderleyden, J. & Rouxhet, P. G. 1996. Direct evidence for the involvement of extracellular proteins in the attachment of *Azospirillum brasilense*. *Microbiology*, **142**, 855–65.

Flint, S. H., Brooks, J. D. & Bremer, P. J. 1997. The influence of cell surface properties of thermophilic streptococci on attachment to stainless steel. *Journal of Applied Microbiology*, **83**, 508–17.

Flint, S. H., Ward, L. J. H. & Brooks, J. D. 1999a. *Streptococcus waius* sp. nov., a thermophilic streptococcus from a biofilm. *International Journal of Systematic Bacteriology*, **49**, 759–67.

Flint, S. H., Van Den Elzen, H., Brooks, J. D. & Bremer, P. J. 1999b. Removal and inactivation of thermo-resistant streptococci colonising stainless steel. *International Dairy Journal*, **9**, 429–36

Flint, S. H., Brooks, J. D., Bremer, P. J. Walker, K. & Hauseman, E. 2002. The heat resistance of thermo-resistant streptococci attached to stainless steel in the presence of milk. *Journal of Industrial Microbiology and Biotechnology*, **28**, 134–6.

Frank, J. F. & Koffi, R. A. 1990 Surface-adherent growth of *Listeria monocytogenes* is associated with increased resistance to surfactant sanitizers and heat. *Journal of Food Protection*, **53**, 550–4.

Gopal, P. K. & Reilly, K. 1995. Molecular architecture of the lactococcal cell surface as it relates to important industrial properties. *International Dairy Journal*, **5**, 1095–111.

Heckels, J. E. & Virji, M. 1988. Separation and purification of surface components. In: Hancock, I. C. & Poxton, I. R. (eds) Bacterial Cell Surface Techniques. John Wiley and Sons, New York, NY, pp. 67–136.
Herald, P. J. & Zottola, E. A. 1989 Effect of various agents upon the attachment of *Pseudomonas fragi* to stainless steel. *Journal of Food Science*, **54**, 461–4.
Hood, S. K. & Zottola, E. A. 1995. Biofilms in food processing. *Food Control*, **6**, 9–18.
Hup, G., Bangma, A., Stadhouders, J. & Bouman, S. 1979. Growth of thermoresistant streptococci in cheese milk pasteurizers. (1) Some observations in cheese making. *Zuivelzicht*, **71**, 1014–16.
Jameson, M. W., Jenkinson, H. F., Parnell, K., Handley, P. S. 1995. Polypeptides associated with tufts of cell-surface fibrils in an oral *Streptococcus*. *Microbiology*, **141**, 2729–38.
Jenkinson, H. F. 1994. Cell surface protein receptors in oral streptococci. *FEMS Microbiology Letters*, **121**, 133–40.
Jeurnink, T. J. M., Walstra, P. & De Kruif, C. G. 1996. Mechanisms of fouling in dairy processing. *Netherlands Milk and Dairy Journal*, **50**, 407–26.
Johansen, C., Falholt, P. & Gram, L. 1997. Enzymatic removal and disinfection of bacterial biofilms. *Applied and Environmental Microbiology*, **63**, 3724–8.
Josyln, L. 1983. Sterilisation by heat. In: Block, S. S. (ed.) Disinfection, Sterilization and Preservation. Lea and Febiger, Philadelphia, PA, pp. 3–46.
Kirtley, S. A. & Mcguire, J. 1989. On differences in surface constitution of dairy product contact materials. *Journal of Dairy Science*, **72**, 1748–53.
Knight, G. C., Nicol, R. S. & Mcmeekin, T. A. 2004. Temperature step changes: a novel approach to control biofilms of *Streptococcus thermophius* in a pilot-scale cheese-milk pasteurisation plant. *International Journal of Food Microbiology*, **93**, 305–18.
Le Chevallier, M. W., Cawthon, C. D. & Lee, R. G. 1988. Inactivation of biofilm bacteria. *Applied and Environmental Microbiology*, **54**, 2492–9.
Lee, M. W., Knight, G. & Nicol, R. S. 1997. Development of biofilm of thermoduric streptococci. Chemica 97 – The 25th Australian and New Zealand Chemical Engineers' Conference and Exhibition, Rotorua, 29 September–1 October 1997.
Lehmann, F. L., Solomon, L. S., Russell, P. S., Murphy, K. & Hull, R. R. 1990. Thermophilic bacteria in pasteurisers and ultrafiltration plants. International Dairy Congress, Montreal, Canada.
Licitra, G., Ogier, J. C., Parayre, S., Pediliggieri, C., Carnemolla, T. M., Falentin, H., Madec, M. N., Carpino, S. & Lortal, S. 2007. Variability of bacterial biofilms of the 'tina' wood vats used in the ragusano cheese-making process. *Applied and Environmental Microbiology*, **73**, 6980–7.
Marshall, K. C. 1996. Attachment as a strategy for access to nutrients. In: Fletcher, M. (ed.) Bacterial Attachment Molecular and Ecological Diversity. Wiley-Liss, New York, NY, pp. 59–88.
Marshall, K. C., Stout, R. & Mitchell, R. 1971. Mechanism of the initial events in the sorption of marine bacteria to surfaces. *Journal of General Microbiology*, **68**, 337–48.
Mora, D., Flint, S. H., Ward, L. J. H., Kelly, W., Fortina, M. G., Parini, C. & Manchini, P. L. 2002. Comparison between *S. macedonicus* and *S. waius* strains and reclassification of *Streptococcus waius* (Flint *et al.* 1999) as *Streptococcus macedonicus* (Tsakalidu *et al.* 1999). *International Journal of Systematic and Environmental Microbiology*, **52**, 945–51.
Neu, T. R. 1992. Microbial 'footprints' and the general ability of microorganisms to label interfaces. *Canadian Journal of Microbiology*, **38**, 1005–8.
Oliveira, R., Melo, L., Oliveira, A. & Salgueiro, R. 1994. Polysaccharide production and biofilm formation by *Pseudomonas fluorescens*: effects of pH and surface material. *Colloids and Surfaces B: Biointerfaces*, **2**, 41–6.
Paul, J. H. & Jeffrey, W. H. 1985. Evidence for separate attachment mechanisms for hydrophilic and hydrophobic surfaces in *Vibrio proteolytica*. *Applied and Environmental Microbiology*, **50**, 431–7.
Pearce, L. E. & Flint, S. H. 2002. Streptococcus thermophilus. In: Encyclopedia of Dairy Science. Academic Press, UK, pp. 2577–82.

Rademacher, B., Walenta, W. & Kessler, H. G. 1995. Contamination during pasteurisation by biofilms of thermophilic streptococci. Heat Treatments and Alternative Methods – Proceedings of the IDF Symposium, Vienna, Austria, 6–8 September 1995, pp. 26–33.

Settanni, L., Grigoli, A. Di., Tornambe, G., Bellina, V., Francesca, N., Moschetti, G. & Bonanno, A. 2012. Persistence of wild *Streptococcus thermophiles* strains on wooden vat and during the manufacture of a traditional Caciocavallo type cheese. *International Journal of Food Microbiology*, **155**, 73–81.

Tsakalidou, E., Zoidou, E., Pot, B., Wassil, L., Ludwig, W., Devriese, L.A., Kalantzopoulos, G., Schleifer, K.-H. & Kersters, K. 1998. Identification of streptococci from Greek Kasseri cheese and description of *Streptococcus macedonicus* sp. nov. *International Journal of Systematic Bacteriology*, **48**, 519–27.

Van der Mei, H. C., De Vries, J. & Busscher, H. J. 1993. Hydrophobic and electrostatic cell surface properties of thermophilic dairy streptococci. *Applied and Environmental Microbiology*, **59**, 4305–12.

Weerkamp, A. H., Handley, P. S., Baars, A. & Slot, J. W. 1986. Negative staining and immunoelectron microscopy of attachment deficient mutants of *Streptococcus salivarius* reveal that the adhesive proteins are separate classes of cell surface fibril. *Journal of Bacteriology*, **165**, 746–55.

Wiencek, K. M., Klapes, N. A. & Foegeding, P. M. 1991. Attachment of *Bacillus* spores to inanimate materials: effects of substratum and spore hydrophobicity. *Biofouling*, **3**, 139–49.

Windholz, M. 1983. The Merck Index – Encyclopoedia of Chemicals, Drugs and Biologicals. Merck and Co. Inc., New Jersey.

Wirtanen, G. & Mattila-Sandholm, T. 1992. Removal of foodborne biofilms – comparison of surface and suspension tests, Part I. *Lebensmittel-Wissenschaft Technologie*, **25**, 43–9.

Yu, F. P. & Mcfeters, G. A. 1994. Physiological responses of bacteria in biofilms to disinfection. *Applied and Environmental Microbiology*, **60**, 2462–6.

7 Thermophilic Spore-Forming Bacilli in the Dairy Industry

Brent Seale[1], Sara Burgess[2], Steve Flint[2], John Brooks[1], Phil Bremer[3] and Shanthi Parkar[4]

[1]School of Applied Sciences, Faculty of Health and Environmental Sciences, Auckland University of Technology, Auckland, New Zealand
[2]Institute of Food, Nutrition and Human Health, Massey University, Palmerston North, New Zealand
[3]Department of Food Science, University of Otago, Dunedin, New Zealand
[4]Plant & Food Research, Palmerston North, New Zealand

7.1 Introduction

Thermophilic spore-forming bacilli are common contaminants of dairy products. Although nonpathogenic, many thermophilic spore-forming bacilli produce extracellular enzymes, which, if allowed to form, can have a negative impact on product quality. Thermophilic spore-forming bacilli are present at very low levels in raw milk, but their spores can survive thermal treatments, such as pasteurisation, and attach to stainless steel surfaces within processing equipment. Cellular adaptations that enable thermophilic spore-forming bacilli to grow and survive in hot environments, such as heat-stable DNA, proteins and membranes, also enable these microorganisms to grow and survive within heated dairy processing equipment. Under favourable conditions, biofilms of thermophilic spore-forming bacilli will develop and vegetative cells and spores will detach from surfaces, enter the product stream and contaminate the final product. This chapter describes the problems associated with thermophilic spore-forming bacilli, our current knowledge on the subject and the control strategies employed by the dairy industry to combat contamination of dairy products.

7.2 Thermophilic spore-forming bacilli of importance to the dairy industry

Thermophilic bacteria are defined in the dairy industry as those bacteria capable of growing on plate count or milk plate count agar (MPCA) during incubation at 55 °C for 48 hours. More generally, they are defined as bacteria capable of growth at temperatures between 45

Biofilms in the Dairy Industry, First Edition.
Edited by Koon Hoong Teh, Steve Flint, John Brooks and Geoff Knight.

and 70 °C. Many species of thermophilic bacteria, belonging to the genus *Bacillus* and related genera, also produce spores and are responsible for causing product quality issues for dairy manufacturers.

The dairy products with the most significant concerns with thermophilic bacteria are milk powders, although thermophilic bacteria do grow in other manufacturing processes. *Anoxybacillus flavithermus* and *Geobacillus stearothermophilus* are the dominant thermophilic spore-forming bacilli found in milk powders manufactured around the world (Ronimus *et al.*, 2003; Rückert *et al.*, 2004; Yuan *et al.*, 2012). Other thermophilic spore-forming bacilli found in milk powder include *Bacillus licheniformis*, *Bacillus coagulans* and *Bacillus subtilis* (Ronimus *et al.*, 2003; Yuan *et al.*, 2012). These latter species are considered to be facultative thermophiles, as they have optimal growth temperatures of less than 50 °C and don't grow at temperatures above 60 °C. Contamination of milk products by thermophilic spore-forming bacilli is not new, as they have been isolated from milk powder recovered from supplies used in an Antarctic expedition in 1907 (Ronimus *et al.*, 2006).

7.2.1 Geobacillus

The genus *Geobacillus* was derived from Group 5 of the genus *Bacillus* as defined in the study of Ash *et al.* (1991). All *Geobacillus* species are thermophilic, with optimal growth temperatures of greater than 50 °C. They produce subterminal and terminal endospores. All species of this genus are closely related, with similarity levels of 16S rDNA sequences in the range 96.0–99.4% (Nazina *et al.*, 2001; Coorevits *et al.*, 2012). Isolates of *Geobacillus* spp. have been obtained from temperate soils, hot springs, oilfields, deep sea sediments, sugar beet juice and dairy products (Nazina *et al.*, 2001; Ronimus *et al.*, 2003, Banat *et al.*, 2004; Tai *et al.*, 2004; Zeigler, 2014).

The predominant species of *Geobacillus* isolated from milk powder is *G. stearothermophilus* (Stadhouders *et al.*, 1982; Flint *et al.*, 2001b; Ronimus *et al.*, 2003; Rückert *et al.*, 2004). Another is *G. thermoglucosidans*, obtained from processing lines and milk powder in the Netherlands (Zhao *et al.*, 2012).

G. stearothermophilus (which includes strains known at various times as *B. calidolactis* and *B stearothermophilus* var. *calidolactis*) has been associated with dairy products since at least the 1950s, when it was found to cause contamination issues with ultra-high-temperature (UHT)-treated dairy products (Galesloot & Labots, 1959a,b). Strains of *G. stearothermophilus* associated with dairy products cannot be differentiated from the type strain of *G. stearothermophilus* (ATCC 12980 = DSM 22) based on 16S rDNA sequencing (Burgess *et al.*, 2014), but they can be differentiated when analysed by molecular biology-based techniques, such as random amplified polymorphic DNA (RAPD) and the internal transcribed spacer (ITS) region, and by phenotypic characterisation (Flint *et al.*, 2001b; Ronimus *et al.*, 2003). Notable phenotypic properties of dairy strains of *G. stearothermophilus* include the ability to utilise lactose and to grow under anaerobic conditions (Flint *et al.*, 2001b; Ronimus *et al.*, 2003). The optimal growth temperature of strains of *G. stearothermophilus* isolated from milk powder is approximately 63 °C (G. Knight, unpublished). Particular strains of *Geobacillus* spp. can produce highly heat-resistant spores that can survive UHT treatment and retorting (Hill & Smythe, 1994).

7.2.2 Anoxybacillus flavithermus

A. flavithermus (formerly *B. flavothermus*) was first isolated from a hot spring in New Zealand and was described as a Gram-positive, facultative anaerobic, motile, rod-shaped spore-forming bacterium (Heinen *et al.*, 1982). The G+C content was 41.6 mol%, the temperature range for growth was between 30 and 70 °C and the optimum growth temperature, under aerobic conditions, was 60 °C (Heinen *et al.*, 1982; Pikuta *et al.*, 2000). This species was transferred to the genus *Anoxybacillus* (and its epithet corrected to flavithermus), alongside the newly described species, *A. pushchinoensis* (Pikuta *et al.*, 2000). *A. pushchinoensis* was initially described as a strict anaerobe, but this has since been revised to aerotolerant anaerobe, and the genus is now considered to contain aerotolerant anaerobes and facultative anaerobes (Pikuta *et al.*, 2003).

Strains of *A. flavithermus* have subsequently been isolated from gelatin and milk powder (Flint *et al.*, 2001b; Ronimus *et al.*, 2003; De Clerck *et al.*, 2004). Isolates from milk powder have an optimum growth temperature of approximately 57 °C (G. Knight, unpublished). The genome sequences of several strains of *A. flavithermus* are available as either draft assemblies or completed genomes (Saw *et al.*, 2008; Matsutani *et al.*, 2013; Wang *et al.*, 2014), with one sequenced strain isolated from a dairy processing environment (Caspers *et al.*, 2013).

7.2.3 Bacillus licheniformis

B. licheniformis is a facultative thermophile, with a growth range of 35–55 °C, and belongs to the *B. subtilis* group. It is a soil microorganism that is commonly found in a range of dairy products, but at low to moderate numbers (Crielly *et al.*, 1994; Cook & Sandman, 2000; Ronimus *et al.*, 2003). *B. licheniformis* is generally considered to be nonpathogenic, but toxigenic strains have been linked to food poisoning outbreaks associated with raw milk and processed baby foods (Salkinoja-Salonen *et al.*, 1999). In addition, some strains have been associated with bovine abortion, septicemia and other infections (Logan, 1988). Spores of *B. licheniformis* can survive pasteurisation but don't appear to germinate and grow in processing lines, and therefore they are not viewed as so much of a concern as spores from *Geobacillus* spp. or *A. flavithermus*.

7.3 Spoilage by thermophilic bacilli

Thermophilic spore-forming bacilli are considered spoilage microorganisms, due to the ability of some strains to produce enzymes, such as proteases and lipases, and to produce acid during growth (Basappa, 1974; Chopra & Mathur, 1984; Cosentino *et al.*, 1997; Chen *et al.*, 2004; Gundogan & Arik, 2004; Murugan & Villi, 2009).

The real potential for the obligate thermophiles to spoil dairy products is thought to be low, since the products are generally stored at temperatures below 37 °C, which is below the lower temperature limit for their growth. The water activity in milk powders is also too low for germination and growth of spores, unless reconstituted milk is temperature-abused. Production of spoilage enzymes may occur in reconstituted and heated products such as cream- and milk-based sauces if thermophiles are present. *Geobacillus* strains also produce amylases, which may degrade starches used as thickening agents for sauces. Finally, *G. stearothermophilus* has also been associated with 'flat-sour' spoilage of evaporated milk (Kalogridou-Vassiliadou, 1992).

In the case of facultative thermophiles, strains of *B. licheniformis* are capable of growth at ambient temperatures and can spoil dairy products if spores survive processing treatments. These strains are also capable of producing a slimy extracellular substance that can affect the quality of pasteurised milk and cream (Gilmour & Rowe, 1990). *B. subtilis* has been associated with ropiness in raw and pasteurised milk, as well as with the spoilage of UHT and canned milk products (Heyndrickx & Scheldeman, 2002). *B. coagulans* has been connected to the spoilage of UHT and canned milk products via the production of lactic acid (Gilmour & Rowe, 1990).

7.4 Bacterial endospores

7.4.1 *Spore structure and resistance*

Bacterial endospores, or more simply spores, are metabolically dormant cell forms that enable microorganisms to survive adverse conditions. Contamination by spores is recognised in many food industries around the world as a major issue affecting food safety and quality (Andersson *et al.*, 1995; Faille *et al.*, 2001, 2014). This is due to the innate resistance of spores to many of the techniques, such as thermal processing and the addition of antimicrobial compounds, employed by food manufacturers to inactivate microorganisms and increase the microbial stability of foods (Chandler *et al.*, 2001; Cortezzo & Setlow, 2005; Jones *et al.*, 2005; Scheldeman *et al.*, 2006). A number of spore structural properties contribute to this resistance.

Starting from the inside and working outwards (Figure 7.1), the innermost structure is the spore core, which contains the genetic material (DNA) of the spore, calcium dipicolinate (CaDPA, making up 5–15% by dry weight of the spore; Powell, 1953) and small acid-soluble proteins (SASPs, which bind to and protect spore DNA from damage; Setlow & Setlow, 1979). CaDPA levels differ between bacterial species and even within the same species (Huang *et al.*, 2007). It is believed that CaDPA replaces much of the water present in the spore core and contributes to resistance to wet and dry heat (Paidhungat *et al.*, 2000). Mineralisation has also been shown to be important in heat resistance (Bender & Marquis, 1985; Beaman *et al.*, 1988). While CaDPA contributes to the resistance of spores to both wet and dry heat (Setlow *et al.*, 2006), it also sensitises spore DNA to damage caused by ultraviolet (UV) irradiation (Douki *et al.*, 2005). SASPs of types α and β have been shown to have an important role in protecting spore DNA from DNA-damaging agents (Setlow, 2007). SASPs bind to GC-rich regions of spore DNA and form a tightly packed assembly (Gerhard & Marquis, 1989; Frenkiel-Krispin *et al.*, 2004; Setlow *et al.*, 2006). Mutants of *B. subtilis* that lack SASPs are more sensitive to UV irradiation, desiccation and wet and dry heat.

Surrounding the core is a membrane, which is itself surrounded by the cortex (Warth & Strominger, 1969). The relative impermeability of the membrane is important to maintenance of conditions in the spore core (Swerdlow *et al.*, 1981). The cortex consists of two peptidoglycan layers: that adjacent to the inner membrane has the same structure as that of the cell wall

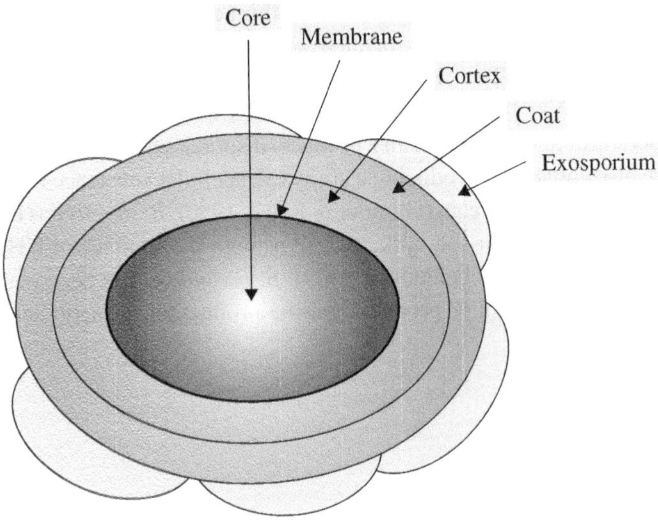

Figure 7.1 Spore structure. The core is surrounded by a membrane, cortex and coat. Certain species contain an additional layer, known as an exosporium.

(referred to as 'primordial cell wall'), while the outer peptidoglycan layer is the main layer of peptidoglycan and has a slightly different structure to the primordial cell wall, with the main difference being the amount of de-*N*-acetylation of an amino sugar glucosamine (Atrih & Foster, 2001).

The spore coat is a highly cross-linked protein layer that occupies most of the spore's volume and contains 70–80% of the spore's total protein (Aronson & Fitz-James, 1976; Driks, 1999; Henriques & Moran, 2000). It contains large amounts of crosslinked cysteine and tyrosine, creating a rigid structure (Aronson & Fitz-James, 1976; Pandey & Aronson, 1979). It is made up of two layers, an inner, laminated layer and an outer, electron-dense layer. The spore coat provides resistance to hydrolytic enzymes, such as lysosyme and trypsin, but remains permeable to small molecules and water. Removal of the spore coat renders a spore sensitive to lysosyme, as the peptidoglycan that makes up the cortex (the site for action of this enzyme), is exposed. It also provides protection from mechanical disruption, UV irradiation and chemicals, such as hydrogen peroxide (Gould & Hitchins, 1963; Riesenman & Nicholson, 2000; Hullo *et al.*, 2001).

In some species, the spore coat is the outermost layer of the spore. However, spores of bacilli such as *B. cereus* and *B. anthracis* have an additional layer, called the exosporium (Gerhardt & Ribi, 1964). The exosporium consists of two layers: a hexagonal crystal lattice structure and an outer 'hair-like' nap of filaments (Gerhardt & Ribi, 1964). The chemical composition of the exosporium of *B. cereus* consists of protein, lipids and polysaccharides, such as glucose, glucosamine and rhamnose (Matz *et al.*, 1970). Specific glycoproteins and carbohydrates (Fox *et al.*, 1993; Sylvestre *et al.*, 2002), as well as proteins (Redmond *et al.*, 2004), have been characterised from the exosporium of *B. anthracis*.

Currently, little is known about the function of the exosporium. It has been shown to protect spores of *B. anthracis* upon ingestion by a macrophage (Weave *et al.*, 2007). Spores of *B. cereus* contain long appendages, constructed from hydrophobic residues and carbohydrate (Stalheim & Granum, 2001), which have been shown to influence the initial attachment of *B. cereus* spores to stainless steel (Klavenes *et al.*, 2002; Tauveron *et al.*, 2006). Transmission electron microscopy (TEM) of thin sections of *Geobacillus* spores isolated from milk powder production lines has revealed an exosporium (Seale *et al.*, 2010). However, the role of this structure for *Geobacillus* is unknown.

The resistance properties of spores are influenced by environmental conditions (such as temperature, pH and media composition) during sporulation (Palop *et al.*, 1999). Spores produced at temperatures greater than their optimal growth temperature tend to be more heat resistant (Beaman & Gerhard, 1986). The metal ion content of the growth medium can also influence their resistance properties (Cazemier *et al.*, 2001). Spores grown on nutrient media supplemented with calcium and magnesium are more heat resistant than those grown in media fortified with manganese only (Cazemier *et al.*, 2001). Spores attached to stainless steel surfaces are also reported to be more heat resistant than spores in suspension (Simmonds *et al.*, 2003), while spores produced during milk powder production are more heat resistant than those grown under laboratory conditions (Hill & Smythe, 2004). Furthermore, spores of *G. stearothermophilus* have a higher heat resistance when suspended in milk than in water (Yildiz & Westhoff, 1989).

7.4.2 *Sporulation*

The life cycle of sporulation, dormancy and germination of bacilli, first observed by Cohn (1876), is shown in Figure 7.2. Many factors are reported to initiate sporulation, including nutrient limitation, population and oxidative stress. These all act through the phosphorylation cascade (Hoch, 1993) to activate a specific set of sigma factors (Driks, 1999) and a master transcription factor, *spo0A* (Piggot & Hilbert, 2004). The first step involves the cell undergoing asymmetric division and inward folding of the cytoplasmic membrane to form a septum. This septum separates the mother cell from the daughter cell (also referred to as the forespore). The forespore is engulfed by the mother cell, and then the cortex is formed, followed by the spore coat. The spore becomes denser through the uptake of CaDPA, dehydration occurs and resistance develops. The mother cell finally undergoes lysis and the spore is released. The entire process takes about 8 hours in *B. subtilis* (Driks, 1999).

The presence of spores in milk powder is a serious concern, due to their resistance to high temperature. It currently remains unclear what triggers sporulation during milk powder production, since there is a constant flow of nutrients present in the milk. Perhaps the nutrients are unable to reach particular cells within a biofilm and therefore trigger sporulation, or perhaps sporulation within a biofilm is part of a complex life cycle of biofilm development. Recent research has shown that sporulation of thermophilic dairy isolates requires optimal growth conditions (Scott *et al.*, 2007; Seale *et al.*, 2008).

7.4.3 *Germination*

Germination occurs in three steps: activation, germination and outgrowth (Dring & Gould, 1971). Activation is a reversible process in which the spore is prepared for germination but retains many of its properties, such as heat resistance. If conditions are favourable, the spore

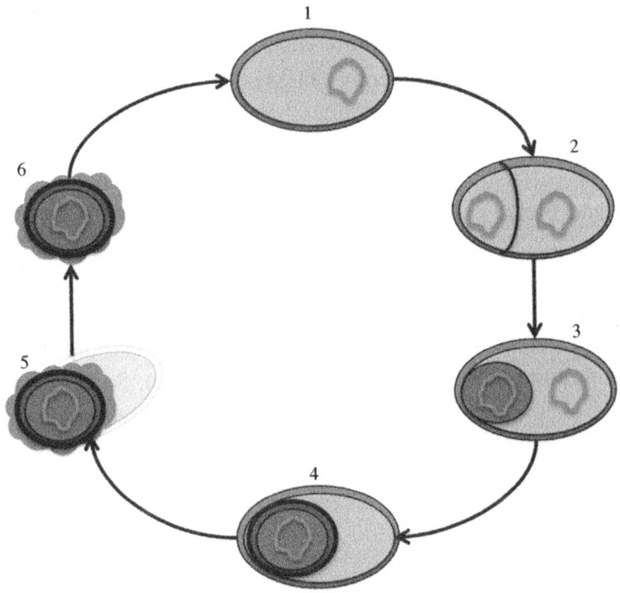

Figure 7.2 The main steps in sporulation by endospore-forming bacteria. Vegetative cells (1) undergo asymmetrical division (2) and the mother cell engulfs the daughter cell (3). The cortex and spore coats are then synthesised (4) and the exosporium (present in some species) forms (5). Finally, the mother cell lyses and the mature endospore is released (6).

will then undergo the irreversible process of germination. During this time, a cascade of events occurs: H^+ ions are secreted, thereby raising the pH of the spore core from 6.5 to 7.7; CaDPA is lost and is replaced by water; and the cell loses dormancy and becomes metabolically active. The next step involves hydrolysis of the peptidoglycan in the cortex and further swelling of the spore core due to the ingress of water and expansion of the germ cell wall. There have been many reports of different signal substances that can induce activation and germination. Heat is commonly used for thermophilic spores (Beaman *et al.*, 1988), while amino acids such as L-alanine are used for mesophilic species (Donnellan *et al.*, 1963). Low pH has also been shown to activate spores, but this does not necessarily lead to germination (Issahary *et al.*, 1970). The spore coat has been shown to be important in the germination of *B. cereus* spores, as components of the coat are used as nutrients by the germinating organism (Kutima & Foegeding, 1987).

7.5 Enumeration of thermophilic bacilli

The enumeration of vegetative cells and spores of thermophilic bacilli is very important in the dairy industry, as it helps ensure that manufactured products meet specifications. Traditional viable plate-counting techniques, used to determine both thermophile and thermophilic spore counts, can take a long time to obtain a result, which can delay the release of dairy products. As a result, there is a focus on developing novel rapid enumeration methods to enable more rapid release of products.

7.5.1 *Viable plate counts*

Currently there is no standard enumeration technique for either vegetative cells or spores of thermophilic bacilli. The traditional methods are the total thermophile plate count (TPC) and the thermophilic spore count (TSC). The TPC method involves reconstitution of product, followed by the transfer of 1 ml of sample and dilutions into separate Petri dishes containing MPCA and incubation at 55 °C for 48 hours (Frank & Yousef, 2004).

In TSC, the sample is first heat treated for 30 minutes at 100 °C to inactivate vegetative cells. The sample and dilutions are then pour plated with MPCA supplemented with 0.2% starch and incubated at 55 °C for 48 hours. This heat treatment is higher than in previously published methods for thermophilic spores (such as 80 °C for 10 minutes, Coorevits *et al.*, 2008; 80 °C for 20 minutes, McGuiggan *et al.*, 2002; or 100 °C for 10 minutes, Rückert *et al.*, 2004, 2005b). The higher temperature over a longer of period of time ensures that the method selects for spores that would survive the higher processing temperatures used during dairy manufacturing.

Recently, a new method has been developed to enumerate highly heat-resistant spores in milk powder that are to be further processed for UHT or retort treatment. This method involves heat treating at 106 °C for 30 minutes; this higher temperature selects for spores of specific *Geobacillus* spp. and destroys spores of *A. flavithermus* (Hill & Smythe, 2004).

7.5.2 *Rapid methods*

Rapid methods have the potential to reduce both labour costs and the time required to obtain results for thermophile and thermophilic spore counts. Unfortunately, rapid methods typically require access to expensive equipment, as well as specialised training. Two rapid methods have recently been developed for the enumeration of thermophilic bacteria in milk powder, one using flow cytometry and the other using real-time polymerase chain reaction (RT-PCR).

A rapid method of determining viable mesophilic bacterial cell numbers in milk powder (equivalent to a standard plate count) was developed by Flint *et al.* (2006) using a BactiFlow flow cytometer and a fluorescent substrate that can detect esterase activity. This method showed promise and was modified to enumerate thermophilic bacteria in milk powder by adding a 55 °C incubation step (Flint *et al.*, 2007). It showed good correlation with TPC data during the development phase, but did not always correlate well during routine use in a manufacturing context. In addition, the detection limit was not low enough for some milk powders.

An RT-PCR assay was developed by Rückert *et al.* (2005a,b) to enumerate total viable vegetative cells and spores of *A. flavithermus*, *B. licheniformis* and *B. megaterium* in milk powder. *Geobacillus* strains were not included in the study. The assay targeted the 16S rDNA gene, and it should be noted that this gene can have a variable copy number, which may have influenced the results. A similar RT-PCR assay was developed which targeted the *spo0A* gene (Rückert *et al.*, 2006); this assay amplified DNA from a variety of the thermophilic bacilli (the targeted strains), as well as a number of nontargeted strains, including *B. cereus* and *B. smithii*. The assay was rapid and provided a result within 1 hour. However, RT-PCR assays are costly to perform, require technical expertise and are not very sensitive. In addition, RT-PCR assays require incorporation of a reverse transcriptase step in order to target viable bacterial cells.

In summary, while there has been progress in the development of rapid methods for the detection and enumeration of thermophilic bacilli, these methods require specialised equipment and training. Further research is also required to improve the sensitivity, specificity and robustness of these techniques before they can be applied in the dairy industry.

7.6 Characterisation and identification of thermophilic bacilli

The identification and typing of thermophilic bacilli is very important in tracing sources of contamination during manufacturing processes and in demonstrating that cleaning procedures are effective at eliminating thermophilic bacilli from equipment. Characterisation of thermophilic bacilli can be achieved using traditional taxonomic approaches or using modern molecular biology-based approaches.

Biochemical test kits (e.g. API CHB kits) are commonly employed to characterise Gram-positive spore-forming bacilli. Such kits are useful for many mesophilic species, such as *B. subtilis* and *B. cereus*, but are not very reliable for the identification of thermophilic bacilli that are commonly found in dairy products. Despite this, phenotypic testing, using biochemical test kits and traditional test methods, is still valuable in characterising thermophilic bacilli. Such testing can identify characters that can be used to easily differentiate between species and can reveal properties that are relevant to growth in dairy processing equipment, such as lactose utilisation and growth under anaerobic conditions (which is important, as the oxygen content of milk during evaporation is low). Such testing can also reveal the spoilage potential of strains or species, including the ability to produce extracellular enzymes such as amylases, proteases and lipases.

The current recommendations for the delineation of new bacterial species include obtaining 16S rDNA gene sequence data, performing DNA–DNA hybridisation with closely related bacteria and determining phenotypic and chemotaxonomic characteristics (Stackebrandt *et al.*, 2002). The distinction between species of *Geobacillus* based on 16S rDNA sequence data is not always clear. Alternatives to 16S rDNA sequencing, and in particular DNA–DNA hybridisation, for delineation of species are currently being investigated. Many of these approaches are based around phylogenies created from sequencing of housekeeping genes.

Recently, a 16S rDNA method was developed by Chauhan *et al.* (2013) which can rapidly identify a number of dairy bacilli, including *Geobacillus* spp., *A. flavithermus* and *B. licheniformis*. This method uses primers to amplify two separate variable regions within the 16S rDNA gene, using PCR. The products then undergo a high-resolution melt analysis (HRMA) using DNA-binding fluorescent dyes and a PCR machine with a highly precise temperature control. While this method is excellent for identifying a wide range of bacilli from a range of different dairy products, it is unable to differentiate between different species of *Geobacillus*, due to the close similarity in the 16S rDNA sequence.

A method for identifying dairy *Bacillus* spp. and *Paenibacillus* spp. has been suggested based on sequence analysis of the housekeeping gene *rpoB* (Durak *et al.*, 2006). However, this may not be a suitable for *Geobacillus* spp. as the *rpoB* gene is highly conserved in this genus. Studies have shown that sequencing of the variable regions within the *rpoB* gene could replace 16S rDNA sequencing in *Geobacillus* spp. as a species-level identification method (Meintanis *et al.*, 2008; Weng *et al.*, 2009). Other gene targets for the typing of *Geobacillus* spp. include *recA* (Weng *et al.*, 2009), *spo0A* (Kuisiene *et al.*, 2009), *recN* (Zeigler, 2005) and the 16S-23S IST region (Flint *et al.*, 2001b; Kuisiene *et al.*, 2007). While these targets are effective at discriminating between different thermophilic bacilli, there has been limited success with *Geobacillus* spp.

7.6.1 *Molecular-based typing methods*

The typing of microorganisms can provide information on sources of contamination within a dairy manufacturing plant. A number of different methods are available for this, including pulsed field gel electrophoresis (PFGE), restriction fragment length polymorphism (RFLP), multilocus sequence typing (MLST), RAPD, multiparametric TaqMan RT-PCR and multilocus variable number of tandem repeat analysis (MLVA).

RAPD-PCR profile analysis (Williams *et al.*, 1990) has previously been used to identify and type *Geobacillus* spp., *A. flavithermus* and *B. licheniformis* isolates obtained from milk powder from New Zealand (Ronimus *et al.*, 2003) and around the world (Rückert *et al.*, 2004). This technique employs PCR amplification of nonspecific regions of the genome using short arbitrary primers. The PCR products are then run through gel electrophoresis, resulting in different banding patterns between isolates, which can be compared between known controls and samples. RAPD-PCR profiling requires no genome sequence information and is quick and easy. However, this technique is known to have poor reproducibility between laboratories and interpretation of the banding patterns can be difficult due to weak bands in an isolate's profile, resulting from varying efficiency of the PCR reaction and mismatches between the primer and the DNA template.

Recently, a multiparametric TaqMan RT-PCR assay was developed that can discriminate between 38 different species of spore-forming bacilli, including psychrotrophic, mesophilic and thermophilic aerobic bacilli, as well as members of the genus *Clostridia*. Sensitivity is high if a pre-enrichment step is used, and the detection limit is one spore of *B. cereus* in a 25 g food sample (Postollec *et al.*, 2010, 2012). Unfortunately, this method has a relatively low sample throughput. Fernández-No *et al.* (2011) reported a quantitative TaqMan-probe assay for *B. cereus*, *B. licheniformis* and *B. subtilis* directly from foods that did not include a pre-enrichment step. However, this method is unable to differentiate between the three species.

Recently, a new typing method has been developed, based on MLVA. Length polymorphisms arise in a variable number of tandem-repeat VNTRs due to the variabile copy number of tandem repeats found within genes or noncoding regions of a genome, which can be analysed using gel electrophoresis or high-resolution melting analysis (MLV-HRMA) (Keim *et al.*, 2004; Vogler *et al.*, 2007; Reyes & Tanaka, 2010). MLVA has been used to genotype a number of different pathogens, including *Salmonella enterica* (Boxrud *et al.*, 2007), *Clostridium difficile* (Broukhanski *et al.*, 2011), *Mycobacterium tuberculosis* (Le Fleche *et al.*, 2002), *Listeria monocytogenes* (Murphy *et al.*, 2007) and *B. anthracis* (Keim *et al.*, 2000). In most cases, the MLVA methods provide discrimination values equal to or greater than the gold-standard genotyping methods (PFGE and MLST) currently used for these microorganisms.

MLV-HRMA techniques have been developed for the typing of *Geobacillus* spp. and *B. licheniformis* isolates obtained from milk powders manufactured in Australia (Seale *et al.*,

2012; Dhakal *et al.*, 2013). The study by Seale *et al.* (2012) demonstrated that three types of *Geobacillus* spp. could coexist during a single processing run, and that specific types were associated with high-spore-count powders. However, there was no correlation of specific types with particular dairy manufacturing plants. Another interesting finding was that isolates obtained in 1995 were of the same type as those obtained in 2012, indicating that the same types remained prominent over 17 years. The study by Dhakal *et al.* (2013) showed that isolates of *B. licheniformis* were more heterogeneous, across multiple product runs and milk powders, than previously thought, and no correlation could be drawn between prominent types and specific dairy manufacturing plants. These studies showed that MLV-HRMA was more discriminative and reproducible than the RAPD method previously used to type thermophilic bacilli.

A microarray-based identification and typing method using 130 genomic markers has been shown to discriminate 34 different strains from six *Bacillus* species and four species of *Geobacillus* isolated from a variety of food products (Caspers *et al.*, 2011). This method looks at differences between core and accessory genome markers across *Bacillus* and related genera. A majority of the core genome markers do not hybridise between species, resulting in discrimination at the species level, while the accessory genome markers can result in high-resolution discrimination between individual isolates of a single species.

Future developments in the typing of thermophilic spore-forming bacilli will arise out of whole-genome sequencing as the technique becomes more readily available and more economical. A number of thermophilic bacilli isolated from dairy products and dairy manufacturing plants have had their genomes sequenced, including isolates of *G. thermoglucosidans* (Zhao *et al.*, 2012), *A. flavithermus* (Caspers *et al.*, 2013) and *B. licheniformis* (Dhakal *et al.*, 2014). These genomes will provide targets that might serve as the basis for typing techniques and might provide some insight into how these microorganisms persist within dairy manufacturing environments.

7.7 Biofilm formation by thermophilic bacilli

A biofilm is a community of microorganisms that are attached to and actively growing on a surface in an aqueous environment. Once attached to a surface, and if conditions are favourable, the microorganisms can replicate and secrete extracellular polymeric substances (EPS), which irreversibly binds them to the surface. As the biofilm develops, cells and spores can detach from the surface and enter the aqueous phase. This section discusses some of the relevant stages of biofilm formation, with a particular focus on thermophilic spore-forming bacilli and dairy manufacturing processes.

7.7.1 *Attachment of cells and spores to surfaces*

A large number of factors influence the attachment of cells and spores to a surface, including interactions between the microorganism and the conditioning film and physicochemical interactions between the microorganism and the surface.

A conditioning film forms on a surface almost immediately after it comes into contact with an aqueous solution. The conditioning film comprises organic and inorganic molecules, which are transported from the aqueous phase to the solid–liquid interface via diffusion or fluid dynamic forces. The conditioning film can alter the physicochemical properties of the surface, such as the surface charge and the hydrophobicity, and molecules adsorbed to the surface can occupy binding sites for bacteria. Molecules in the conditioning film may also serve as a nutrient source for microorganisms actively growing on the surface.

Milk is a complex colloid suspension consisting of proteins, fats and salts. Studies investigating the influence of the presence of milk proteins adsorbed to a surface on the attachment of thermophilic bacilli have produced conflicting results. Some have shown that milk proteins inhibit attachment of thermophilic bacilli to surfaces (Parkar *et al.*, 2001; Han *et al.*, 2011), but Flint *et al.* (2001a) found that the attachment of vegetative cells of *Geobacillus* spp. to stainless steel increased 10–100-fold with the presence of milk fouling. The same result is supported by the work of Hinton *et al.* (2002), who found a milk fouling layer enhanced accumulation of *G. stearothermophilus* on stainless steel. The conflicting results of these studies may be explained by the different characteristics of the milk protein layers (thickness, native versus denatured proteins) and how these influence attachment.

A number of studies have looked at the attachment of spores to surfaces (Husmark & Ronner, 1990, 1992; Ronner *et al.*, 1990; Ronner & Husmark, 1992; Faille *et al.*, 2002; Seale *et al.*, 2008). These studies have shown that hydrophobicity plays an important role in the attachment of spores to surfaces, with the general rule that the more hydrophobic a microorganism or surface, the greater the attachment. This rule is supported by the observation that spores of *Geobacillus* spp. suspended in simple saline solutions attached in greater numbers to surfaces with greater hydrophobicity, such as polystyrene, than to hydrophilic surfaces, such as glass (Seale, 2009). However, when spores were suspended in skim milk and exposed to the same surfaces, no differences in the number of spores attaching to surfaces was observed. This may be due to the milk proteins, adsorbed to the substrate (i.e. the conditioning film) and spore surfaces, masking the original surface properties of both the spores and the substrate.

The influence of the hydrophobic nature of the spore surface is not as clear. Seale *et al.* (2008) found that the spores of *Geobacillus* spp. isolates that were the most hydrophilic attached in greater numbers to stainless steel than those which were more hydrophobic. Another study by Parkar *et al.* (2001) found there was no correlation between spore hydrophobicity and the adhesion of spores from thermophilic bacilli. From these studies, it can be assumed that hydrophobic/hydrophilic interactions are not the only factors involved in attachment of spores of thermophilic bacilli to surfaces.

The outer layers of spores have been shown to be involved in attachment. A study by Faille *et al.* (2007) demonstrated a slight reduction in the attachment of *B. cereus* spores to model food processing surfaces after removal of the exosporium. In contrast, removal of the outer protein coats of *Geobacillus* spp. and *A. flavithermus* spores did not result in any changes in the number of spores that attached to surfaces (Parkar *et al.*, 2001).

7.7.2 *Biofilm development*

Biofilm development by thermophilic bacilli has been analysed extensively in laboratory settings (Flint *et al.*, 2001a; Parkar *et al.*, 2003; Burgess *et al.*, 2009). Under favourable conditions (temperature, pH and water activity), attached spores germinate, grow and form

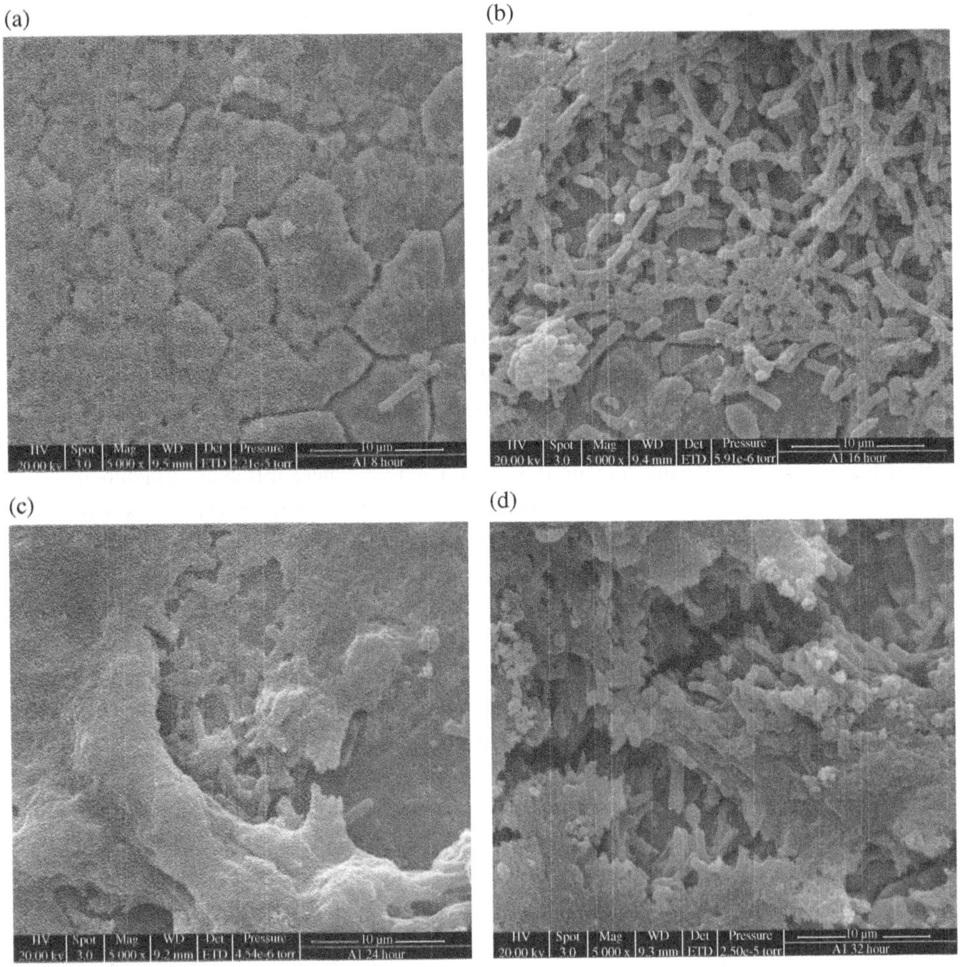

Figure 7.3 Scanning electron micrographs displaying the development of a *Geobacillus* spp. biofilm on stainless steel in skim milk at 55 °C after (a) 8, (b) 16, (c) 24 and (d) 32 hours.

a biofilm (Flint *et al.*, 2001a). Figure 7.3 shows the development of a biofilm of a strain of *Geobacillus* spp. on stainless steel, in the presence of reconstituted skim milk, over 32 hours. Burgess *et al.* (2009) demonstrated that the biofilms created by *A. flavithermus* are capable of being initiated from either vegetative cells or spores. Strains of both *A. flavithermus* and *Geobacillus* spp. reached a biofilm cell density of 6–7 $\log_{10}$ CFU/cm^2 after 6 hours (Flint *et al.*, 2001a; Parkar *et al.*, 2003; Burgess *et al.*, 2009). Parkar *et al.* (2003) also found that uninoculated pasteurised milk with low levels of an unknown thermophilic bacterium (~1 CFU/ml) produced a biofilm after 18 hours. This is more representative of what actually happens in a milk powder manufacturing plant, since thermophiles are typically found in very low numbers in raw milk (<10 CFU/ml).

It is likely that biofilms that develop in milk powder production equipment will contain multiple species of thermophilic bacilli. Seale *et al.* (2012) demonstrated that multiple

Geobacillus spp. types were present during a single production run. A study has shown that *G. thermoglucosidans*, isolated from dairy products, is unable to grow and produce a biofilm in milk in pure culture but can do so in the presence of other strains of thermophilic bacilli (Zhao *et al.*, 2013). Biofilms of thermophiles within processing lines are thought to develop as a monolayer, due to the thin boundary layer created by the high shear rates of turbulent flow (Beyenal & Lewandowski, 2002). However, multilayer biofilms may occur in locations where the flow rate is low, such as underneath distribution plates in evaporators. More research is required in order to better understand the composition and structure of thermophilic biofilms growing *in situ* on a dairy manufacturing plant processing line.

7.7.3 *Spore development within biofilms*

A number of studies have examined spore production within biofilms (Storgårds *et al.*, 2006; Shi & Zhu, 2009; Shaheen *et al.*, 2010). Lindsay *et al.* (2005) demonstrated that *B. subtilis* biofilms produced spores when placed under nutrient-limiting conditions. A recent study by Faille *et al.* (2014) analysed spore production for both mono- and mixed-species biofilms. It found that biofilms allowed to develop on stainless steel surfaces for 48 hours consisted of 90% spores. It also demonstrated that spores were easily transferred by direct contact with agar surfaces, a procedure used to mimic the transfer of spores from equipment surfaces to food. Burgess *et al.* (2009) demonstrated that *A. flavithermus* produced spores when grown in a continuous-flow reactor, and that they were released into the milk flowing through the reactor. Interestingly, spores were produced when the system was operated at 55 and 60 °C, but not when it operated at 48 °C. Currently, little is known about the extent of spore formation by thermophilic bacilli growing in biofilms within dairy manufacturing plants.

7.8 Thermophilic bacilli in dairy manufacturing

7.8.1 *Thermophilic bacilli in raw milk*

Levels of thermophilic bacteria in raw milk are generally very low (<10 CFU/ml) (Hill & Smythe, 1994; McGuiggan *et al.*, 2002). High thermophile levels (>100 CFU/ml) are generally associated with the presence of facultative thermophiles, such as *B. licheniformis* and *B. coagulans*. The groups of thermophilic bacilli that are often found at high levels in milk powder (i.e. *A. flavithermus* and *G. stearothermophilus*) are rarely isolated from raw milk.

7.8.2 *Milk powder manufacturing*

Most of our knowledge about thermophilic bacilli in dairy manufacturing has come from studies focusing on the production of milk powder (Murphy *et al.*, 1999; Ronimus *et al.*, 2003; Scott *et al.*, 2007; Seale *et al.*, 2012; Zhao *et al.*, 2013; Burgess *et al.*, 2014). Manufacture of milk powder involves the removal of water from milk and typically employs

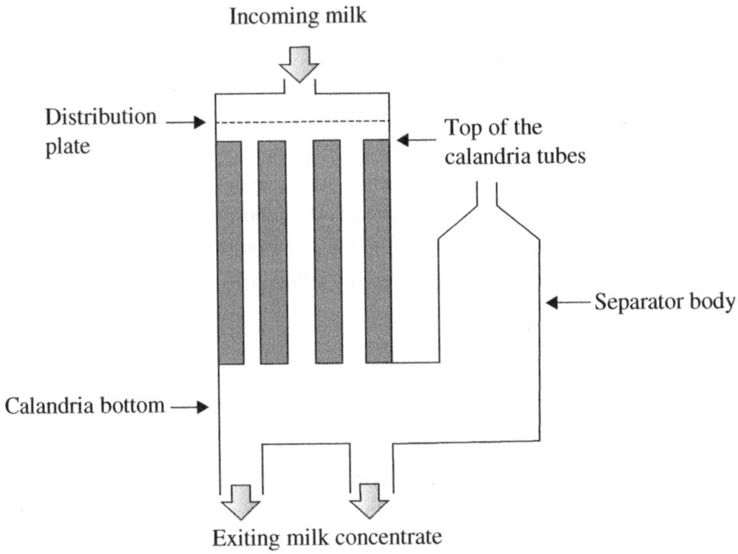

Figure 7.4 Schematic diagram of an evaporator.

an evaporation step, in which water is evaporated from milk to produce a milk concentrate, and a drying step, in which most of the remaining water is removed by spray drying.

The preheating sections of evaporators are particularly vulnerable to colonisation by thermophilic bacteria, due to the large surface area held at temperatures suitable for their growth. The preheating section is rarely opened and examined, so it is difficult to determine the true nature and extent of biofilms formed there. Anecdotal evidence suggests that the use of dual-plate heat exchangers or preheating equipment that has a reduced surface area, such as direct steam injectors (DSIs), can reduce thermophile growth.

Evaporators also operate at temperatures that support the growth of thermophilic bacteria (approximately 45–68 °C). Again, evaporators are rarely opened and examined, so it is difficult to ascertain the full extent of biofilm development in such equipment. However, there is evidence that thermophilic bacteria are associated with foulant that accumulates around distribution plates and at the tops and the bottoms of calandria tubes (Figure 7.4). Fouling underneath distribution plates can be caused by recirculation of milk underneath the plate and/or foaming, which can result in a fluffy foulant that contains high numbers of thermophiles (10^5–10^6 CFU/g) (Scott *et al.*, 2007). Incorrect alignment between the distribution plate and the calandria tubes can result in poor distribution of milk at the top of the calandria tubes and can ultimately lead to blockage of the calandria tubes. Blocked calandria tubes can be a significant source of thermophile contamination.

7.8.3 *Thermophilic bacilli in other dairy processes*

Thermophiles are likely to grow as biofilms in any dairy manufacturing plant processing equipment where temperatures are suitable (Lindsay & Flint, 2009). While much of the work conducted on thermophiles has focused on milk powder manufacturing, other processes in which thermophile contamination may be an issue include plate heat exchangers, separators and ultrafiltration (UF) equipment operating in the temperature range at which thermophiles grow (45–70 °C) (Lehmann, 1995; Scott *et al.*, 2007). In addition, recycle loops, further processing of ingredients (e.g. buttermilk, milk powders) containing thermophiles, milk solids recovery, dead ends, holding tanks at thermophile growth temperatures and damaged seals or gaskets can all contribute to increased levels of thermophiles. Cheese, buttermilk, whey, pasteurised milk and cream are among the other dairy products that have been associated with thermophilic bacteria (Langeveld *et al.*, 1990; Cosentino *et al.*, 1997; Murphy *et al.*, 1999; Scott *et al.*, 2007; Burgess *et al.*, 2010).

Milk separators operate best at warm temperatures and have a large surface area to support the growth of thermophilic bacteria. Where multiple separators are used in a manufacturing process, such as in the manufacture of anhydrous milk fat, the aqueous streams in particular are likely to contain thermophilic spore-forming bacteria. It is reasonable to expect some bacteria to be carried through with the fat phase in each subsequent separation process, which may result in high levels of contamination by thermophilic bacteria as they continue to multiply.

B. licheniformis has been reported as the dominant isolate from UF plants used in the dairy industry (Lehmann, 1995). This thermophilic bacterium has long been associated with the dairy industry but there are few reports on biofilms containing it. This may be due its being outgrown by faster-growing thermophilic bacteria such as *G. stearothermophilus* and *A. flavithermus*. How *B. licheniformis* interacts with other thermophilic bacteria in a biofilm is unknown and requires investigation.

7.9 Control of thermophilic bacilli

Current practices employed in the dairy industry to reduce contamination by thermophilic bacilli include short production times, increased cleaning frequency and the use of sanitisers. Recently, focus has turned to the development of novel control techniques such as temperature cycling, reduction of the surface area of equipment in the optimal temperature growth zone and duplication of equipment.

7.9.1 *Cleaning-in-place*

At the end of every production run, processing equipment must be cleaned using an appropriate cleaning-in-place (CIP) regime (Romney, 1990; Christi, 1999). Using a CIP regime means performing cleaning without having to dismantle the processing equipment. A typical CIP regime, explained in more detail in Section 4.5.2, consists of the following steps: a warm water rinse, an alkaline wash, a water rinse, a nitric acid wash and a final water rinse. The alkaline wash is designed to remove organic matter, such as fats and proteins. Nitric acid is a strong oxidiser and removes inorganic material, such as calcium phosphate and other salts. In some cases, a sanitiser may be applied at the end of CIP, to inactivate any microorganisms that might remain in equipment.

The ability of CIP procedures and sanitisers to remove biofilms and spores from processing equipment is still subject to debate. Microorganisms may remain on surfaces following CIP, even though they appear clean (Watkinson, 2008). Parkar *et al.* (2004) demonstrated that the sequential application of a 2% sodium hydroxide solution (75 °C for 30 minutes) and a 1.8% nitric acid solution (75 °C for 30 minutes) removed biofilms of *A. flavithermus* from stainless steel surfaces. However, changing the temperature and/or the concentrations of the sodium hydroxide and nitric acid solutions reduced the ability of the cleaning procedure to remove biofilm cells. The sodium hydroxide and nitric acid treatments employed here were sporicidal (Knight & Weeks, 2008; Seale *et al.*, 2011).

It is important to monitor and control the chemical concentrations of cleaning solutions and the temperatures employed during cleaning, as both affect the sporicidal activity of cleaning solutions (Knight & Weeks, 2008). Lindsay *et al.* (2000) were able to isolate viable spores of *Bacillus* spp., in particular *B. cereus*, from alkaline cleaning solutions that had been used for dairy CIP procedures, while Seale *et al.* (2011) demonstrated that exposure to a sodium hydroxide solution enhanced the ability of spores of *Geobacillus* spp. to attach to stainless steel. These findings suggest that circulation of sodium hydroxide cleaning solutions during CIP could potentially spread viable spores around processing equipment. This also has serious implications for the practice of reusing sodium hydroxide cleaning solutions. It therefore becomes very important to design a cleaning regime to ensure that spores are removed from the surfaces of equipment and that spores suspended in cleaning solutions are inactivated.

7.9.2 *Other control methods*

The growth of thermophilic bacteria in dairy processing equipment essentially comes down to a time–temperature relationship. Control can be achieved by either limiting the production runtime, which limits the time available for the growth of thermophilic bacteria, or operating processing equipment at temperatures at which the growth rates of thermophilic bacteria are reduced.

It is very common for dairy manufacturers to reduce production runtimes for centrifugal separators and plate heat exchanger equipment to between 6 and 8 hours, in order to limit the growth of thermophilic bacteria. Similarly, production runtimes for the manufacture of milk powder can be limited to between 18 and 24 hours, or to less than 10 hours when manufacturing ‘high-spec’ milk powders, which have very strict limits on counts for thermophilic spores. It is also common to reduce the operating temperature (e.g. to between 15 and 30 °C) of processing equipment, such as centrifugal separators and UF plants, to prevent thermophile growth.

Another approach that uses temperature to control biofilm development is the implementation of temperature step changes, which have been shown to control the development of biofilms of thermoresistant streptococci in cheese-milk pasteurisation equipment (Knight *et al.*, 2004). Temperature step changes are discussed in more detail in Chapter 6. A modification of this method may be feasible as a way of controlling the development of biofilms and sporulation by thermophilic bacilli: Burgess *et al.* (2009) demonstrated that lowering the temperature of the growth environment from 55 to 48 °C prevented the formation of spores by *A. flavithermus* in biofilms.

Reducing the surface area of processing equipment that is at a temperature that allows the growth of thermophilic bacteria may also reduce thermophile growth in milk powder manufacturing plants (Refstrup, 2000). This can be achieved by using a direct-contact heating system, such as a DSI unit. Heating in a DSI unit is achieved by injecting steam directly into the milk stream, rapidly increasing the temperature of the milk. A DSI unit can be used to heat milk from 45 up to 70 °C, for example, and thus virtually eliminate surfaces that are at a temperature that supports the growth of thermophilic bacteria. Such a system is more expensive to operate than an indirect heating system, such as a plate heat exchanger, due to the requirement for additional steam. The use of a DSI unit also results in dilution of the milk, due to the added water (as steam), so flash evaporation is required downstream.

Finally, it is also possible to use a dual preheating system, in which milk is directed from one preheater to another after 8–12 hours of processing. This allows the first preheater to undergo a CIP procedure without disrupting the manufacturing cycle.

References

Andersson, A., Ronner, U. & Granum, P. E. 1995. What problems does the food industry have with the spore-forming pathogens *Bacillus cereus* and *Clostridium perfringens*? *International Journal of Food Microbiology*, **28**, 145–55.

Aronson, A. I. & Fitz-James, P. 1976. Structure and morphogenesis of the bacterial spore coat. *Bacteriological Reviews*, **40**, 360–402.

Ash, C., Farrow, J. A. E., Wallbanks, S. & Collins, M. D. 1991. Phylogenetic heterogeneity of the genus *Bacillus* revealed by comparative analysis of small-subunit-ribosomal RNA. *Letters in Applied Microbiology*, **13**, 202–6.

Atrih, A. & Foster, S. J. 2001. Analysis of the role of bacterial endospore cortex structure in resistance properties and demonstration of its conservation amongst species. *Journal of applied microbiology*, **91**, 364–72.

Banat, I. M., Marchant, R. & Rahman, T. J. 2004. *Geobacillus debilis* sp. nov., a novel obligately thermophilic bacterium isolated from a cool soil environment, and reassignment of *Bacillus pallidus* to *Geobacillus pallidus* comb. nov. *International Journal of Systematic Evolutionary Microbiology*, **54**, 2197–201.

Basappa, P. 1974. Growth characteristics of thermophilic micro-organisms in milk. *Indian Journal of Dairy Science*, **27**, 103–8.

Beaman, T. C. & Gerhard, P. 1986. Heat resistance of bacterial spores correlated with protoplast dehydration, minerlization, and thermal adaption. *Applied and Environmental Microbiology*, **52**, 1242–6.

Beaman, T. C., Pankratz, H. S. & Gerhard, P. 1988. Heat shock affects permeability and resistance of *Bacillus stearothermophilus* spores. *Applied and Environmental Microbiology*, **54**, 2515–20.

Bender, G. R. & Marquis, R. E. 1985. Spore heat resistance and specific mineralization. *Applied and Environmental Microbiology*, **50**, 1414–21.

Beyenal, H. & Lewandowski, Z. 2002. Internal and external mass transfer in biofilms grown at various flow velocities. *Biotechnology Progress*, **18**, 55–61.

Boxrud, D., Pederson-Gulrud, K., Wotton, J., Medus, C., Lyszkowicz, E., Besser, J. & Bartkus, J. M. 2007. Comparison of multiple-locus variable-number tandem repeat analysis, pulsed-field gel electrophoresis, and phage typing for subtype analysis of *Salmonella enterica* serotype enteritidis. *Journal of Clinical Microbiology*, **45**, 536–43.

Broukhanski, G., Low, D. E. & Pillai, D. R. 2011. Modified (M) MLVA for rapid identification and typing of *Clostridium difficile* during institutional outbreaks. *Journal of Clinical Microbiology*, JCM.02359-10.

Burgess, S. A., Brooks, J. D., Rakonjac, J., Walker, K. M. & Flint, S. H. 2009. The formation of spores in biofilms of *Anoxybacillus flavithermus*. *Journal of Applied Microbiology*, **107**, 1012–18.

Burgess, S. A., Lindsay, D. & Flint, S. H. 2010. Thermophilic bacilli and their importance in dairy processing. *International Journal of Food Microbiology*, **144**, 215–25.

Burgess, S. A., Flint, S. H. & Lindsay, D. 2014. Characterization of thermophilic bacilli from a milk powder processing plant. *Journal of Applied Microbiology*, **11**, 350–9.

Caspers, M. P. M., Schuren, F. H. J., Van Zuijlen, A. C. M., Brul, S., Montijn, R. C., Abee, T. & Kort, R. 2011. A mixed-species microarray for identification of food spoilage bacilli. *Food Microbiology*, **28**, 245–51.

Caspers, M. P. M., Boekhorst, J., Abee, T., Siezen, R. J. & Kort, R. 2013. Complete genome sequence of *Anoxybacillus flavithermus* TNO-09.006, a thermophilic sporeformer associated with a dairy-processing environment. *Genome Announcements*, **1**, e00010-13.

Cazemier, A. E., M., W. S. F. & Ter Steeg, P. F. 2001. Effect of sporulation and recovery medium on the heat resistance and amount of injury of spores from spoilage bacilli. *Journal of Applied Microbiology*, **90**, 761–70.

Chandler, D. P., Brown, J., Bruckner-Lea, C. J., Olson, L., Posakony, G. J., Stults, J. R., Valentine, N. B. & Bond, L. J. 2001. Continuous spore disruption using radially focused, high-frequency ultrasound. *Analytical Chemistry*, **73**, 3784–9.

Chauhan, K., Dhakal, R., Seale, R. B., Deeth, H. C., Pillidge, C. J., Powell, I. B., Craven, H. & Turner, M. S. 2013. Rapid identification of dairy mesophilic and thermophilic sporeforming bacteria using DNA high resolution melt analysis of variable 16S rDNA regions. *International Journal of Food Microbiology*, **165**, 175–83.

Chen, L., Coolbear, T. & Daniel, R. M. 2004. Characteristics of proteinases and lipases produced by seven *Bacillus* sp. isolated from milk powder production lines. *International Dairy Journal*, **14**, 495–504.

Chopra, A. K. & Mathur, D. K. 1984. Isolation, screening and characterization of thermophilic Bacillus species isolated from dairy products. *Journal of Applied Microbiology*, **57**, 263–71.

Christi, Y. 1999. Modern systems of plant cleaning. *Encyclopedia of Food Microbiology: Process Hygiene*, **3**, 1806–15.

Cohn, F. 1876. Untersuchungen über Bakterien. IV. Beiträge zur Biologie der Bacillen. *Beiträge zur Biologie der Pflanzen*, **2**, 249–76.

Cook, G. M. & Sandman, R. M. 2000. Sources and characterisation of spore-forming bacteria in raw milk. *Australian Journal of Dairy Technology*, **55**, 119–26.

Coorevits, A., De Jonghe, V., Vandroemme, J., Reekmans, R., Heyrman, J., Messens, W., De Vos, P. & Heyndrickx, M. 2008. Comparative analysis of the diversity of aerobic spore-forming bacteria in raw milk from organic and conventional dairy farms. *Systematic and Applied Microbiology*, **31**, 126–40.

Coorevits, A., Dinsdale, A. E., Halket, G., Lebbe, L., De Vos, P., Van Landschoot, A. & Logan, N. A. 2012 Taxonomic revision of the genus *Geobacillus*: emendation of *Geobacillus*, *G. stearothermophilus*, *G. jurassicus*, *G. toebii*, *G. thermodenitrificans* and *G. thermoglucosidans* (nom. corrig., formerly '*thermoglucosidasius*'); transfer of *Bacillus thermantarcticus* to the genus as *G. thermantarcticus* comb. nov.; proposal of *Caldibacillus debilis* gen. nov., comb. nov.; transfer of *G. tepidamans* to *Anoxybacillus* as *A. tepidamans* comb. nov.; and proposal of *Anoxybacillus caldiproteolyticus* sp. nov. *International Journal of Systematic and Evolutionary Microbiology*, **62**, 1470–85.

Cortezzo, D. E. & Setlow, P. 2005. Analysis of factors that influence the sensitivity of spores of *Bacillus subtilis* to DNA damaging chemicals. *Journal of Applied Microbiology*, **98**, 606–17.

Cosentino, S., Mulargia, A. F., Pisano, B., Tuveri, P. & Palmas, F. 1997. Incidence and biochemical characteristics of *Bacillus* flora in Sardinian dairy products. *International Journal of Food Microbiology*, **38**, 235–8.

Crielly, E. M., Logan, N. A. & Anderton, A. 1994. Studies on the Bacillus flora of milk and milk products. *Journal of Applied Bacteriology*, **77**, 256–63.

De Clerck, E., Vanhoutte, T., Hebb, T., Geerinck, J., Devos, J. & De Vos, P. 2004. Isolation, characterization, and identification of bacterial contaminants in semifinal gelatin extracts. *Applied and Environmental Microbiology*, **70**, 3664–72.

Dhakal, R., Chauhan, K., Seale, R. B., Deeth, H. C., Pillidge, C. J., Powell, I. B., Craven, H. & Turner, M. S. 2013. Genotyping of dairy *Bacillus licheniformis* isolates by high resolution melt analysis of multiple variable number tandem repeat loci. *Food Microbiology*, **34**, 344–51.

Dhakal, R., Seale, R. B., Deeth, H. C., Craven, H. & Turner, M. S. 2014. Draft genome comparison of representatives of the three dominant genotype groups of dairy *Bacillus licheniformis* strains. *Applied Environmental Microbiology*, **80**, 3453–62.

Donnellan, J. E., Nags Jr., E. H. & Levinson, H. S. 1963. Chemically defined, synthetic media for sporulation and for germination and growth of *Bacillus subtilis*. *Journal of Bacteriology*, **87**, 332–6.

Douki, T., Setlow, B. & Setlow, P. 2005. Photosensitization of DNA by dipicolinic acid, a major component of spores of *Bacillus* species. *Photochemical and Photobiological Sciences*, **4**, 591–7.

Driks, A. 1999. *Bacillus subtilis* spore coat. *Microbiology and Molecular Biology Reviews*, **63**, 1–20.

Dring, G. J. & Gould, G. W. 1971. Sequence of events during rapid germination of spores of *Bacillus cereus*. *Journal of General Microbiology*, **65**, 101–4.

Durak, M. Z., Fromm, H. I., Huck, J. R., Zadoks, R. N. & Boor, K. J. 2006. Development of molecular typing methods for *Bacillus* spp. and *Paenibacillus* spp. Isolated from fluid milk products. *Journal of Food Science*, **71**, M50–6.

Faille, C., Fontaine, F. & Benezech, T. 2001. Potential occurrence of adhering living Bacillus spores in milk product processing lines. *Journal of Applied Microbiology*, **90**, 892–900.

Faille, C., Jullien, C., Fontaine, F., Bellon-Fontaine, M. N., Slomianny, C. & Benezech, T. 2002. Adhesion of Bacillus spores and *Escherichia coli* cells to inert surfaces: role of surface hydrophobicity. *Canadian Journal of Microbiology*, **48**, 728–38.

Faille, C., Tauveron, G., Gentil-Lelievre, L. C. & Slomianny, C. 2007. Occurrence of *Bacillus cereus* spores with a damaged exosporium: consequences on the spore adhesion on surfaces of food processing lines. *Journal of Food Protection*, **70**, 2346–53.

Faille, C., Bénézech, T., Midelet-Bourdin, G., Lequette, Y., Clarisse, M., Ronse, G., Ronse, A. & Slomianny, C. 2014. Sporulation of *Bacillus* spp. within biofilms: a potential source of contamination in food processing environments. *Food Microbiology*, **40**, 64–74.

Fernández-No, I. C., Guarddon, M., Böhme, K., Cepeda, A., Calo-Mata, P. & Barros-Velázquez, J. 2011. Detection and quantification of spoilage and pathogenic *Bacillus cereus*, *Bacillus subtilis* and *Bacillus licheniformis* by real-time PCR. *Food Microbiology*, **28**, 605–10.

Flint, S., Palmer, J., Bloemen, K., Brooks, J. & Crawford, R. 2001a. The growth of *Bacillus stearothermophilus* on stainless steel. *Journal of Applied Microbiology*, **90**, 151–7.

Flint, S., Ward, L. J. H. & Walker, K. 2001b. Functional grouping of thermophilic *Bacillus* strains using amplification profiles of the 16S–23S internal spacer region. *Systematic and Applied Microbiology*, **24**, 539–48.

Flint, S., Drocourt, J.-L., Walker, K., Stevenson, B., Dwyer, M., Clarke, I. & Mcgill, D. 2006. A rapid, two-hour method for the enumeration of total viable bacteria in samples from commercial milk powder and whey protein concentrate powder manufacturing plants. *International Dairy Journal*, **16**, 379–84.

Flint, S., Walker, K., Waters, B. & Crawford, R. 2007. Description and validation of a rapid (1 h) flow cytometry test for enumerating thermophilic bacteria in milk powders. *Journal of Applied Microbiology*, **102**, 909–15.

Fox, A., Black, G. E., Fox, K. & Rostovtseva, S. 1993. Determination of carbohydrate profiles of *Bacillus anthracis* and *Bacillus cereus* including identification of 0-Methyl methylpentoses by using gas chromatography-mass spectrometry. *Journal of Clinical Microbiology*, **31**, 887–94.

Frank, J. F. & Yousef, A. E. 2004. Thermophilic bacteria. In: Wehr, M. H. & Frank, J. F. (eds) Standard Methods for the Examination of Dairy Products. American Public Health Association, Washington, DC.

Frenkiel-Krispin, D., Sack, R., Englander, J., Shimoni, E., Eisenstein, M., Bullitt, E., Horowitz-Scherer, R., Hayes, C. S., Setlow, P., Minsky, A. & Wolf, S. G. 2004. Structure of the DNA-SspC complex: implications for DNA packaging, protection, and repair in bacterial spores. *Journal of Bacteriology*, **186**, 3525–30.

Galesloot, T. E. & Labots, H. 1959a. Concerning the bacteriology of sterilized milk and milk products. *Netherlands Milk and Dairy Journal*, **13**, 16–37.

Galesloot, T. E. & Labots, H. 1959b. Thermophilic bacilli in milk with special reference to the making of sterilised milk and chocolate milk. *Netherlands Milk and Dairy Journal*, **13**, 155–79.

Gerhard, P. & Marquis, R. E. 1989. Spore thermoresistance mechanisms. In: Smith, I., Slepecky, R. A. & Setlow, P. (eds) Regulation of Prokaryotic Development. American Society for Microbiology, Washingon, DC.

Gerhardt, P. & Ribi, E. 1964. Ultrastructure of the exosporium enveloping spores of *Bacillus cereus*. *Journal of Bacteriology*, **88**, 1774–89.

Gilmour, A. & Rowe, M. T. 1990. Micro-organisms associated with milk. In: Robinson, R. K. (ed.) Dairy Microbiology. Elsevier Applied Science, London.

Gould, G. W. & Hitchins, A. D. 1963. Sensitization of bacterial spores to lysozyme and to hydrogen peroxide with agents which rupture disulphide bonds. *Journal of General Microbiology*, **33**, 413–23.

Gundogan, N. & Arik, M. T. 2004. Comparison of the protease activity of psychrotrophic and thermophilic *Bacilli* spp. isolated from raw milk samples. *The Indian Veterinary Journal*, **81**, 1013–15.

Han, J., Seale, R. B., Silcock, P., McQuillan, A. J. & Bremer, P. J. 2011. The physico-chemical characterization of casein-modified surfaces and their influence on the adhesion of spores from a Geobacillus species. *Biofouling*, **27**, 459–66.

Heinen, W., Lauwers, A. M. & Mulders, J. W. 1982. *Bacillus licheniformis*, a newly isolated facultative thermophile. *Antonie van Leeuwenhoek*, **48**, 265–72.

Henriques, A. O. & Moran, C. P. 2000. Structure and assembly of the bacterial endospore coat. *Methods*, **20**, 95–110.

Heyndrickx, M. & Scheldeman, P. 2002. Bacilli associated with spoilage in dairy products and other food. In: Berkeley, R. C. W., Heyndrickx, M., Logan, N. & De Vos, P. (eds.) *Applications and Systematics of Bacillus and Relatives*. Blackwell Science, Oxford.

Hill, B. M. & Smythe, B. W. 1994. Progress in understanding the behaviour of thermophilic bacteria during milk powder manufacture. In: Chen, X. D. (ed.) Milk Powders for the Future II. Dunmore Press, Palmerston North.

Hill, B. M. & Smythe, B. W. 2004. Thermophilic spores in milk powder destined for use in UHT and retort sterilisation processes. Proceedings of the New Zealand Microbiological Society Conference, 2004, Palmerston North, New Zealand.

Hinton, A. R., Trinh, K. T., Brooks, J. D. & Manderson, G. J. 2002. Thermophile survival in milk fouling and on stainless steel during cleaning. *Transactions of the Institution of Chemical Engineers*, **80**, 299–304.

Hoch, J. A. 1993. Regulation of the phosphorelay and the initiation of sporulation in *Bacillus subtilis*. *Annual Review of Microbiology*, **47**, 87–99.

Huang, S., Chen, D., Pelczar, P. L., Vepachedu, V. R., Setlow, P. & Li, Y. 2007. Levels of Ca^2-dipicolinic acid in individual *Bacillus* spores determined using microfluidic raman tweezers. *Journal of Bacteriology*, **189**, 4681–7.

Hullo, M.-F., Moszer, I., Danchin, A. & Martin-Verstraete, I. 2001. CotA of *Bacillus subtilis* is a copper-dependent laccase. *Journal of Bacteriology*, **183**, 5426–30.

Husmark, U. & Ronner, U. 1990. Forces involved in adhesion of *Bacillus cereus* spores to solid surfaces under different environmental conditions. *Journal of Applied Microbiology*, **69**, 557–62.

Husmark, U. & Ronner, U. 1992. The influence of hydrophobic, electrostatic and morphological properties on the adhesion of Bacillus spores. *Biofouling*, **5**, 335–44.

Issahary, G., Evenchik, Z. & Keynan, A. 1970. Low-pH activation of *Bacillus cereus* spores. *Journal of Bacteriology*, **101**, 418–22.

Jones, C. A., Padula, N. L. & Setlow, P. 2005. Effect of mechanical abrasion on the viability, disruption and germination of spores of *Bacillus subtilis*. *Journal of Applied Microbiology*, **99**, 1484–94.

Kalogridou-Vassiliadou, D. 1992. Biochemical activities of *Bacillus* species isolated from flat sour evaporated milk. *Journal of Dairy Science*, **75**, 2681–6.

Keim, P., Price, L. B., Klevytska, A. M., Smith, K. L., Schupp, J. M., Okinaka, R., Jackson, P. J. & Hugh-Jones, M. E. 2000. Multiple-locus variable-number tandem repeat analysis reveals genetic relationships within *Bacillus anthracis*. *Journal of Bacteriology*, **182**, 2928–36.

Keim, P., Van Ert, M. N., Pearson, T., Vogler, A. J., Huynh, L. Y. & Wagner, D. M. 2004. Anthrax molecular epidemiology and forensics: using the appropriate marker for different evolutionary scales. *Infection, Genetics and Evolution*, **4**, 205–13.

Klavenes, A., Stalheim, T., Sjøvold, O., Josefsen, K. & Granum, P. E. 2002. Attachment of *Bacillus cereus* spores with and without appendages to stainless steel surfaces. *Food and Bioproducts Processing*, **80**, 312–18.

Knight, G. C. & Weeks, M. G. 2008. Conditions for inactivation of thermophilic spores in NaOH solutions for reuse applications. *Australian Journal of Dairy Technology*, **63**, 82–6.

Knight, G. C., Nicol, R. S. & Mcmeekin, T. A. 2004. Temperature step changes: a novel approach to control biofilms of *Streptococcus thermophilus* in a pilot plant-scale cheese-milk pasteurisation plant. *International Journal of Food Microbiology*, **93**, 305–18.

Kuisiene, N., Raugalas, J., Stuknyte, M. & Chitavichius, D. 2007. Identification of the genus *Geobacillus* using genus-specific primers, based on the 16S–23S rRNA gene internal transcribed spacer. *FEMS Microbiology Letters*, **277**, 165–72.

Kuisiene, N., Raugalas, J. & Chitavichius, D. 2009. Phylogenetic, inter, and intraspecific sequence analysis of spo0A gene of the genus *Geobacillus*. *Current Microbiology*, **58**, 547–53.

Kutima, P. M. & Foegeding, P. M. 1987. Involvement of the spore coat in germination of *Bacilus cereus* T spores. *Applied and Environmental Microbiology*, **53**, 47–52.

Langeveld, A. N., Van Der Klijn, W. & Van Der Waals, C. B. 1990. Bacteriegroei in indampers voor melk en wei. *Voedingamiddelentechnologie*, **23**, 13–17.

Le Fleche, P., Fabre, M., Denoeud, F., Koeck, J. & Vergnaud, G. 2002. High resolution, on-line identification of strains from the *Mycobacterium tuberculosis* complex based on tandem repeat typing. *BMC Microbiology*, **2**, 37.

Lehmann, F. L. 1995. Recontamination of industrial ultrafiltration units and pasteurisers by thermoduric-thermophilic bacteria. International Dairy Federation Symposium on Heat Treatments and Alternative Methods, 1995, Vienna.

Lindsay, D. & Flint, S. 2009. Biofilm formation by spore-forming bacteria in food processing environments. In: Fratmico, B., Annous, A. & Guenther, N. W. (eds.) Biofilms in the Food and Beverage Industries. Woodhead, Cambridge.

Lindsay, D., Brözel, V. S., Mostert, J. F. & Von Holy, A. 2000. Physiology of dairy-associated *Bacillus* spp. over a wide pH range. *International Journal of Food Microbiology*, **54**, 49–62.

Lindsay, D., Brözel, V. S. & Von Holy, A. 2005. Spore formation in *Bacillus subtilis* biofilms. *Journal of Food Protection*, **68**, 860–5.

Logan, N. A. 1988. Bacillus species of medical and veterinary importance. *Journal of Medical Microbiology*, **25**, 157–65.

Matsutani, M., Shirakihara, Y., Imada, K., Yakushi, T. & Matsushita, K. 2013. Draft genome sequence of a thermophilic member of the Bacillaceae, Anoxybacillus flavithermus strain Kn10, isolated from the Kan-nawa Hot Spring in Japan. *Genome announcements*, **1**, e00311-13.

Matz, L. L., Beaman, T. C. & Gerhardt, P. 1970. Chemical composition of exosporium from spores of *Bacillus cereus*. *Journal of Bacteriology*, **101**, 196–201.

McGuiggan, J. T. M., Mccleery, D. R., Hannan, A. & Gilmour, A. 2002. Aerobic spore-forming bacteria in bulk raw milk: factors influencing the numbers of psychrotrophic, mesophilic and thermophilic *Bacillus* spores. *International Journal of Dairy Technology*, **55**, 100–7.

Meintanis, C., Chalkou, K. I., Kormas, K. A., Lymperopoulou, D. S., Katsifas, E. A., Hatzinikolaou, D. G. & Karagouni, A. D. 2008. Application of *rpoB* sequence similarity analysis, REP-PCR and BOX-PCR for the differentiation of species within the genus *Geobacillus*. *Letters in Applied Microbiology*, **46**, 395–401.

Murphy, P. M., Lynch, D. & Kelly, P. M. 1999. Growth of thermophilic spore forming bacilli in milk during the manufacture of low heat powders. *International Journal of Dairy Technology*, **52**, 45–50.

Murphy, M., Corcoran, D., Buckley, J. F., O'Mahony, M., Whyte, P. & Fanning, S. 2007. Development and application of multiple-locus variable number of tandem repeat analysis (MLVA) to subtype a collection of *Listeria monocytogenes*. *International Journal of Food Microbiology*, **115**, 187–94.

Murugan, B. & Villi, R. A. 2009. Lipolytic activity of *Bacillus* species isolated from milk and dairy products. *The Indian Veterinary Journal*, **86**, 80–1.

Nazina, T. N., Tourova, T. P., Poltaraus, A. B., Novikova, E. V., Grigoryan, A. A., Ivanova, A. E., Lysenko, A. M., Petrunyaka, V. V., Osipov, G. A., Belyaev, S. S. & Ivanov, M. V. 2001. Taxonomic study of aerobic thermophilic bacilli: descriptions of *Geobacillus subterraneus* gen. nov., sp. nov. and *Geobacillus uzenensis* sp. nov. from petroleum reservoirs and transfer of *Bacillus stearothermophilus*, *Bacillus thermocatenulatus*, *Bacillus thermoleovorans*, *Bacillus kaustophilus*, *Bacillus thermoglucosidasius* and *Bacillus thermodenitrificans* to Geobacillus as the new combinations *G. stearothermophilus*, *G. thermocatenulatus*, *G. thermoleovorans*, *G. kaustophilus*, *G. thermoglucosidasius* and *G. thermodenitrificans*. *International Journal of Systematic Evolutionary Microbiology*, **51**, 433–46.

Paidhungat, M., Setlow, B., Driks, A. & Setlow, P. 2000. Characterization of spores of *Bacillus subtilis* which lack dipicolinic acid. *Journal of Bacteriology*, **182**, 5505–12.

Palop, A., Mañas, P. & Condón, S. 1999. Sporulation temperature and heat resistance of *Bacillus* spores: A review. *Journal of Food Safety*, **19**, 57–72.

Pandey, N. K. & Aronson, A. I. 1979. Properties of the *Bacillus subtilis* spore coat. *Journal of Bacteriology*, **137**, 1208–18.

Parkar, S. G., Flint, S. H., Palmer, J. S. & Brooks, J. D. 2001. Factors influencing attachment of thermophilic bacilli to stainless steel. *Journal of Applied Microbiology*, **90**, 901–8.

Parkar, S. G., Flint, S. H. & Brooks, J. D. 2003. Physiology of biofilms of thermophilic bacilli-potential consequences for cleaning. *Journal of Industrial Microbiology and Biotechnology*, **30**, 553–60.

Parkar, S. G., Flint, S. H. & Brooks, J. D. 2004. Evaluation of the effect of cleaning regimes on biofilms of thermophilic bacilli on stainless steel. *Journal of Applied Microbiology*, **96**, 110–16.

Piggot, P. J. & Hilbert, D. W. 2004. Sporulation of *Bacillus subtilis*. *Current Opinions in Microbiology*, **7**, 579–86.

Pikuta, E., Lysenko, A., Chuvilskaya, N., Mendrock, U., Hippe, H., Suzina, N., Nikitin, D., Osipov, G. & Laurinavichius, K. 2000. *Anoxybacillus pushchinoensis gen. nov., sp no.*, a novel anaerobic, alkaliphilic, moderately thermophilic bacterium from manure, and the decription of *Anoxybacillus flavithermus comb nov. International Journal of Systematic and Evolutionary Microbiology*, **50**, 2109–17.

Pikuta, E. Cleland, D. & Tang, J. 2003. Aerobic growth of *Anoxybacillus pushchinoensis* K1^T: emended descriptions of *A. pushchinoensis* and the genus *Anoxybacillus*. *International Journal of Systematic and Evolutionary Microbiology*, **53**, 1561–2.

Postollec, F., Bonilla, S., Baron, F., Jan, S., Gautier, M., Mathot, A. G., Hallier-Soulier, S., Pavan, S. & Sohier, D. 2010. A multiparametric PCR-based tool for fast detection and identification of spore-forming bacteria in food. *International Journal of Food Microbiology*, **142**, 78–88.

Postollec, F., Mathot, A. G., Bernard, M., Divanac'h, M. L., Pavan, S. & Sohier, D. 2012. Tracking spore-forming bacteria in food: from natural biodiversity to selection by processes. *International Journal of Food Microbiology*, **158**, 1–8.

Powell, J. F. 1953. Isolation of dipicolinic acid (pyridine-2:6-dicarboxylic acid) from spores of *Bacillus megaterium*. *Biochemical Journal*, **54**, 210–11.

Redmond, C., Baillie, L. W. J., Hibbs, S., Moir, A. J. G. & Moir, A. 2004. Identification of proteins in the exosporium of *Bacillus anthracis*. *Microbiology*, **150**, 355–63.

Refstrup, E. 2000. Evaporation and drying technology developments. *International Journal of Dairy Technology*, **53**, 163–7.

Reyes, J. F. & Tanaka, M. M. 2010. Mutation rates of spoligotypes and variable numbers of tandem repeat loci in *Mycobacterium tuberculosis*. *Infection, Genetics and Evolution*, **10**, 1046–51.

Riesenman, P. J. & Nicholson, W. L. 2000. Role of the spore coat layers in *Bacillus subtilis* spore resistance to hydrogen peroxide, artificial UV-C, UV-B and solar UV radiation. *Applied Environmental Microbiology*, **66**, 620–6.

Romney, A. J. D. 1990. CIP: Cleaning in Place. The Society for Dairy Technology, Cambridge.

Ronimus, R. S., Parker, L. E., Turner, N., Poudel, S., Rückert, A. & Morgan, H. W. 2003. A RAPD-based comparison of thermophilic bacilli from milk powders. *International Journal of Food Microbiology*, **85**, 45–61.

Ronimus, R. S., Rückert, A. & Morgan, H. W. 2006. Survival of thermophilic spore-forming bacteria in a 90+ year old milk powder from Ernest Shackleton's Cape Royds Hut in Antarctica. *Journal of Dairy Research*, **73**, 235–43.

Ronner, U. & Husmark, U. 1992. Adhesion of *Bacillus cereus* spores – a hazard to the dairy industry. In: Melo, L. F. (ed.) Biofilms: Science and Technology. Kluwer Academic Publishers, The Netherlands.

Ronner, U., Husmark, U. & Henriksson, A. 1990. Adhesion of bacillus spores in relation to hydrophobicity. *Journal of Applied Bacteriology*, **69**, 550–6.

Rückert, A., Ronimus, R. S. & Morgan, H. W. 2004. A RAPD-based survey of thermophilic bacilli in milk powders from different countries. *International Journal of Food Microbiology*, **96**, 263–72.

Rückert, A., Ronimus, R. S. & Morgan, H. W. 2005a. Development of a rapid detection and enumeration method for thermophilic bacilli in milk powders. *Journal of Microbiological Methods*, **60**, 155–67.

Rückert, A., Ronimus, R. S. & Morgan, H. W. 2005b. Rapid differentiation and enumeration of the total, viable vegetative cell and spore content of thermophilic bacilli in milk powders with reference to *Anoxybacillus flavithermus*. *Journal of Applied Microbiology*, **99**, 1246–55.

Rückert, A., Ronimus, R. S. & Morgan, H. W. 2006. Development of a real time PCR assay targeting the sporulation gene, *spo0A*, for the enumeration of thermophilic bacilli in milk powder. *Food Microbiology*, **23**, 220–30.

Salkinoja-Salonen, M. S., Vuorio, R., Andersson, M. A., Kampfer, P., Andersson, M. C., Honkanen-Buzalski, T. & Scoging, A. C. 1999. Toxigenic strains of *Bacillus licheniformis* related to food poisoning. *Applied and Environmental Microbiology*, **65**, 4637–45.

Saw, J. H., Mountain, B. W., Feng, L., Omelchenko, M. V., Hou, S., Saito, J. A., Stott, M. B., Li, D., Zhao, G. & Wu, J. 2008. Encapsulated in silica: genome, proteome and physiology of the thermophilic bacterium *Anoxybacillus flavithermus* WK1. *Genome Biology*, **9**, R161.

Scheldeman, P., Herman, L., Foster, S. & Heyndrickx, M. 2006. *Bacillus sporothermodurans* and other highly heat-resistant spore formers in milk. *Journal of Applied Microbiology*, **101**, 542–55.

Scott, S. A., Brooks, J. D., Rakonjac, J., Walker, K. M. R. & Flint, S. H. 2007. The formation of thermophilic spores during the manufacture of whole milk powder. *International Journal of Dairy Technology*, **60**, 109–17.

Seale, R. B. 2009. The surface characteristics of spores from thermophilic bacilli isolated from a milk powder production line and their influence on adhesion to surfaces. PhD thesis, University of Otago, New Zealand.

Seale, R. B., Flint, S. H., Mcquillan, A. J. & Bremer, P. J. 2008. Recovery of spores from thermophilic dairy bacilli and effects of their surface characteristics on attachment to different surfaces. *Applied and Environmental Microbiology*, **74**, 731–7.

Seale, R. B., Bremer, P. J., Flint, S. H. & Mcquillan, A. J. 2010. Characterization of spore surfaces from a *Geobacillus* sp. isolate by pH dependence of surface charge and infrared spectra. *Journal of Applied Microbiology*, **109**, 1339–48.

Seale, R. B., Flint, S. H., Mcquillan, A. J. & Bremer, P. J. 2011. Effect of NaOH (caustic wash) on the viability, surface characteristics and adhesion of spores of a *Geobacillus* sp. isolated from a milk powder production line. *Letters in Applied Microbiology*, **52**, 104–8.

Seale, R. B., Dhakal, R., Chauhan, K., Craven, H. M., Deeth, H. C., Pillidge, C. J., Powell, I. B. & Turner, M. S. 2012. Genotyping of present-day and historical *Geobacillus* species isolates from milk powders by high-resolution melt analysis of multiple variable-number tandem-repeat loci. *Applied and Environmental Microbiology*, **78**, 7090–7.

Setlow, P. 2007. I will survive: DNA protection in bacterial spores. *Trends in Microbiology*, **15**, 172–80.

Setlow, B. & Setlow, P. 1979. Localization of low-molecular-weight proteins in *Bacillus megaterium* spores by cross-linking with ultraviolet light. *Journal of Bacteriology*, **139**, 486–94.

Setlow, B., Atluri, S., Kitchel, R., Koziol-Dube, K. & Setlow, P. 2006. Role of dipicolinic acid in resistance and stability of spores of *Bacillus subtilis* with or without DNA-protective alpha/beta-type small acid-soluble proteins. *Journal of Bacteriology*, **188**, 3740–7.

Shaheen, R., Svensson, B., Andersson, M. A., Christiansson, A. & Salkinoja-Salonen, M. 2010. Persistence strategies of *Bacillus cereus* spores isolated from dairy silo tanks. *Food Microbiology*, **27**, 347–55.

Shi, X. & Zhu, X. 2009. Biofilm formation and food safety in food industries. *Trends in Food Science and Technology*, **20**, 407–13.

Simmonds, P., Mossel, B. L., Intaraphan, T. & Deeth, H. C. 2003. Heat resistance of *Bacillus* spores when adhered to stainless steel and its relationship to spore hydrophobicity. *Journal of Food Protection*, **66**, 2070–5.

Stackebrandt, E., Frederiksen, W., Garrity, G. M., Grimont, P. A. D., Kämpfer, P., Maiden, M. C. J., Nesme, X., Rosselló-Mora, R., Swings, J., Trüper, H. G., Vauterin, L., Ward, A. C. and Whitman, W. B. 2002. Report of the ad hoc committee for the re-evaluation of the species definition in bacteriology. *International Journal of Systematic Evolutionary Microbiology*, **52**, 1043–7.

Stadhouders, J., Hup, G. & Hassing, F. 1982. The conceptions index and indicator organisms discussed on the basis of the bacteriology of spray-dried milk powder. *Netherlands Milk and Dairy Journal*, **36**, 231–60.

Stalheim, T. & Granum, P. E. 2001. Characterization of spore appendages from *Bacillus cereus* strains. *Journal of Applied Microbiology*, **91**, 839–45.

Storgårds, E., Tapani, K., Hartwall, P., Saleva, R. & Suihko, M. L. 2006. Microbial attachment and biofilm formation in brewery bottling plants. *Journal of the American Society of Brewing Chemists*, **64**, 8–15.

Swerdlow, B. M., Setlow, B. & Setlow, P. 1981. Levels of H+ and other monovalent cations in dormant and germinating spores of *Bacillus megaterium*. *Journal of Bacteriology*, **148**, 20–9.

Sylvestre, P., Couture-Tosi, E. & Mock, M. 2002. A collagen-like surface glycoprotein is a structural component of the *Bacillus anthracis* exosporium. *Molecular Microbiology*, **45**, 169–78.

Tai, S. K., Lin, H. P., Kuo, J. & Liu, J. K. 2004. Isolation and characterization of a cellulolytic *Geobacillus thermoleovorans* T4 strain from a sugar refinery wastewater. *Extremophiles*, **8**, 345–9.

Tauveron, G., Slomianny, C., Henry, C. & Faille, C. 2006. Variability among *Bacillus cereus* strains in spore surface properties and influence on their ability to contaminate food surface equipment. *International Journal of Food Microbiology*, **110**, 254–62.

Vogler, A. J., Keys, C. E., Allender, C., Bailey, I., Girard, J., Pearson, T., Smith, K. L., Wagner, D. M. & Keim, P. 2007. Mutations, mutation rates, and evolution at the hypervariable VNTR loci of *Yersinia pestis*. *Mutation Research/Fundamental and Molecular Mechanisms of Mutagenesis*, **616**, 145–58.

Wang, Y., Zheng, Y., Wang, M., Gao, Y., Xiao, Y. & Hui, P. 2014. Non-contiguous finished genome sequence of *Anoxybacillus flavithermus* subsp. yunnanensis type strain (E13T), a strictly thermophilic and organic solvent-tolerant bacterium. *Standards in Genomic Sciences*, **9**, 735–43.

Warth, A. D. & Strominger, J. L. 1969. Structure of the peptidoglycan of bacterial spores: occurrence of the lactam of muramic acid. *PNAS*, **64**, 528–35.

Watkinson, W. J. 2008. Chemistry of detergents and disinfectants. In: Tamime, A. Y. (ed.) Cleaning-in-Place: Dairy, Food and Beverage Operations. Wiley-Blackwell, Hoboken, NJ, pp. 56–79.

Weave, J., Kang, T. J., Raines, K. W., Cao, G. L., Hibbs, S., Tsai, P., Baillie, L., Rosen, G. M. & Cross, A. S. 2007. Protective role of *Bacillus anthracis* exosporium in macrophage-mediated killing by nitric oxide. *Infection and Immunity*, **75**, 3894–901.

Weng, F. Y., Chiou, C. S., Lin, P. H. P. & Yang, S. S. 2009. Application of *recA* and *rpoB* sequence analysis on phylogeny and molecular identification of *Geobacillus* species. *Journal of Applied Microbiology*, **107**, 452–64.

Williams, J. G. K., Kubelik, A. R., Livak, K. J., Rafallski, J. A. & Tingey, S. V. 1990. DNA polymorphisms amplified by abitrary primers are useful as genetic markers. *Nucleic Acids Research*, **18**, 6531–5.

Yuan, D., Liu, G., Ren, D., Zhang, D., Zhao, L., Kan, C., Yang, Y., Ma, W., Li, Y. & Zhang, L. 2012. A survey on occurrence of thermophilic bacilli in commercial milk powders in China. *Food Control*, **25**, 752–7.

Yildiz, F. & Westhoff, D. C. 1989. Sporulation and thermal resistance of *Bacillus stearothermophilus* spores in milk. *Food Microbiology*, **6**, 245–50.

Zeigler, D. R. 2005. Application of a recN sequence similarity analysis to the identification of species within the bacterial genus *Geobacillus*. *International Journal of Systematic and Evolutionary Microbiology*, **55**, 1171–9.

Zeigler, D. R. 2014. The *Geobacillus* paradox: why is a thermophilic bacterial genus so prevalent on a mesophilic planet? *Microbiology*, **160**, 1–11.

Zhao, Y., Caspers, M. P., Abee, T., Siezen, R. J. & Kort, R. 2012. Complete genome sequence of *Geobacillus thermoglucosidans* TNO-09.020, a thermophilic sporeformer associated with a dairy-processing environment. *Journal of Bacteriology*, **194**, 4118.

Zhao, Y., Caspers, M. P. M., Metselaar, K. I., De Boer, P., Roeselers, G., Moezelaar, R., Groot, M. N., Montijn, R. C., Abee, T. & Kort, R. 2013. Abiotic and microbiotic factors controlling biofilm gormation by thermophilic sporeformers. *Applied and Environmental Microbiology*, **79**, 5652–60.

8 Biofilm Contamination of Ultrafiltration and Reverse Osmosis Plants

Xuemei Tang[1], Steve Flint[1], Rod Bennett[1], John Brooks[2] and Siti Norbaizura Md Zain[1]

[1]Institute of Food, Nutrition and Human Health, Massey University, Palmerston North, New Zealand
[2]School of Applied Sciences, Faculty of Health and Environmental Sciences, Auckland University of Technology, Auckland, New Zealand

8.1 Introduction

Ultrafiltration (UF) and reverse osmosis (RO) membranes are used in several different dairy processes, from milk concentration to whey processing and water purification. Membrane systems represent a large surface area of synthetic material that is prone to biofilm contamination. Control of this biofilm is generally achieved through cleaning-in-place (CIP) systems. The original membrane systems used in the dairy industry could not be cleaned with standard cleaning chemicals as their structure was sensitive to acid and alkaline conditions. Enzyme cleaners were used, but these were more expensive than the standard caustic and acid cleaning systems used on modern UF and RO membrane plants. Current cleaning systems are similar to those used in the cleaning of stainless steel surfaces in other parts of the dairy manufacturing plant. However, the build-up of fouling and biofilm development is a problem. The high percentage of surface area compared with the rest of the manufacturing plant provides the greatest opportunity for biofilm development in a membrane processing plant, which is only enhanced by the temperatures of around 50 °C used for maximum filtration efficiency, which encourage the growth of thermophilic bacteria. Modern dairy UF and RO plants operate at lower temperatures – generally around 10 °C – which limits microbial growth to slow growing psychrotrophic bacteria. This extends the time between cleans and avoids product contamination by the spore-forming bacteria that used to grow in the filtration systems, which causing minimal changes in filtration efficiency compared with plants that use warmer temperatures.

A biofilm is a population of microbial cells growing on a surface and enclosed in an amorphous extracellular matrix (Donlan *et al.*, 2002). Biofilm growth is the predominant form of microbial growth in most environments; it can consist of either single or multiple species (O'Toole *et al.*, 2000). Biofilm can develop on any surface exposed to an aqueous environment (Flint *et al.*, 1997a). In the dairy and food industries, serious problems caused by biofilms

Biofilms in the Dairy Industry, First Edition.
Edited by Koon Hoong Teh, Steve Flint, John Brooks and Geoff Knight.

include interference with the flow of heat across the surface (Criado *et al.*, 1994) and increases in the fluid frictional resistance (Kumar & Anand, 1998) and the corrosion rate at the surface (Liu *et al.*, 2007). In addition, microorganisms growing in biofilms are more difficult to eliminate than free-floating bacterial cells (Flint *et al.*, 1997a), and thus cross-contamination and post-processing contamination may occur once biofilms have become established in a manufacturing plant (Kumar & Anand, 1998), leading to reduced product shelf life (Zottola, 1994). Such microbial contamination is the major cause of poor-quality dairy products (Flint *et al.*, 1997a).

Membrane processes in the dairy industry are severely limited by the problem of fouling, mainly by protein, as just a small degree of adsorption causes membrane pore blockage (Cheryan & Mehaia, 1986). Biofilm formation is enhanced by fouling of the membrane (Kumar & Anand, 1998) and will eventually lead to blockage of the membrane pores, preventing further manufacture (Flint *et al.*, 1997b). A mature biofilm on a membrane surface can also change the zones at which filtration can occur and the surface properties of the filter (Cogan & Chellam, 2008). Biofilm can release bacteria into the retentate stream, contaminating the product with potential spoilage issues.

8.2 Ultrafiltration and reverse osmosis membranes

UF and RO membranes are both semipermeable membranes that contain many tiny pores. Smaller molecules pass through the membranes, while larger molecules are retained. The feed stock will generally be split into two streams; materials that pass

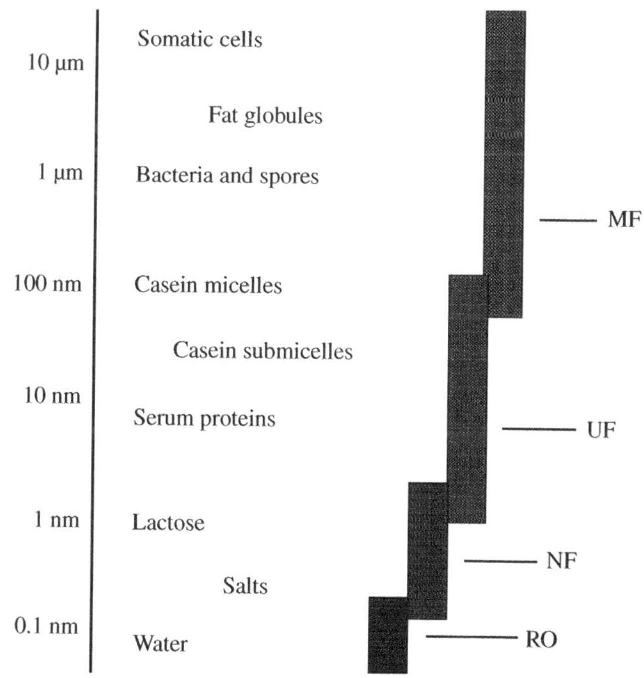

Figure 8.1 Permeability of membranes in dairy manufacture. MF, microfiltration; UF, ultrafiltration; NF, nanofiltration; RO, reverse osmosis. (From Brans *et al.*, 2004; used with permission from Elsevier.)

through the membrane are called 'permeates' and those that are retained by the membrane are called 'retentates' (Bird, 1996).

UF is widely used in the dairy industry (Daufin *et al.*, 2001). The pore size of UF membranes (10–100 nm) is larger than that of RO membranes (0.1–1.0 nm) (Figure 8.1), which allows protein and fat to be retained and permits water, lactose and ash to pass through. UF membrane applications in the dairy industry include the manufacture of whey protein concentrates (WPCs) and milk protein concentrates (MPCs), milk standardisation before cheese manufacture, liquid milk concentration for market milk product and clarification of cheese brine (Bird, 1996).

RO is a high-pressure membrane separation process that operates at between 25 and 40 bar (Hiddink *et al.*, 1980; Bird, 1996) and allows only water to pass through (Figure 8.1). Applications of RO membranes in the dairy industry include concentration of UF permeates for lactose manufacture, milk standardisation, lactose fermentation, recovery of proteins and lactose from casein whey wash waters, recovery of CIP water from UF and concentration of whey prior to transportation (Bird, 1996).

8.3 Membrane configuration and materials

In membrane applications today, the most common configuration is spiral-wound (Ridgway *et al.*, 1983; Woodhams, 2014), due to its high membrane surface area-to-volume ratio and its convenience of replacement and purchase (Bodalo-Santoyo *et al.*, 2004). However, this configuration has extreme susceptibility to fouling, due to the close spacing of the membrane leaves (Cartwright, 2003). In spiral-wound membrane modules, feed is separated by membrane layers. Retentates are collected from the sides of the layers, and permeates enter the central tube through permeate collection holes. Other configurations include plate and frame, tubular and hollow fibre (Maubois, 1980).

The most common materials used for fabrication of spiral-wound membranes in the dairy industry are polyethersulphone (PES) and polysulphone (PS) (D'Souza & Mawson, 2005; Pearce, 2007a). PES membranes have good strength and high permeability, and their properties can be modified through a polymer blend (Pearce, 2008). Membranes are usually modified to have a hydrophilic surface, which gives the advantages of being easily wetted and resisting fouling (Pearce, 2007b). Polyvinylidene fluoride (PVDF) has begun to be used for membranes since the 1990s (Pearce, 2008). Both PES and PVDF are now important materials for the membrane market (Pearce, 2007b). PVDF is stronger and more flexible than PES and has excellent chemical resistance (Boributha *et al.*, 2009). Thus, PVDF membranes tend to have a longer life (Pearce, 2007b). However, since the hydrophobic surface of a PVDF membrane is difficult to modify (Fontananova *et al.*, 2006), it is more susceptible to fouling than other materials (Lozier *et al.*, 2006; Pearce, 2007b).

8.4 Crossflow and biofouling

Membrane filtration in the dairy industry is almost exclusively operated in a crossflow mode (Figure 8.2), especially for the more difficult feeds such as whole milk (Pearce, 2008). The circulation in crossflow filtration is parallel to the membrane (Anon., 2007). The consistent turbulent

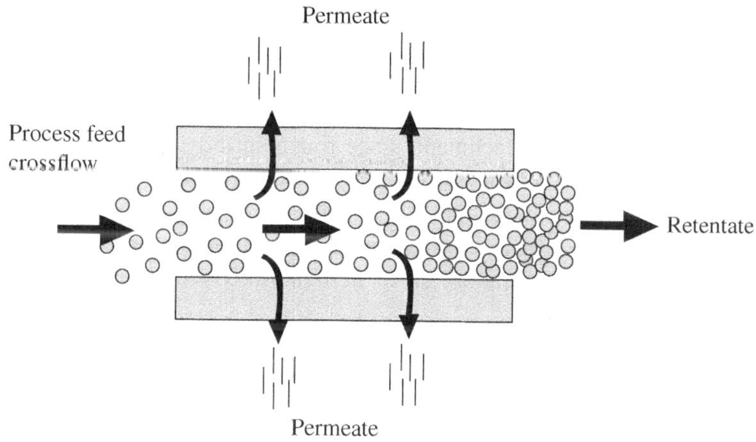

Figure 8.2 Crossflow filtration. (From Caridis & Papathanasiou, 1997; used with permission from Springer.)

flow (Anon., 2007) creates the shearing effect of the fluid as it passes over the membrane to remove any particles that may have accumulated at the membrane surface (Caridis & Papathanasiou, 1997). This helps to maintain a relatively steady flux through the membrane.

Crossflow filtration is a pressure-driven process and is profoundly influenced by the applied pressure differential between the retentate and the permeate (Caridis & Papathanasiou, 1997). During the filtration of protein solutions (e.g. whey suspension), increased transmembrane pressure (TMP) results in accumulation of a stronger fouling layer on the membrane surface (Karasu *et al.*, 2009). This preconditioning layer will influence the subsequent biofilm formation. Results from a modelling study of UF of whey determined that higher feed flow rate caused a larger volume of particles to be removed from the fouling layer (Karasu *et al.*, 2009). Therefore, very high crossflow velocities may be necessary to control fouling (Pearce, 2008).

8.5 Biofilm development

8.5.1 *Membrane surface characteristics and biofilm formation*

The surface properties of membranes are believed to be important in biofilm formation (Pasmore *et al.*, 2001). Bacterial attachment is regulated by the physicochemical nature of both the bacterial cell and the polymer membrane surface (Ridgway, 1991). This includes hydrophobicity and surface charge. In addition, the surface roughness will also affect biofilm formation (Herzberg *et al.*, 2009).

Surface roughness

Surface roughness refers to the steepness, evenness and topology of peaks and valleys on the surface of the membrane material (Lee *et al.*, 2010b). Membrane surface roughness is an important surface property for biofilm formation (Characklis, 1990; Elimelech *et al.*, 1997; Vrijenhoek *et al.*, 2001), affecting the development of younger biofilms more than that of mature ones (Pang *et al.*, 2005). Pasmore *et al.* (2001) concluded that bacterial attachment was affected by surface roughness through two primary means: (i) the roughness disrupts fluid flow by creating surface areas where the shear rate and the forces that might remove attached bacteria are significantly reduced; and (ii) the increased roughness increases the available surface area for cell attachment, since rough surfaces have more contours and valleys (Pasmore *et al.*, 2001). They also observed an increase in biofilm formation by *Pseudomonas aeruginosa* on a rougher UF membrane surface (Pasmore *et al.*, 2001). Similarly, it was found that the degree of roughness had a strong linear relationship with the maximum adhered cell concentration of *P. aeruginosa* PAO1 on nanofiltration (NF) membranes (Myint *et al.*, 2010).

Using atomic force microscopy (AFM) combined with scanning electron microscopy (SEM), Pang *et al.* (2005) observed that both roughness and depression areas of RO membranes made up of cellulose acetate (CA), polyamide (PA) and thin film composites (TFCs) increased when membranes were in a hydrated form. They compared the roughness of these three types of membrane and concluded that the CA membrane had the lowest, while the PA membrane had the largest depression areas (18 888 nm^2 for dry membrane (72.5 nm deep) and 33 416 nm^2 for hydrated membrane (133 nm deep). Microorganism entrapment is relatively easy in depression areas, and, therefore, PA membrane is more likely to promote biofilm formation (Pang *et al.*, 2005). While the depth of the depressions is not large enough to hide a whole cell, it will provide an area in which cells can become trapped. Similar observations were also reported by Campbell *et al.* (1999), who studied the attachment of *Mycobacterium sp.* on to PA and CA membranes in batch assays.

Hydrophobicity

The hydrophobicity of inanimate substrata influences the strength and kinetics of microbial adhesion and early biofouling (Ridgway *et al.*, 1999). It has been proposed that a hydrophobic substratum attracts bacteria with a hydrophobic surface and a hydrophilic substratum attracts bacteria with a hydrophilic surface (An & Friedman, 1998; Katsikogianni & Missirlis, 2004). Lee *et al.* (2010a) found that a relatively hydrophilic NF membrane had a higher potential for biofouling by hydrophilic bacteria than a hydrophobic UF membrane. Pasmore *et al.* (2001) found that biofilm initiation by a *P. aeruginosa* strain increased as a UF membrane surface became more hydrophobic. Similarly, Lee *et al.* (2010b) observed that the adhered cell concentration of *P. aeruginosa* PAO1 increased proportionally to the RO membrane hydrophobicity.

Surface charge

Most polymer materials used for fabrication of membranes possess some degree of surface charge, due to the presence of trace quantities of free carboxylate or sulphonate groups (Ridgway *et al.*, 1999). Surface charge can affect the attractive and repulsive forces that act between the bacterial cells and the substrate (Pasmore *et al.*, 2001). Charge attraction has even been suggested to have a stronger effect than hydrophobicity on attachment of cells to surfaces (Koo *et al.*, 2002). Under physiologically relevant pH values (~7), the polymer materials used for RO membranes tend to be negatively charged (Elimelech *et al.*, 1997; Vrijenhoek *et al.*, 2001). Negative membrane surface charge can reduce bacterial attachment due to electrostatic repulsion of negatively charged bacterial surfaces (Her *et al.*, 2000). However, other studies have observed that the ability to recover performance upon washing is higher for membranes with chemically neutral surfaces than for charged membranes (Pasmore *et al.*, 2001; Kochkodan *et al.*, 2006). This may be due to the absence of opposite charges generating a strong bond.

8.5.2 *Other factors*

Biofilm formation is an extremely complicated process that is affected by a number of factors. In addition to those just described, it is also influenced by environmental parameters such as flow conditions, the level of nutrients, the concentration of electrolytes and the pH (Lee *et al.*, 2010b).

Flow rate is considered a dominant factor that strongly influences bacterial attachment (Isberg & Barnes, 2002) and biofilm structure (Stoodley *et al.*, 1999b). Higher shear rates result in higher detachment forces, which decrease the number of attached cells (Katsikogianni & Missirlis, 2004). However, studies show that a high flow rate will not prevent bacterial attachment nor completely remove existing biofilm (Dreeszen, 2003), although it will make the biofilm denser and thinner (Chang *et al.*, 2002). This may be due to the lower growth yield obtained when the shear rate is increased (Katsikogianni & Missirlis, 2004), which may result from the biofilm bacteria putting energy into producing more extracellular polymeric substances (EPS) rather than cells to resist the shear forces.

Bacteria require certain nutrients for growth and replication. Limiting the nutrients will limit bacterial growth. However, biofilm will reach a certain equilibrium thickness according to both shear force and available nutrient levels (Dreeszen, 2003). For example, Ivnitsky *et al.* (2005) observed a bacterial count of approximately 10^7 CFU/cm^2 in biofilm on an NF membrane surface regardless of the feed applied. This suggests that nutrient levels used in these trials are not the limiting factor and that sufficient nutrients are available for the biofilm to reach equilibrium. There is a general assumption that nutrients are more concentrated at a substrate interface anyway, so even in an environment with minimal nutrients, there may often be sufficient nutrients at the substrate surface to sustain good biofilm growth.

Ionic strength and pH influence bacterial attachment by changing the surface characteristics of both the bacteria and the substrate, resulting in changing interactions between bacteria and substrates (Katsikogianni & Missirlis, 2004). Bunt *et al.* (1993) found that pH and ionic strength influenced the cell surface hydrophobicity and charge. The highest adhesion to hydrophobic surfaces was found at pHs in the range of the isoelectric point when bacteria were uncharged (Bunt *et al.*, 1993). In the range pH 3–9, an increase of the pH of the environment above the isoelectric point of the surface (PA membrane) resulted in an increased negative charge and increased repulsion of the bacteria from the surface (Bellona & Drewes, 2005). The chemicals adsorbed to the membrane surface are responsible for most of the changes in surface properties (Pasmore *et al.*, 2001). Studies have shown that positively charged ions such as sodium, calcium, magnesium and cationic surfactants can bind to the negatively charged membrane surface, resulting in a reduced negative surface charge (Bellona & Drewes, 2005).

8.6 Biofilm structure

SEM and confocal laser scanning microscopy (CLSM) are widely used to visualise and investigate biofilm structure. A membrane sample carrying biofilm can be fixed and dyed with suitable stains for examination using a CLSM or can be examined directly using SEM without dying (Camargo *et al.*, 2005). Useful parameters such as biovolume and substratum coverage can be measured (Pang *et al.*, 2005).

8.6.1 *Models and bioreactors for biofilm study*

Flemming (2003) and others have proposed molecular modelling techniques for the exploration and delineation of some of the theoretical mechanisms underlying primary bacterial adhesion to synthetic membrane materials. Such techniques may provide information on the structures and conformations of the adhesive biopolymers and membrane materials, and their dynamic interactions in different chemical environments. However, accurate modelling requires proper software tools (Flemming, 2003).

A recent investigation of biofilm formation on membrane surfaces was conducted by Pang *et al.* (2005), using a continuous flow system (Figure 8.3). Unfortunately, with this system,

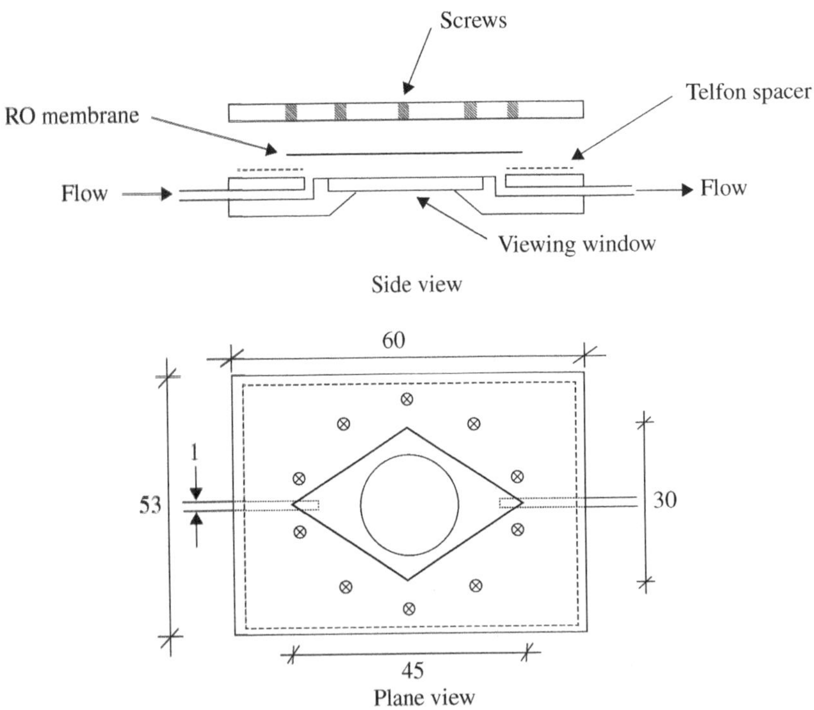

Figure 8.3 Schematic representation of the flow cell used in monitoring biofilm development. The channel depth is set by the thickness of the Teflon spacer (1 mm). All dimensions are given in mm. (Reprinted with permission from Pang *et al.*, 2005; copyright 2005 American Chemical Society.)

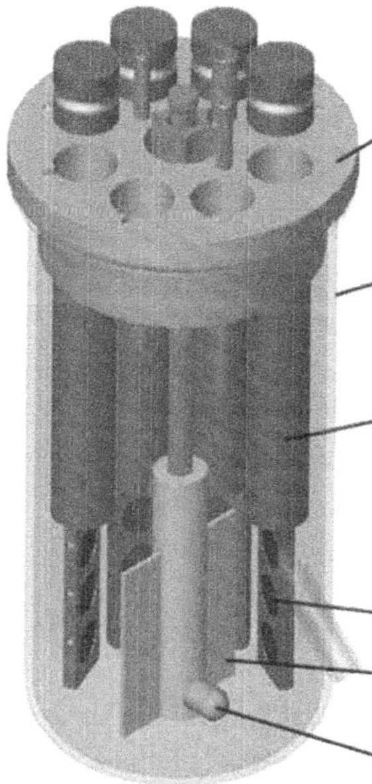

Figure 8.4 CBR 90 biofilm reactor. (From http://www.biofilms.biz/biofilm-reactors, last accessed 12 March 2015; used with permission from BioSurface Technologies Inc.)

only one biofilm sample could be obtained for each run. This model examines only flow parallel to the membrane and not through the pores. Laminar or turbulent flow in glass flow cell biofilm reactors can be achieved by adjusting flow velocity (Stoodley *et al.*, 1999a).

A CBR 90 biofilm reactor (BioSurface Technologies, Bozeman, USA) (Figure 8.4) that can generate up to 24 coupon samples was tested by Goeres *et al.* (2005). Unfortunately, the surface material investigated was polystyrene, which makes it difficult to compare results with membrane surface materials. Other materials can also be tested using the CBR 90 biofilm reactor.

8.7 Investigation of persistent biofilms on UF membranes

We conducated a detailed study of the microbial population of dairy UF membranes (Tang *et al.*, 2009a,b, 2010), examining the attachment, growth and detachment of isolates obtained from dairy UF membranes following CIP. The purpose of these trials was to determine the microflora remaining on the membrane surfaces of a whey UF processing plant after standard cleaning. This would allow the biofilms of most concern in whey processing, which persist following cleaning, to be identified, enabling the risk to product quality to be assessed and providing a focus for the development of an improved cleaning system.

The spiral-wound UF and RO membranes were obtained from dairy manufacturing plants in New Zealand (Table 8.1). All membranes had been in routine use in manufacturing plants processing milk, whey or whey permeate. All operated at either 15–20 °C or

Table 8.1 Details of membrane samples from a New Zealand dairy manufacturing plant.

Manufacturing plant	Sample details
A	Polyethersulphone (PES) RO membrane used for processing whey at 15–20 °C
B	Four PES UF membranes used for whey processing under various temperatures of 10–50 °C; four different stages of the plant were labeled as 1–4
C	PES RO membrane used for processing of casein whey permeates at 15–20 °C
D	Two PES RO membranes used for milk permeate treatment at 15–20 °C; loop 1 was the first stage of the membrane processing, while loop 4 was the last stage
E	PES RO membrane from a pilot plant used for processing of milk protein concentrate at 15–20 °C
F	PES UF membrane used for whey processing at 55 °C
G	PES RO membrane used for whey processing at 55 °C

55 °C under turbulent flow at pH 4.6–6.2. Specific details of the shear rate and flux were not provided, although all manufacturers aim to operate their plants according to the membrane manufacturer's guidelines. Membranes had been cleaned, using the standard caustic-based CIP system in the plant, before being removed, sealed in plastic bags to retain moisture and sent by courier to our research laboratory.

Microbiological analysis of the bacteria recovered from the membranes produced a variety of isolates, many of which were surprisingly Gram-negative bacteria (Table 8.2). Gram-negative bacteria would be killed during the heat treatment of whey before UF, so these bacteria found on the membrane surfaces most likely originate from the water used in the diafiltration or washing of the plant. Gram-negative bacteria produce strong biofilms, and these survived standard dairy cleaning. The predominant isolate was *Klebsiella*, so this was used in subsequent trials.

8.7.1 *Attachment of* Klebsiella *isolates to UF membranes*

Studies of the initial attachment of bacterial cells to surfaces are essential in any programme aimed at elimination or control of biofilms (Dang & Lovell, 2000). Three *Klebsiella* strains (TR002, B001 and B006) isolated from the UF membranes of a whey processing plant readily attached to surfaces in a model microtitre plate system and to membrane surfaces. A further *Klebsiella* strain (EL4019) with poor ability to attach also originated from a whey manufacturing plant. There was no indication that the isolates with the greatest attachment were specific to any manufacturing plant. The microtitre plate assay was found to be a useful tool with which to screen for the attachment of cells to polysulphone membrane surfaces.

The increase in the attachment of two mixed strains (*P. fluorescens.* TR001 with *K.* TR002 or B001 or B006) compared with the attachment of each individual strain indicates an interaction between these strains in the initiation of a biofilm (Tang *et al.*, 2009a). Biofilms in many environments are multispecies, rather than single-species (Kawarai *et al.*, 2007; Macleod & Stickler, 2007). It is well known that *Pseudomonas*

Table 8.2 Bacteria isolated from dairy plant membrane surfaces.

Strain	Species	Dairy plant	Type of plant	Membrane side
WL001	*Chryseobacterium indologenes*	A	Ultrafiltration (whey)	Retentate
WL004	*Bacillus firmus*	A	Ultrafiltration (whey)	Retentate
WL008	*Lactococcus lactis* ssp *cremoris*	A	Ultrafiltration (whey)	Retentate
B001	*Klebsiella oxytoca*	A	Ultrafiltration (whey)	Permeate
B003	*Cronobacter sakazakii*	A	Ultrafiltration (whey)	Permeate
B006	*Klebsiella oxytoca*	A	Ultrafiltration (whey)	Permeate
WA001	*Lactobacillus*	B	Ultrafiltration (whey)	Permeate
WA002	*Bacillus licheniformis*	B	Ultrafiltration (whey)	Retentate
TR001	*Pseudomonas fluorescens*	C	RO (casein whey permeate)	Retentate
TR002	*Klebsiella oxytoca*	C	RO (casein whey permeate)	Retentate
TR004	*Bacillus licheniformis*	C	RO (casein whey permeate)	Retentate
H1	*Blastoschizomyces capitatus*	C	RO (casein whey permeate)	Retentate
EL4019	*Klebsiella oxytoca*	D	RO (milk permeate)	Retentate

are often the primary colonising organisms of surfaces. They have been shown to enhance the attachment of others to surfaces (Zottola, 1994), and their coexistence with *Klebsiella* has been documented (Stewart *et al.*, 1997). In Tang *et al.* (2009a), 10 of the 13 strains showed no ability to attach from pure culture, which suggests that either the majority of isolates did not form biofilm and were trapped in the accumulation of protein and biofilm on the membranes or the required conditions were not present in our experiments (e.g. combination with other microorganisms or specific environmental conditions required for attachment).

Whey and whey permeate were found to increase the attachment of most of the strains compared with phosphate-buffered saline (PBS) at pH 6.5 (Tang *et al.*, 2009a). Therefore, further details of the effects of whey components were investigated using three *Klebsiella* strains in the microtitre plate assay. Four whey components – α-lactalbumin, β-lactoglobulin, glycomacropeptide (GMP) and bovine albumin (BA) – were used. These experiments did not show which component played a major role in increasing attachment. It can be concluded that all components of whey may enhance bacterial attachment.

Whey protein concentration, membrane type (including membrane material and age), strain type and the interactions between different microorganisms are all important factors for biofilm development on UF membrane surfaces (Tang *et al.*, 2009a). Strains varied in their ability to form biofilm as individual strains, but dual strains produced a higher biofilm density than single strains. Biofilm density tended to increase with increased whey protein concentration. The saturated biofilm was approximately $8\log_{10}$ CFU/cm^2. PES membranes appeared to support biofilm growth less readily than did PVDF membranes; they may therefore be better suited to use as UF membranes, in order to reduce problems with microbial colonisation (Figure 8.5). Used membranes were more susceptible to colonisation with

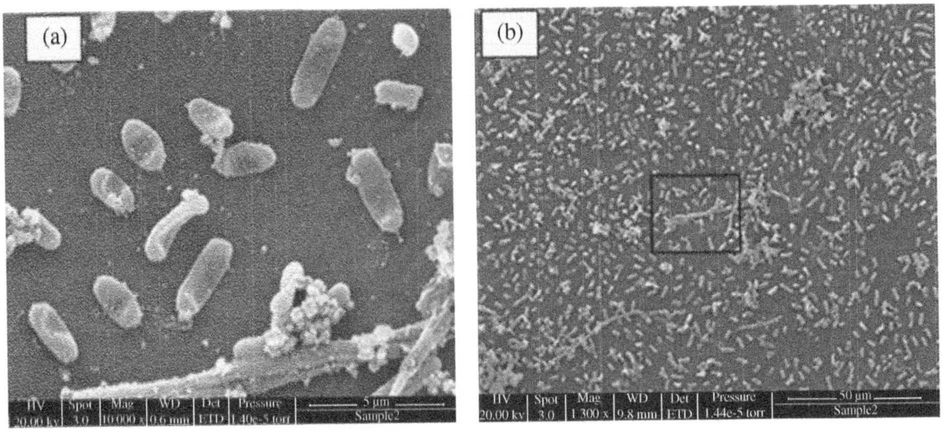

Figure 8.5 SEM of biofilm of *K. oxytoca* B006 on used PES membranes after 24 hours' incubation with 5% whey; (a) shows a magnification of the rectangular area in (b). (Reprinted with permission from Tang *et al.*, 2009; used with permission from Springer)

biofilm than were new membranes. Therefore, selecting a membrane type and monitoring membrane age will help manage biofilm development during UF.

8.7.2 *Removal of* Klebsiella *biofilms from membranes*

The use of sanitisers following a standard dairy industry caustic/acid CIP procedure reduced the number of culturable bacterial cells on membrane surfaces (Tang *et al.*, 2010). The most effective sanitiser in this study was the MIOX EW anolyte (120 ppm FAC, pH 6.8), as compared with the control CIP. MIOX EW is activated water, often described as a mixed oxidant cleaner/sanitiser, produced from the electrolysis of sodium chloride. Sodium hypochlorite and Perform (hydrogen peroxide/peracetic acid) functioned equally well when combined with Reflux E1000 (Protease). This study indicated that, if a dairy processor were to use a standard CIP (such as the control) on membrane systems, a further flush with MIOX EW anolyte would further reduce residual attached microbial populations. In addition, using protease followed by a sanitation (sodium hypochlorite, Perform or an anolyte of MIOX EW) produced the best clean, based on a >2 log reduction in residual cells, and left no culturable and viable cells at a detection limit of 0.1 $\log_{10}$ CFU/cm^2.

8.8 Other isolates from WPCs

A recent study profiling the microflora in WPCs showed *Bacillus licheniformis* to be a predominant contaminant (Zain, unpublished data) (Table 8.3). This suggests that areas other than UF membrane surfaces may be more important as a source of biofilms contaminating WPCs, as most membrane plants, including the ones used in this study, operate at temperatures of 10–15 °C, at which these isolates cannot grow. This suggests the preheater plate heat exchanger and evaporators are the most likely sites for biofilm development.

Table 8.3 Bacteria isolated from six batches of WPC.

No.	Bacterial identity	Frequency
1	*Bacillus licheniformis*	34
2	*Bacillus cereus*	9
3	*Bacillus thuringensis*	2
4	*Bacillus subtilis*	2
5	*Bacillus pumilus*	2
6	*Paenibacillus glucanolyticus*	1
7	*Lactobacillus plantarum*	3
8	*Staphyloccocus spp*	2

8.9 Conclusion

Biofilm formation is a major impediment to the use of filtration membranes in crossflow processes in dairy plants. Membrane cleaning strategies require improvements for the effective control of biofilms.

The main effects of biofilm on membranes are: (i) reduction of membrane flux and productivity; (ii) biodegradation of the membrane material; (iii) an increase in power consumption for the raising of operation pressure, (iv) potential contamination and spoilage of the product; and (v) an increase in the cost of cleaning and even consequent replacement of membrane modules.

The initiation of biofilm formation on membrane surfaces depends not only on the physical and chemical characteristics of the membranes, but also on the characteristics of early adhering bacteria and the operating conditions inside the membrane system. A suitable laboratory-scale biofilm reactor must be developed that can closely mimic the conditions in the dairy membrane plant in order to enable further study of the factors affecting biofilm formation and membrane cleaning (e.g. membrane material, strains, feed, flow rate, pH and temperature).

Dairy manufacturers have focused on the control of biofilm formed by *Pseudomonas* species and foodborne pathogens (Flint *et al.*, 1997a). Detailed studies of membrane biofilms need to understand the microbial community that exists in membrane plants. For example, if membranes are predominantly colonised by mixed-species biofilms, this will have an impact on the ability to clean. The biofilm developed by mixed cultures is more complicated than that of pure cultures. A map showing where control should be focused can be generated only when the mechanisms of biofilm formation by the true biofilm formers are explored. This requires setting up a microbe library for specific membrane plants before further study can be carried out.

The control of membrane biofilms in the dairy industry has to date been dependent upon frequent CIP with chemicals, enzymes or disinfectants/sanitisers commonly used in cleaning systems in food manufacturing plants. A study of improved control strategies should focus on both the membranes (e.g. selection of membrane materials with modifications), to lower bacterial attachment, and improvements in membrane cleaning methods, to eliminate the biofilm and prevent regrowth.

A study of *Klebsiella oxytoca* biofilm formation on UF and RO membranes indicates that the growth of this bacteria is significantly affected by strain type, medium concentration (whey protein concentration) and membrane type (membrane material and age). The use of enzymatic detergent in a CIP procedure, combined with an extra sanitation step using an EW anolyte, would improve membrane cleaning by removing biofilms from membrane surfaces.

References

An, Y. H. & Friedman, R. J. 1998. Concise review of mechanisms of bacterial adhesion to biomaterial surfaces. *Journal of Biomedical Materials Research*, **43**, 338–48.

Anon. 2007. Crossflow filtration: a new approach to wine clarification. *Filtration and Separation*, **44**, 36–9.

Bellona, C. & Drewes, J. E. 2005. The role of membrane surface charge and solute physic-chemical properties in the rejection of organic acids by NF membranes. *Journal of Membrane Science*, **249**, 227–34.

Bird, J. 1996. The application of membrane systems in the dairy industry. *Journal of the Society of Dairy Technology*, **49**, 16–23.

Bodalo-Santoyo, A., Gomez-Carasco, J. L., Gomez-Gomez, E., Maximo-Martin, M. F. & Hidalgo-Montesinos, A. M. 2004. Spiral wound membrane reverse osmosis and the treatment of industrial effluents. *Desalination*, **160**, 151–8.

Boributha, S., Chanachaia, A. & Jiraratananon, R. 2009. Modification of PVDF membrane by chitosan solution for reducing protein fouling. *Journal of Membrane Science*, **342**, 97–104.

Brans, C., Schroën, C. G. P. H., Van Der Sman, R. G. M. & Boom, R. M. 2004. Membrane fractionation of milk: state of the art and challenges. *Journal of Membrane Science*, **243**, 263–72.

Bunt, C. R., Jones, D. S. & Tucker, I. G. 1993. The effects of pH, ionic-strength and organic-phase on the bacterial adhesion to hydrocarbons (Bath) test. *International Journal of Pharmaceutics*, **99**, 93–8.

Camargo, G. M. P. A., Pizzolitto, A. C. & Pizzolitto, E. L. 2005. Biofilm formation on catheters used after cesarean section as observed by scanning electron microscopy. *International Journal of Gynecology and Obstetrics*, **90**, 148–9.

Campbell, P., Srinivasan, R., Knoell, T., Phipps, D., Ishida, K., Safarik, J. & Cormack, T. 1999. Quantitative structure-activity relationship (QSAR) analysis of surfactants influencing attachment of a *Mycobacterium* spp to cellulose acetate and aromatic polyamide reverse osmosis membranes. *Biotechnology and Bioengineering*, **64**, 527–44.

Caridis, K. A. & Papathanasiou, T. D. 1997. Pressure effects in cross-flow microfiltration of suspensions of whole bacterial cells. *Bioprocess Engineering*, **16**, 199–208.

Cartwright, P. S. 2003. Guides to selecting membrane separation technologies. *Industrial Water World*, **4**, 9–13.

Chang, I. S., Le Clech, P., Jefferson, B. & Judd, S. 2002. Membrane fouling in membrane bioreactors for wastewater treatment. *Journal of Environmental Engineering – ASCE*, **128**, 1018–29.

Characklis, W. G. 1990. Biofilm processes. In W. G. Characklis & K. C. Marshall (eds.) Biofilms, 1st edn. John Wiley & Sons, New York, pp. 195–231.

Cheryan, M. & Mehaia, M. A. 1986. Membrane bioreactors. *Chemtech*, **16**, 676–81.

Cogan, N. G. & Chellam, S. 2008. Regularized Stokeslets solution for 2-D flow in dead-end microfiltration: application to bacterial deposition and fouling. *Journal of Membrane Science*, **318**, 379–86.

Criado, M. T., Suarez, B. & Ferreiros, C. M. 1994. The importance of bacterial adhesion in the dairy-industry. *Food Technology*, **48**, 123–6.

D'souza, N. M. & Mawson, A. J. 2005. Membrane cleaning in the dairy industry: a review. *Critical Reviews in Food Science and Nutrition*, **45**, 125–34.

Dang, H. Y. & Lovell, C. R. 2000. Bacterial primary colonization and early succession on surfaces in marine waters as determined by amplified rRNA gene restriction analysis and sequence analysis of 16S rRNA genes. *Applied and Environmental Microbiology*, **66**, 467–75.

Daufin, G., Escudier, J. P., Carrere, H., Berot, S., Fillaudeau, L. & Decloux, M. 2001. Recent and emerging applications of membrane processes in the food and dairy industry. *Food and Bioproducts Processing*, **79**, 89–102.

Donlan, R. M., Murga, R., Carpenter, J., Brown, E., Besser, R. & Fields, B. 2002. Monochloramine disinfection of biofilm-associated *Legionella pneumophila* in a potable water model system. In Marre, R., Kwaik, Y. A., Bartlett, C., Cianciotto, N. P., Fields, B. S., Frosch, M., Nacker J. & Luck P. C. (eds) Legionella, 1st edn. American Society of Microbiology Press, Washington, DC, pp. 406–10.

Dreeszen, P. H. 2003. Biofilm: The Key to Understanding and Controlling Bacterial Growth in Automated Drinking Water Systems, 2nd edn. Edstrom Industries, Inc., Waterford, WI.

Elimelech, M., Zhu, X. H., Childress, A. E. & Hong, S. K. 1997. Role of membrane surface morphology in colloidal fouling of cellulose acetate and composite aromatic polyamide reverse osmosis membranes. *Journal of Membrane Science*, **127**, 101–9.

Flemming, H. C. 2003. Role and levels of real-time monitoring for successful anti-fouling strategies – an overview. *Water Science and Technology*, **47**, 1–8.

Flint, S. H., Bremer, P. J. & Brooks, J. D. 1997a. Biofilms in dairy manufacturing plant – description, current concerns and methods of control. *Biofouling*, **11**, 81–97.

Flint, S. H., Brooks, J. D. & Bremer, P. J. 1997b. The influence of cell surface properties of thermophilic *Streptococci* on attachment to stainless steel. *Journal of Applied Microbiology*, **83**, 508–17.

Fontananova, E., Jansen, J. C., Cristiano, A., Curcio, E. & Drioli, E. 2006. Effect of additives in the casting solution on the formation of PVDF membranes. *Desalination*, **192**, 190–7.

Goeres, D. M., Loetterle, L. R., Hamilton, M. A., Murga, R., Kirby, D. W. & Donlan, R. M. 2005. Statistical assessment of a laboratory method for growing biofilms. *Microbiology*, **151**, 757–62.

Her, N., Amy, G. & Jarusutthirak, C. 2000. Seasonal variations of nanofiltration (NF) foulants: identification and control. *Desalination*, **132**, 143–60.

Herzberg, M., Kang, S. & Elimelech, M. 2009. Role of extracellular polymeric substances (EPS) in biofouling of reverse osmosis membranes. *Environmental Science and Technology*, **43**, 4393–8.

Hiddink, J., Deboer, R. & Nooy, P. F. C. 1980. Reverse-osmosis of dairy liquids. *Journal of Dairy Science*, **63**, 204–14.

Isberg, R. R. & Barnes, P. 2002. Dancing with the host: flow-dependent bacterial adhesion. *Cell*, **110**, 1–4.

Ivnitsky, H., Katz, I., Minz, D., Shimoni, E., Chen, Y., Tarchitzky, J., Semiat, R. & Dosoretz, C. G. 2005. Characterization of membrane biofouling in nanofiltration processes of wastewater treatment. *Desalination*, **185**, 255–68.

Karasu, K., Yoshikawa, S., Kentish, S. E. & Stevens, G. W. 2009. A model for cross-flow ultrafiltration of dairy whey based on the rheology of the compressible cake. *Journal of Membrane Science*, **341**, 252–60.

Katsikogianni, M. & Missirlis, Y. F. 2004. Concise review of mechanisms of bacterial adhesion to biomaterials and of techniques used in estimating bacteria-material interactions. *European Cells and Materials Journal*, **8**, 37–57.

Kawarai, T., Furukawa, S., Ogihara, H. & Yamasaki, M. 2007. Mixed-species biofilm formation by lactic acid bacteria and rice wine yeasts. *Applied and Environmental Microbiology*, **73**, 4673–6.

Kochkodan, V. M., Hilal, N., Goncharuk, V. V., Al-Khatib, L. & Levadna, T. I. 2006. Effect of the surface modification of polymer membranes on their microbiological fouling. *Colloid Journal*, **68**, 267–73.

Koo, J. Y., Hong, S. P., Kang, J. W., Kim, J. E., Hyung, H., Kim, Y. H., Yoon, S. & Kim, S. S. 2002. Fouling resistant reverse osmosis membranes. *American Water Works Association Membrane Technology Conference*, **223**, 512–22.

Kumar, C. G. & Anand, S. K. 1998. Significance of microbial biofilms in food industry: a review. *International Journal of Food Microbiology*, **42**, 9–27.

Lee, E., Shon, H. K. & Cho, J. 2010a. Biofouling characteristics using flow field-flow fractionation: effect of bacteria and membrane properties. *Bioresource Technology*, **101**, 1487–93.

Lee, W., Ahn, C. H., Hong, S., Kim, S., Lee, S., Baek, Y. & Yoon, J. 2010b. Evaluation of surface properties of reverse osmosis membranes on the initial biofouling stages under no filtration condition. *Journal of Membrane Science*, **351**, 112–22.

Liu, H., Huang, L., Huang, Z. & Zheng, J. 2007. Specification of sulfate reducing bacteria biofilms accumulation effects on corrosion initiation. *Materials and Corrosion*, **58**, 44–8.

Lozier, J., Amy, G., Jacangelo, J., Mysore, C. & Heijmann, B. 2006. Natural organic matter fouling of low-pressure membrane systems. Proceedings of NWRI Microfiltration 4 Conference, Canada, pp. 19–27.

Macleod, S. M. & Stickler, D. J. 2007. Species interactions in mixed-community crystalline biofilms on urinary catheters. *Journal of Medical Microbiology*, **56**, 1549–57.

Maubois, J. L. 1980. Ultrafiltration of whey. *Journal of the Society of Dairy Technology*, **33**, 55–8.

Myint, A. A., Lee, W., Mun, S., Ahn, C. H., Lee, S. & Yoon, J. 2010. Influence of membrane surface properties on the behavior of initial bacterial adhesion and biofilm development onto nanofiltration membranes. *Biofouling*, **26**, 313–21.

O'Toole, G., Kaplan, H. B. & Kolter, R. 2000. Biofilm formation as microbial development. *Annual Review of Microbiology*, **54**, 49–79.

Pang, C. M., Hong, P. Y., Guo, H. L. & Liu, W. T. 2005. Biofilm formation characteristics of bacterial isolates retrieved from a reverse osmosis membrane. *Environmental Science and Technology*, **39**, 7541–50.

Pasmore, M., Todd, P., Smith, S., Baker, D., Silverstein, J., Coons, D. & Bowman, C. N. 2001. Effects of ultrafiltration membrane surface properties on *Pseudomonas aeruginosa* biofilm initiation for the purpose of reducing biofouling. *Journal of Membrane Science*, **194**, 15–32.

Pearce, G. 2007a. Introduction to membranes: manufacturers' comparison – part 2. *Filtration and Separation*, **44**, 28–31.

Pearce, G. 2007b. Introduction to membranes: membrane selection. *Filtration and Separation*, **44**, 35–37.

Pearce, G. 2008. Introduction to membranes-MBRs: Manufacturers' comparison: part 1. *Filtration and Separation*, **45**, 28–31.

Ridgway, H. F. 1991. Bacteria and membranes – ending a bad relationship. *Desalination*, **83**, 53.

Ridgway, H.F., Kelly, A., Justice, C. & Olson, B. H. 1983. Microbial fouling of reverse-osmosis membranes used in advanced wastewater treatment technology – chemical, bacteriological and ultrastructural analysis. *Applied and Environmental Microbiology*, **45**, 1066–84.

Ridgway, H., Ishida, K., Rodriguez, G., Safarik, J., Knoell, T. & Bold, R. 1999. Biofouling of membranes: membrane preparation, characterization, and analysis of bacterial adhesion. *Biofilms*, **310**, 463–94.

Stewart, P. S., Camper, A. K., Handran, S. D., Huang, C.-T. & Warnecke, W. 1997. Spatial distribution and coexistence of *Klebsiella pneumoniae* and *Pseudomonas aeruginosa* in biofilms. *Microbial Ecology*, **33**, 2–10.

Stoodley, P., Dodds, I., Boyle, J. D. & Lappin-Scott, H. M. 1999a. Influence of hydrodynamics and nutrients on biofilm structure. *Journal of Applied Microbiology*, **85**, 19S–28S.

Stoodley, P., Lewandowwski, Z., Boyle, J. D. & Lappin-Scott, H. M. 1999b. Structural deformation of bacterial biofilms caused by short-term fluctuations in fluid shear: an in situ investigation of biofilm rheology. *Biotechnology and Bioengineering*, **65**, 83–92.

Tang, X., Flint, S. H., Brooks, J. D. & Bennett, R. J. 2009a. Factors affecting the attachment of micro-organisms isolated from ultrafiltration and reverse osmosis membranes in dairy processing plants. *Journal of Applied Microbiology*, **107**, 443–51.

Tang, X., Flint, S. H., Bennett, R. J., Brooks, J. D. & Morton, R. H. 2009b. Biofilm growth of individual and dual strains of *Klebsiella oxytoca* from the dairy industry on ultrafiltration membranes. *Journal of Industrial Microbiology and Biotechnology*, **36**, 1491–7.

Tang, X., Flint, S. H., Brooks, J. D. & Bennett, R. J. 2010. The efficacy of different cleaners and sanitisers in cleaning biofilms on UF membranes used in the dairy industry. *Journal of Membrane Science*, **352**, 71–75.

Vrijenhoek, E. M., Hong, S. & Elimelech, M. 2001. Influence of membrane surface properties on initial rate of colloidal fouling of reverse osmosis and nanofiltration membranes. *Journal of Membrane Science*, **188**, 115–28.

Woodhams, D. 2014. Membrane filtration technology. In: MacGibbon, J. (ed.) Whey To Go. Ngaio Press, Martinborough, New Zealand, pp. 200–17.

Zottola, E. A. 1994. Microbial attachment and biofilm formation – a new problem for the food industry. *Food Technology*, **48**, 107–14.

9 Pathogen Contamination in Dairy Manufacturing Environments

Shanthi Parkar[1], Steve Flint[2], Koon Hoong Teh[2], John Brooks[3] and Phil Bremer[4]

[1]Plant & Food Research, Palmerston North, New Zealand
[2]Institute of Food, Nutrition and Human Health, Massey University, Palmerston North, New Zealand
[3]School of Applied Sciences, Faculty of Health and Environmental Sciences, Auckland University of Technology, Auckland, New Zealand
[4]Department of Food Science, University of Otago, Dunedin, New Zealand

9.1 Introduction

Milk is consumed extensively throughout the world as a dietary staple or supplement because of its high nutritional value, being rich in proteins, fats, carbohydrates, vitamins and minerals. Milk's nutrient density, coupled with its near-neutral pH and high water activity, means that it is also a good growth medium for a range of pathogenic or spoilage bacteria, yeasts and moulds (Quigley *et al.*, 2013). These nutrients act as substrates for bacteria, such as lactic acid bacteria (LAB), that utilise lactose, releasing metabolites that other microorganisms can use. Raw milk analysed in the United Kingdom (Neaves, 2013) and New Zealand (Hill *et al.*, 2012) had a low number of bacteria detected by a total viable count method, but still contained pathogens at a low frequency. Other risks associated with the consumption of raw milk include the transmission of multidrug-resistant microorganisms (Chandra *et al.*, 2010). Cold storage of raw milk favours the growth of psychrotrophic bacteria, such as *Pseudomonas* and *Acinetobacter* species. Furthermore, infections in the dairy herd have been found to contaminate raw milk with increased levels of opportunistic pathogens (Bhatt *et al.*, 2012).

The presence of pathogens in milk-based beverages and foods has the potential to cause illness or death to all consumers, but especially groups such as the sick or elderly who consume dairy-based nutritional supplements and those with developing or fragile immune health, such as neonates who consume powdered infant formula (PIF). Recalls of contaminated product can also have a significant economic impact. In order to provide consumers with safe food, dairy industries all over the world manage their manufacturing practices to control the safety and quality of dairy products. It is necessary to determine critical control points in

Biofilms in the Dairy Industry, First Edition.
Edited by Koon Hoong Teh, Steve Flint, John Brooks and Geoff Knight.

all factories and to regulate processes and automate systems in order to eliminate or at least minimise the risk of contamination.

There are two separate environments that impact on the safety and quality of dairy products, namely the primary production environment (dairy farm and milking facilities), which is largely influenced by feed, sick cows, contamination in the bulk raw milk storage tank and so on; and the dairy product manufacturing environment, which includes raw milk reception, pasteurisation and other processes used for the manufacture of specific dairy products, as well as processing and packaging environments, cleaning systems and refrigerated storage facilities.

Pathogens that can be present in milk include viruses and bacteria, some of which may survive by forming spores, which are resistant to conditions that would normally inactivate vegetative cells. This chapter reviews pathogenic contaminants in the dairy manufacturing environment are discusses the major ones in terms of their impact on humans (especially high-risk populations), their growth characteristics and their responses to environmental stress. Also discussed are the mechanisms of contamination and persistence in the dairy processing environments and processing lines; detection of pathogens in dairy products using traditional and novel techniques; and the control of pathogens in the dairy industry, including established and alternative methods.

9.2 Pathogenic bacteria

Bacterial growth can be a major problem for dairy manufacturing plants, the extent of which depends on their required growth conditions and their survival in the different environments from farm to the factory. We discuss here the pathogens that are at the greatest risk of contaminating dairy manufacturing plants, and describe their growth characteristics, their mode of contamination, the control measures required to contain their dissemination and their detection.

9.2.1 Cronobacter *species (formerly* Enterobacter sakazakii*)*

Cronobacter is a recently described genus comprising six species (Lehner, 2010) that are of emerging importance as foodborne pathogens (Norberg *et al.*, 2012; Fakruddin *et al.*, 2013; Hunter & Bean, 2013; Lu *et al.*, 2014). They are opportunistic pathogens that can contaminate PIF and cause life-threatening infections in neonates. The International Commission on Microbiological Specification for Foods (2002) has classified *Cronobacter* as a severe hazard for restricted populations (Adekunte *et al.*, 2010). *Cronobacter* infection may cause symptoms such as necrotising enterocolitis, bacteraemia and meningitis, and up to 80% of cases are fatal (Hunter & Bean, 2013). Three species, *C. sakazakii*, *C. malonaticus* and *C. turicensis*, are known to invade human intestinal cells, replicate in macrophages and invade the blood–brain barrier (Kucerova *et al.*, 2011). Infections caused by these organisms have also been reported in immunocompromised individuals, such as the elderly (See *et al.*, 2007).

These species are all potentially pathogenic, and while the minimum lethal dose is not known for oral routes of infection, PIF contamination at counts of 0.36–66.0 CFU/100 g has been suggested to be potentially infectious to the immunologically susceptible neonatal population (Pagotto & Farber, 2009).

Growth characteristics and response to environmental stress

Mathematical modelling has revealed that the growth rate for *Cronobacter* in reconstituted PIF is maximal at 37 °C (0.924 CFU/ml/h), as compared to 25 °C (0.384 CFU/ml/h) and 8 °C (0.027 CFU/ml/h) (Pina-Perez *et al.*, 2012), indicating the potential impact of heat abuse (e.g. prolonged storage at room temperature) in PIF post-preparation. Hence, it is extremely important to detect *Cronobacter* in milk powder formulations and comply with health directives for the preparation of PIF.

Cronobacter species are highly tolerant to heat treatment (Walsh *et al.*, 2011), desiccation and osmotic stress (Osaili & Forsythe, 2009) and thrive in the harsh conditions of a processing environment. The osmoprotection has been attributed to their ability to accumulate molecules such as trehalose, glycine and betaine to counteract high osmotic pressure (Osaili & Forsythe, 2009).

Mode of contamination

Cronobacter species present in milk powder have been identified by epidemiological approaches as a cause of PIF-borne infection (Lehner *et al.*, 2010). However, *Cronobacter* species have not been detected in raw milk, even with a large sample size of 875 bulk milk samples from large milk processing companies (Baumgartner & Niederhauser, 2010). Thus, an analysis of 867 samples collected from PIF manufacture, with contaminants traced back to dry processing environments (Reich *et al.*, 2010), suggests that the major route of contamination to milk powder is from the factory environment.

Cronobacter species have been detected in microbiological surveillance studies of milk processing plants (Kandhai *et al.*, 2004; Lehner & Stephan, 2004). They were detected post-pasteurisation, and a high percentage of positive samples (28%) came from vacuum cleaners used in the packaging areas (Reich *et al.*, 2010); this indicates some risk of recontamination when the product is being filled and packaged (Songzhe *et al.*, 2011), given the persistence of the bacteria at room temperature. The packaging area was also found to contain other members of the Enterobacteriaceae, which is an indicator of poor factory hygiene.

The protein components of milk powder – whey and casein – appear to be critical to the formation of exopolysaccharide-rich *Cronobacter* biofilms, which confer further protection in the processing environment (Dancer *et al.*, 2009; Hartmann *et al.*, 2010). *Cronobacter* species may be disseminated into production lines/products via two routes: post-pasteurisation from soil/dust contaminating a product at a point after drying, or from other equipment in the factory; and dry ingredients/vitamins/supplements added to the product without an additional heating step (Lehner, 2010; Larsen *et al.*, 2014). Indeed, the ability of these species to survive and thrive may be due to the protecting layer of capsular polysaccharides, which are also important in the attachment of the bacteria to surfaces in their natural environment and to manufacturing surfaces such as silicone, latex and polycarbonate (Kucerova *et al.*, 2011). The genes involved in synthesis of curli fimbriae-mediated adhesion in urinary catheters are also the ones involved in attachment to abiotic surfaces in processing plants (Zogaj *et al.*, 2003).

Other genes involved in mediating adhesion and biofilm formation on abiotic surfaces include those associated with the uptake of ions such as divalent magnesium and those involved in the biosynthesis of cellulose (Hartmann *et al.*, 2010).

The formation of biofilms in the processing environment has long been a major concern in the dairy industry because of the greater resistance of bacteria in biofilms to cleaning conditions and the ability of bacteria in biofilms to persist after cleaning (Flint *et al.*, 1997). Experimental model biofilms of *C. sakazakii* formed on manufacturing substrates dipped in reconstituted PIF had microbial counts of up to 7.96, 7.91 and 6.99 $\log_{10}$ CFU/cm^2 on silicone, polycarbonate and stainless steel surfaces, respectively (Jo *et al.*, 2010). Further, biofilms of C. *sakazakii* (and *Staphylococcus aureus*) were recalcitrant, as they showed little reduction in numbers after storage at relative humidities (RH) of 23, 43, 68 and 85% for 5 days, with the greatest resistance being at 100% RH (Bae *et al.*, 2012). Sanitation with at least 70% alcohol was found to decrease the microbial levels (Jo *et al.*, 2010), implying that alcohol-based sanitisers may provide effective control; this was confirmed with a biofilm model of *Escherichia coli* (Bae *et al.*, 2012). Thus, alcohol-based sanitisers may effectively inactivate pathogenic bacteria attached to or present in biofilms on the surfaces of utensils and cooking equipment used in the dairy industry.

Detection in dairy

Cronobacter species have been detected in PIF using a range of techniques, from culture-dependent techniques using growth media with enrichment steps to selective identification of the organism using biochemical growth characteristics (Norberg *et al.*, 2012). Molecular biology-based tools are also used to identify *Cronobacter* in final products (such as PIF) and the production environment, and thereby control their dissemination. These include protein profiling by matrix-assisted laser desorption/ionisation time-of-flight mass spectrometry (MALDI-TOF MS), biochemical identification using the API 20E system (bioMerieux, Marcy l'Etoile, France) and genotypic profiling by ribotype analysis. Biochemical profiling and genotyping have been used more routinely as accurate methods for identification and characterisation of *Cronobacter* species (Lu *et al.*, 2014). Pulsed field gel electrophoresis (PFGE) typing has been used to locate sources of contamination by tracing back clonal populations to different factory areas, such as tanker bays, evaporator rooms, an employee's shoes and external roofs (Craven *et al.*, 2010). Multilocus sequence typing (MLST) is also being investigated as a sensitive method by which to identify strains that dominate the dry processing ecosystems of dairy manufacturing plants; 21 out of the 72 strains of C. *sakazakii* isolated in one study belonged to sequence type 4 (ST4), the clonal complex associated with neonatal meningitis (Sonbol *et al.*, 2013).

Control of Cronobacter

Heat treatment of whole milk at 68 °C for 16 seconds has been shown to be very effective at inactivating *C. sakazakii* in a high-temperature short-time pilot-scale pasteuriser (Nazarowec-White *et al.*, 1999). Inactivation of *C. sakazakii* was also studied under conditions simulating contamination of stainless steel equipment surfaces, glass window panes and Teflon machinery parts. The treatment temperature was critical, as the bacteria were able to survive exposure to temperatures of up to 70 °C for 2 hours. At room temperature (25–30 °C), *C. sakazakii* survived on stainless steel and glass surfaces for up to 6 days, indicating the potential for persistence and crosscontamination (Kuo *et al.*, 2013). Ultraviolet (UV) irradiation is often used in dairy factories to kill bacteria. Exposure to 15 W UV irradiation at a distance of 55 cm caused 100% loss of viability of *Cronobacter* species within 5 minutes (Kuo *et al.*, 2013). The authors cautioned that this may not be reproducible in the manufacturing environment, where the machinery is large and has many corners and crevices in which bacteria can shelter.

In view of the serious hazard *Cronobacter*-contaminated PIF poses to neonates and the high tolerance of *Cronobacter* to heat and low water activity, procedures such as filtration and immunomagnetic separation are being investigated to remove contaminating bacteria in the post-pasteurisation stages of PIF manufacture. UV irradiation has proven to be a safe, cost-effective and sensitive method by which to inactivate *Cronobacter* species. Exposure to UV irradiation for 20 minutes has been found to inactivate *Cronobacter species* in dry PIF; this was more effective if followed by hot water treatment at a moderately high temperature of 60 °C (Liu *et al.*, 2012). Post-processing supplementation of PIF with bioactive preservatives has also been considered as a means of inactivating these pathogens. The presence of the major constituents of vanilla extract, such as vanillin (4-hydroxy-3-methoxybenzaldehyde), ethyl vanillin (3-ethoxy-4-hydroxybenzaldehyde) and vanillic acid (4-hydroxy-3-methoxybenzoic acid), may provide antimicrobial properties and decrease the thermal tolerance of *C. sakazakii* in reconstituted PIF. Supplementation of reconstituted PIF with millimolar concentrations of these bioactives has been found to result in the inactivation of *C. sakazakii* following heat treatment at 58 °C for 20 minutes (a treatment known to be sub-lethal at best) and storage at 10 °C for 48 hours or at 21 °C for 24 hours (Yemis *et al.*, 2012).

Considering that contamination by *Cronobacter* species is most likely to occur during drying and filling (Songzhe *et al.*, 2011), and given their high resistance to heat and their ability to persist at room temperature, strict compliance with quality and hygiene protocols is essential to ensure microbiological safety of the product.

9.2.2 Escherichia coli

Escherichia coli are Gram-negative, facultatively anaerobic bacteria that naturally reside in the bovine gastrointestinal tract. While most strains are harmless commensals, the herd may harbour bacteria that are potentially pathogenic when transmitted to humans through dairy products. Because *E. coli* are always present in the intestine, they can serve as indicators of faecal contamination and hygiene status in dairy manufacturing plants. While recent surveillance in Australia indicates that such safety issues are now well managed (Fegan & Desmarchelier, 2010), *E. coli* contamination may depend on the scale of the production environment, with microbial quality being poorest in small-scale manufacturing plants due to manual operations, noncompliance with hygiene practices and minimal documentation (Opiyo *et al.*, 2013).

Escherichia can undergo genetic exchange with other genera of the Enterobacteriaceae that commonly reside in the bovine gut, and this has been known to spread undesirable traits such as multiple-drug resistance. The extrachromosomal genes coding for CTXM-type extended-spectrum β-lactamases (ESBLs) confer resistance against several antibiotics, including penicillins and oxyimino-cephalosporins. Over a span of 2 years, these genes spread to different clones of *E. coli* and other Enterobacteriaceae in Japanese dairy farms (Ohnishi *et al.*, 2013).

There are many pathogenic strains of *E. coli*, but the cause of greatest concern to the dairy industry are the Shiga toxin-producing E. coli (STEC) serotypes, particularly *E. coli* O157:H7, which has an infectious dose of 5–50 cells and causes serious illness, including haemorrhagic diarrhoea (Farrokh *et al.*, 2013). Some of the serotypes produce cytotoxic Shiga toxins (Stx seropathotypes), which cause inhibition of protein synthesis in intestinal cells (Elhadidy & Mohammed, 2013).

Growth characteristics and response to environmental stress

Most *E. coli* grow at between 10 and 46 °C, with the more virulent strains well adapted to temperatures below 15 °C. Maintaining low temperatures during storage and transportation of raw milk is thus vital to preventing the growth of virulent strains of *E. coli* (Farrokh *et al.*, 2013). Pasteurisation of milk at 72 °C for 15 seconds inactivates *E. coli* O157:H7, but a higher microbial load or the proliferation of psychrotrophic bacteria may lower the efficiency of pasteurisation, leading to unacceptable levels of potential pathogens remaining in the milk (Silva *et al.*, 2010).

E. coli O157:H7 have a high tolerance to cold, surviving at subzero temperatures (−18 to −20 °C) in dairy products such as ice cream (Amer *et al.*, 2010). Exposure to mild heat of 43–51 °C enhances their resistance to cell death at the lethal temperature of 53 °C (Nakano *et al.*, 2012). Oxidative stress may also increase their thermotolerance (Blackman *et al.*, 2005), indicating that oxidative sanitisers should be used with caution and that experimentation to determine not just their cleaning potential but also their ability to induce thermotolerance in contaminating bacteria is required. The presence of an acidic environment has been shown to protect against heat inactivation (Buchanan & Edelson, 1999), and this may influence their survival in acidic dairy foods such as yoghurt and cheese.

Contamination of dairy foods

Ruminants harbour STEC as part of their normal gut microflora, and up to 27.5% of cattle harboured the Stx serotypes in one farm study, with the highest occurrence in cows with unweaned calves (Renter *et al.*, 2005). STEC contamination can occur during milking, and an incidence average of 2% has been noted for raw milk, although other serotypes, such as O26:H11, were also detected (Trevisani *et al.*, 2013). STEC may enter raw milk during milking as a result of contamination of teats by faecal material or through cows with subclinical mastitis (Farrokh *et al.*, 2013). In pasteurised milk, contamination may be the result of an ineffective treatment cycle or of supplementation with contaminated additives.

The major dairy foods at high risk of STEC contamination include raw milk, which is often consumed by farm families in the United States (Jayarao *et al.*, 2006), and raw milk cheeses, in which the acid production, mould ripening, thermal stress and osmotic stress all enhance the survival of STEC (Lee *et al.*, 2012; Peng *et al.*, 2012; Elhadidy & Mohammed, 2013). The blue-type cheeses ripened with *Penicillium* have been known to favour the growth of STEC, especially under mild acidic environments (Lee *et al.*, 2012), although

STEC numbers declined after ripening, especially the O157:H7 serotype (Miszczycha *et al.*, 2013). This organism was also inhibited in cooked cheese and lactic cheese, with a long period of coagulation at pH<4.5.

STEC can also persist within biofilms in dairy environments, which can increase their resistance to sanitisers (Sharma *et al.*, 2005). Genes required for curli formation play an important role in their ability to form biofilms and tolerate sanitisers (Wang *et al.*, 2012). Biofilms also serve to further propagate the contaminating seropathotypes as the Stx-encoding genes undergo horizontal gene transfer via Stx-encoding bacteriophages (Solheim *et al.*, 2013).

Detection of STEC in food

Sensitive culture-based methods based on enrichment of the organisms in dairy food and can detect 1–2 CFU/25 g as per ISO 16654:2001 (Anon., 2001). Multiplex polymerase chain reaction (PCR), using specific primers for serotypes such as O157:H7 and O111, has been successful in detecting contamination of milk samples (Alwathnani & Hessain, 2013).

Control

Pasteurisation and other thermal treatments are the best means of controlling this microorganism in dairy products, along with environmental hygiene control measures employed in dairy factories (Elhadidy & Mohammed, 2013). Many of the problems associated with STEC involve the consumption of raw milk and products made with raw milk. High hydrostatic-pressure (HHP) processing may be an applicable nonthermal treatment method for inactivation of pathogenic bacteria that does not affect the sensory or nutritional value of cheese (Martinez-Rodriguez *et al.*, 2012).

9.2.3 Salmonella *species*

Salmonella are Gram-negative facultative anaerobic rods that cause foodborne infections such as self-limiting gastroenteritis (Pui *et al.*, 2011). The infectious dose of *Salmonella* necessary to induce infection is dependent on multiple factors, including the virulence of the serovar and host-specific factors such as age, health and immune response to the bacterial infection (Pui *et al.*, 2011). Following a review of clinical studies performed with participants administered doses of live *Salmonella*, the infectious dose was found to be easily achievable; for example, just 25 cells of *Salmonella* Sofia and *Salmonella* Bovismorbificans to 2×10^9 cells of *Salmonella* Typhimurium, and a dose of 1×10^{10} cells of *Salmonella* Pullorum was necessary to induce salmonellosis in humans (Blaser & Newman, 1982).

In general, food poisoning outbreaks due to *Salmonella* are more common in Australia and New Zealand than in the rest of the world; *Salmonella* is the second most common cause of bacterial food poisoning in New Zealand (Lee, 2014). *Salmonella* serovars have been isolated, at a low frequency, from dairy products manufactured in Victoria, Australia over many years (Eddy *et al.*, 2010). These pathogens may be found in raw milk (probably as a result of faecal contamination of udders) and may also contaminate dairy products post-pasteurisation. Two large consecutive outbreaks involving *Salmonella* Agona affected

141 infants consuming contaminated PIF produced in France (Brouard *et al.*, 2007). An outbreak in Germany was epidemiologically linked to a strain of *S.* Agona contaminating Turkish herbal teas containing aniseed (Koch *et al.*, 2005).

Growth characteristics and survival in environmental stresses

Salmonellae grow optimally at 35–37 °C, under aerobic or anaerobic conditions, but they can grow at much lower temperatures if the incubation time is suitably extended (El-Gazzar & Marth, 1992). *Salmonella* species exhibit many of the survival mechanisms of Enterobacteriaceae, such as resistance to pH, ionic strength, temperature and water activity. They may survive the acid conditions of yoghurt, for example: *Salmonella* Infantis was shown to survive in yoghurt with pH 4.5–4.6 for 10 days (Nassib *et al.*, 2003). *Salmonella* are stable to salt stress, remaining viable for 13 weeks in Domiatis cheeses prepared from highly salted milk. *Salmonella* also survived in ice cream for 4 months, albeit with lowered viability (Nassib *et al.*, 2003). *Salmonella* survived in low-water-activity (a_w < 0.7) dairy foods such as PIF and whey powder for 168 days, although there was a 2–3 $\log_{10}$ CFU/g decrease in cells when water activity was greater than 0.33 (Farrokh *et al.*, 2013). Ageing of artificially inoculated cheeses with different ionic strengths and pH levels demonstrated the survival of *Salmonella* for up to 90 days when stored at 4 or 10 °C and for up to 30 days at 21 °C (Shrestha *et al.*, 2011).

Biofilm formation and persistence of Salmonella

The persistence of *Salmonella* in food production environments has been associated with their ability to form biofilms (Vestby *et al.*, 2009; Nicolay *et al.*, 2011). Indeed, heat-injured *Salmonella* have been shown to resort to biofilm formation as a survival mechanism (Honjoh *et al.*, 2009). *Salmonella* serovars implicated in outbreaks, such as *S.* Typhimurium and *S.* Infantis, have also been shown to attach to food processing surfaces such as stainless steel, glass and rubber (Chia *et al.*, 2009). The biofilm formation depends on the surface energy of the substrate (Chia *et al.*, 2009), but also on the genomic traits of the organisms that help in the initiation of biofilm formation and the production of biofilm matrix components, such as curli fimbriae, cellulose, capsular polysaccharides and lipopolysaccharides (Tabak *et al.*, 2007; Jesudhasan *et al.*, 2010).

The mode of growth of biofilms was found to confer bacteria with protection against sanitisers, although the extent of protection depended on biofilm age, temperature and pH (Hai & Yuk, 2013). Cells within the biofilm matrix are naturally sheltered from harsh events and chemicals, and further protection is provided by pumps that divert chemicals away from cells located deep within biofilms and by increased levels of exopolysaccharide production, which provide resistance to antimicrobials such as triclosan (Tabak *et al.*, 2007).

A salmonellosis outbreak resulting from consumption of contaminated dry cereal in the United States persisted over 10 years, and was finally sourced to *S.* Agona growing in factory wall crevices and released back into the manufacturing environment (Russo *et al.*, 2013). The chemicals in the cleaning-in-place (CIP) regime were unable to remove the bacteria in crevices, due either to insufficient access (due to biofilm formation or a physical inability to enter the crevice) or to insufficient contact time (Korber *et al.*, 1997).

Detection

PFGE has been used extensively for accurate identification of different strains of *Salmonella* and to trace isolates back to the source of contamination (Eddy *et al.*, 2010; Hall *et al.*, 2010; Brichta-Harhay *et al.*, 2011). Recently, new molecular techniques such as MLST and multilocus variable number of tandem repeat analysis (MLVA) have been used (Ricke *et al.*, 2013). High-throughput sequencing techniques have led to the publishing of complete genomes of bacterial contaminants, which are useful in the development of accurate genome-based identification methods (Haley *et al.*, 2014). Phage typing using an international typing scheme has also been used to identify salmonella isolates with a specific source, such as hard farmhouse cheese (Duynhoven *et al.*, 2009).

Control

Biofilms act as a reservoir for the seeding of pathogens into dairy manufacturing lines (Kumar & Anand, 1998). Hence, it is essential to understand the growth and survival characteristics of *Salmonella* in different environments, from farm to factory. Some knowledge is transferable, such as the effect of desiccation on thermal resistance, which is the same in dry chicken litter (Chen *et al.*, 2013) as in milk powders (Farakos *et al.*, 2013). Understanding the survival characteristics of *Salmonella* over long periods of time is essential for pathogen control. Thus, high-temperature inactivation of *Salmonella* is effective in low-moisture foods, but success may depend on the fat content of the food (Farakos *et al.*, 2013).

Natural biocides are now being investigated as a way of overcoming resistance to sanitisers among the bacteria in biofilms, including *Salmonella* species and *Staphylococcus aureus* (Braoudaki & Hilton, 2004; Doyle *et al.*, 2012). For example, carvacrol, an essential oil derived from common herbs, was pulse fed into laboratory models of dual-species biofilms formed by *S. aureus* and *S.* Typhimurium and found to inhibit growth (Knowles *et al.*, 2005).

9.2.4 Campylobacter jejuni

C. jejuni is a major cause of food poisoning (Wysok & Uradzinski, 2009). Food poisoning outbreaks caused by *Campylobacter* have been associated with the consumption of raw milk and cheese made with raw milk in the United States (Anon., 2009; Greig & Ravel, 2009; Castrodale *et al.*, 2013; Longenberger *et al.*, 2013) and Europe (Hauri *et al.*, 2013). Outbreaks in Europe and Australia are more often caused by contaminated meat (Greig & Ravel, 2009; Unicomb *et al.*, 2009). While *Campylobacter* is the most common food poisoning bacteria in New Zealand, the most common sources are meat products. This is because the most common dairy source for *Campylobacter* is raw milk, which can be legally sold in rationed quota but is not widely consumed in New Zealand (Anon., 2014c). Contamination of raw milk may be seasonal in tropical countries, associated with an increased incidence in summer and a reduced incidence in winter (Elango *et al.*, 2010), but seasonality is not linked to temperature in subtropical New Zealand (Spencer *et al.*, 2012).

Growth characteristics and survival in environmental stress

Campylobacter jejuni is a Gram-negative, microaerophilic and thermotrophic spiral rod, adaptable to nutrition depletion (Garcia & Heredia, 2013). It is aerotolerant in the presence of pyruvate or some antioxidants (Verhoeff-Bakkenes, 2012). Due to a strict minimum growth temperature (30 °C), *C. jejuni* is still assumed to be generally unable to grow in many foods. *C. jejuni* can survive for long periods under non-growing conditions, especially at low temperatures (around 4 °C). Thus, chilling foods will prolong the survival of *C. jejuni*. Furthermore, survival is enhanced under low-oxygen conditions, as are often found in packaged foods (Verhoeff-Bakkenes, 2012).

Survival/persistence

Campylobacter species have been reported to form biofilms in the water supplies and plumbing systems of animal husbandry facilities and animal processing plants (Garcia & Heredia, 2013). When grown in a mixed-species biofilm with *Enterococcus faecium*, *C. jejuni* was found to survive incubation at 23 °C for up to 2 days. Under the same conditions, *C. jejuni* cells attached to stainless steel were inactivated within 2 days (Trachoo & Brooks, 2005). Growth of *C. jejuni* in biofilms also increased the survival of *C. jejuni* when exposed to higher temperatures (56 and 63 °C).

Detection

The current standard methods regulated by the US Food and Drug Administration (FDA) employ a combined bacteriological enrichment technique using the modified cefoperazone charcoal deoxycholate agar (mCCD agar) for enrichment of *C. jejuni* from milk samples. *Campylobacter jejuni/Campylobacter coli* chromogenic plating medium was found to be most sensitive in the isolation of low-cell-count milk samples. The quantitative PCR protocol was also more rapid and sensitive, enabling the detection of organisms at counts as low as 5 CFU/25 g of raw silo milk (Gharst *et al.*, 2013).

PCR, based on the amplification of specific DNA fragments, is sensitive, specific and rapid. Detection of *C. jejuni* using PCR was tested in faecal (human/animal) and food samples (Harkanwaldeep *et al.*, 2011); compared with culture-based isolation of organisms and biochemical characterisation, more samples were identified as positive using PCR. The sensitive and rapid diagnosis and high throughput capacity make PCR-based detection ideal for laboratory screening of food samples. Commercial real-time PCR (RT-PCR) kits based on ISO 16140:2003 were successful with a limit of detection (LOD) set at 4 CFU/25 g or ml (Vencia *et al.*, 2014). PFGE and *flaA*-typing (by analysis of flagellin A gene amplicon digests) were used to identify the *Campylobacter* types contaminating 40 dairy farms in Canada, and to trace their transmission in the farm and manufacturing environments (Guevremont *et al.*, 2014), while MLST methods were used to identify new clones of these organisms in waterways and farms in New Zealand (Carter *et al.*, 2009).

Major outer membrane protein (MOMP) typing is a new proteomics-based method for the epidemiological study of campylobacteriosis (Jay-Russell *et al.*, 2013). It uses a comparison between the *porA* gene sequences of isolates from human stools collected during outbreaks and suspect dairy farm strains to trace transmission routes. MOMP typing has been found to correlate with the results from standard PFGE results (Jay-Russell *et al.*, 2013).

Culture-based isolation, followed by MALDI-TOF spectrometric analysis of the colonies, has recently been used to identify the factors leading to *Campylobacter* contamination on dairy farms, and to identify the relative importance of water segregation (between cows and poultry), weaning, use of an individual bucket for each calf, feeding of waste milk and so on in controlling bacterial transmission (Klein *et al.*, 2013).

Control

Since the cow is the major source of *Campylobacter* contamination, most likely as a result of the contamination of the udder by faecal material, milking hygiene is essential in controlling the entry of *Campylobacter* into raw milk (Oermeci & Oezdemir, 2007; Klein *et al.*, 2013). It follows that appropriate storage conditions for raw milk and heat treatment (such as pasteurisation) are the primary factors controlling the spread of *Campylobacter* during the manufacture of dairy products.

9.2.5 Bacillus cereus

Bacillus cereus sensu lato is a group of Gram-positive aerobic spore-forming bacilli that includes six closely related species: *Bacillus cereus sensu stricto*, *Bacillus anthracis*, *Bacillus thuringiensis*, *Bacillus mycoides*, *Bacillus pseudomycoides* and *Bacillus weihenstephanensis* (Carlin *et al.*, 2010).

While *B. cereus* is the major pathogenic spore former found in milk, other *Bacillus* species, such as *Bacillus licheniformis*, *Bacillus amyloliquefaciens* and *Bacillus pumilus*, are also found in milk and may produce toxins that lead to food poisoning (Lucking *et al.*, 2013). These bacteria also cause food spoilage and decrease the shelf life of dairy products. Microbial enzymes, such as proteases, lipases and phospholipases, can break down the food structure, causing 'bitty cream' and 'sweet curdling' defects. Aerobic spore-forming bacilli also cause typical off-flavours, including flat-sour spoilage and bitter, fruity and rancid off-flavours (Lucking *et al.*, 2013).

Growth characteristics and toxin production

B. cereus is ubiquitous and is adapted to survival in diverse ecological niches as vegetative cells and spores (De Jonghe *et al.*, 2010). It can grow at mesophilic temperatures (10–42 °C) and under aerobic and anaerobic conditions. In addition, it produces spores that can survive unfavourable conditions. The bacterium survives in the host by producing virulent enterotoxins that weaken host defences (Pexara & Govaris, 2010).

B. cereus occurs as two main forms: the diarrhoeal type and the emetic type. The diarrhoeal type is pathogenic only on consumption of food contaminated with its heat-resistant spores, not the vegetative cells or the toxin. This is because the vegetative cells do not survive the gastrointestinal transit. Furthermore, its enterotoxin is destroyed by stomach acids and enzymes, such as intestinal proteases (within 30 minutes). After the spores reach the small intestine, they germinate, grow and produce diarrhoeal enterotoxins (Ceuppens *et al.*, 2013). In contrast, consumption of the emetic pathogen's toxin results in emetic food poisoning, as the cereulide toxin is resistant to the acidic conditions and proteases encountered in the gut (De Jonghe *et al.*, 2010). Emesis has often been associated with the consumption of starch-rich foods, such as rice and pasta, while the diarrhoeal poisoning is often linked to the consumption of vegetables, meat and especially milk and dairy products (Pexara & Govaris, 2010).

Consumption of food containing between 10^5 and 10^8 *B. cereus* cells and/or spores will cause disease, although this depends on the food matrix, the growth pattern of the strain, the storage conditions of the food and the immune status of the consumer (Ceuppens *et al.*, 2013). Cheese and milk inoculated with 10^5 cells were found to contain the enterotoxin following storage at 10 °C for 7 days (Sadek *et al.*, 2006). Interestingly, the presence of cheese starter probiotics such as *Lactobacillus reuteri* and *Lactobacillus rhamnosus* decreased the number of viable *B. cereus* and no enterotoxin was detected (Sadek *et al.*, 2006).

Mode of contamination/persistence

A wide and diverse range of species belonging to the genus *Bacillus* can contaminate raw milk via soil, feed and farm equipment. A Romanian study reported a 15% frequency of *Bacillus* contamination, with the pathogenic *B. cereus* making up 5% (Ciotau & Sindilar, 2010). *B. cereus* is mainly found in the soil, where it occurs at 10^5–10^6 cells or spores per gram. *B. cereus* is present in dairy pastures, and contamination of the udders of cows occurs during grazing or via bedding material and feed.

The persistence mechanisms of *B. cereus* were studied under conditions simulating those in operational dairy silos. The spores were studied for attachment to stainless steel at 4 °C, for germination and biofilm formation in milk and for survival during CIP treatments (1.0% sodium hydroxide at pH 13.1, 75 °C and 0.9% nitric acid at pH 0.8, 65 °C) (Shaheen *et al.*, 2010). The spores survived CIP regimes by withstanding the hot alkaline wash treatment ($\leq 1.5 \log_{10}$ CFU/ml inactivation after 15 minutes), remained attached to stainless steel following cold water rinses and germinated, post-rinse, at low temperatures. The spores of a cereulide-producing group germinated slowly in rich medium and remained viable after exposure to heating at 90 °C (Shaheen *et al.*, 2010). Some *B. cereus* strains can grow under refrigerated storage conditions (4 °C) and form biofilms in dairy chilling tanks (Kumari & Sarkar, 2014; Pena *et al.*, 2014).

The identification of psychrotrophic strains of *Bacillus*, especially those of the pathogenic *B. cereus* that grow at 4–10 °C, is of major concern, as they can grow during production processes and produce toxins in milk and cheese (Sadek *et al.*, 2006), pasteurised milk (Haldar & Kuila, 2011) and fermented milk (Hanamant & Bansilal, 2012), and even in extended-shelf-life (ESL) milk (Schmidt *et al.*, 2012). Furthermore, psychrotrophic *B. cereus* growing in ice creams was found to be resistant to common antibiotics such as ampicillin, amoxicillin, streptomycin and neomycin, which would mean that illnesses caused by consumption of such contaminated food would be difficult to treat with antibiotics (Amin & Shaker, 2011). *B. cereus* is emerging as a serious threat to neonates through contamination of PIF (Pinto *et al.*, 2013). In one survey of 100 reconstituted PIF samples, *Cronobacter* species were not detected at all, but *B. cereus* was detected at low levels in 24 samples, even though the total aerobic counts were in the acceptable range. After reconstitution and storage at ≥20 °C for 14 hours, the levels of *B. cereus* for 59 samples were > 10^3 CFU/g. This highlights the need to eliminate spores and not just vegetative cells in high-risk foods, and to conduct appropriate handling of reconstituted PIF (Haughton *et al.*, 2010).

The spores of *B. cereus* strains are hydrophobic and attach to equipment surfaces already coated with organic material from milk, such as proteins and phospholipids. Under favourable conditions, the spores will germinate, grow and initiate biofilm formation (Hornstra *et al.*, 2007). Biofilm formation has been observed on a wide variety of materials commonly encountered in dairy manufacturing plants (Heyndrickx, 2011; Faille *et al.*, 2014). It is stronger at the air–liquid interface, as seen in milking lines, possibly due to the aerobic conditions required by these microorganisms (Wijman *et al.*, 2007). These biofilms increase the persistence of *B. cereus* by encouraging sporulation (up to 90% spores) and providing resistance to sanitisers (Faille *et al.*, 2014).

Detection

PCR-based methods have been used to differentiate and enumerate different *Bacillus* isolates in milk (Oliwa-Stasiak *et al.*, 2011) for simultaneous detection of emetic and enterotoxin-producing strains using primers specific for toxin-producing genes (Kim *et al.*, 2012). However, the sensitivity of detection is reduced by tenfold in the milk matrix as compared with culture medium. Another recent assay based on a similar principle for simultaneous detection of emetic and non-emetic strains had an LOD of 1.91×10^3 spores/ml milk in artificial inoculation experiments, with a mean recovery rate averaging 81% (Dzieciol *et al.*, 2013).

Control

Efficient CIP processes are required to ensure that cells and spores of *B. cereus* present within biofilms are eliminated from dairy processing equipment. Simple biofilm models using different substrates (e.g. stainless steel and gasket materials) and growth conditions (e.g. temperature) may provide experimental set-ups to trial different CIP regimes. Optimal CIP may be assessed against a 24-hour biofilm model using coupons with test strains of *B. cereus* growing in milk at 4 °C. Increasing the length of the alkali cleaning step appeared to increase the cleaning efficiency (Salustiano *et al.*, 2010; Kumari & Sarkar, 2014).

Understanding the effect of CIP on *B. cereus* spores may be a key to controlling this microorganism in dairy manufacturing plants. For example, *B. cereus* spores that were produced in biofilms on surfaces, in the presence of whole milk, were sensitive to hot 0.9% nitric acid, providing a simple CIP treatment to eliminate these spores (Shaheen *et al.*, 2010). The surface chemistry of the spores provides a clue to the temperature, strength and duration of the CIP treatments required, and may be more important than mechanical action or turbulence (Sundberg *et al.*, 2011; Faille *et al.*, 2013). Spores are highly resistant to inactivation and removal by oxidising sanitisers (chlorine, chlorine dioxide and a peroxyacetic acid-based sanitiser) (Ryu & Beuchat, 2005), although these are more effective than other types of santisers (Sundberg *et al.*, 2011).

Another strategy is the use of mild heat or germinants, such as mixtures of l-alanine and inosine, to germinate the spores. This results in an immediate loss of resistance

(Lovdal *et al.*, 2011), so that *B. cereus* spores (now germinated) are more susceptible to CIP treatments (Hornstra *et al.*, 2007).

Plant-derived flavonoids may be used for their antibacterial properties. For example, counts for psychrotrophic strains of *B. cereus* decreased by 2–4 $\log_{10}$ CFU/ml following incubation with kaempferol at 50 µM concentration for 24 hours (Lee *et al.*, 2011). These tests are yet to be validated with milk or milk products, as the food matrix, water activity and other growth conditions will also influence bacterial viability (Desai & Varadaraj, 2013).

9.2.6 Listeria monocytogenes

Listeria monocytogenes is a ubiquitous microorganism found in farm environments and carried by farm animals and humans, often asymptomatically. It is a Gram-positive motile rod-shaped bacterium and is a particular concern in chilled, wet food production environments, such as those often found at dairy manufacturing plants (Knochel, 2010).

Listeria food poisoning may result in listeriosis, which can cause bacteraemia, meningitis, foetal loss and death. Populations that are vulnerable include immunocompromised individuals, the elderly and pregnant women. For example, 1651 cases of listeriosis were reported in the United States during 2009–11 (Anon., 2013). In the 2009–10 European outbreak, 34 people were affected, with 8 fatalities; the source was found to be Quargel cheese contaminated with just two different clones of *L. monocytogenes* of serotype 1/2a (Schoder *et al.*, 2014). Consumption of cheese was linked to a two-state outbreak of listeriosis that led to seven hospitalisations and one death (Anon., 2014b).

Growth characteristics and response to environmental stresses

Listeria is extremely versatile in its growth requirements, with temperature ranging from −1.5 to 50 °C, and pH from 4.3 to 9.6. It is extraordinarily resilient to stresses and has been reported to survive for 132 days at 4 °C in a growth medium containing 25.5% NaCl (Donnelly & Diez-Gonzalez, 2013). The dairy products that are most commonly associated with foodborne outbreaks involving *L. monocytogenes* are soft cheeses. They are particularly at risk because of the ability of strains of *L. monocytogenes* to grow in the acidic environment of cheese, and to grow within soft cheeses under maturation and refrigerated storage conditions. For example, *L. monocytogenes*, inoculated at a level of 3 $\log_{10}$ CFU/g, was found to grow rapidly in Minas Frescal cheese (a Brazilian fresh cheese) during refrigerated storage at 8–10 °C; the level of *L. monocytogenes* increased by 3 $\log_{10}$ CFU/g after 6 days and by a further 2 $\log_{10}$ CFU/g after 12 days (Pimentel-Filho *et al.*, 2014).

L. monocytogenes has also been shown to be resistant to stresses such as pH (5.5) and salt (3.5%), which further enhances its ability to survive in cheese by induction of tolerance responses (Melo *et al.*, 2013). This ability to mount tolerance responses also appears to be responsible for its ability to survive in humans after ingestion of contaminated foods. Indeed, upregulation of genes involved in stress adaptation has been observed in pathogenic strains of *L. monocytogenes* after simulated gastrointestinal transit (Mataragas *et al.*, 2014).

Mode of contamination/persistence

The extraordinary resilience of *L. monocytogenes* to stresses such as acid, temperature, humidity and salt concentration, combined with its ability to form biofilms and its resistance to routine cleaning and disinfection practices, makes it a significant and continuous challenge to the food industry (Todd, 2011). It has been detected in raw milk, dairy products (e.g. cheeses), food contact surfaces (milk reception, production lines, cheese ripening, cheese washing, refrigerated product storage rooms) and non-contact food surfaces (plant entrance and shipping containers) (Almeida *et al.*, 2013).

L. monocytogenes can form biofilms on a wide range of surfaces in the food industry, including the rubber used in gaskets (Ronner & Wong, 1993), stainless steel (Bonsaglia *et al.*, 2014), polytetrafluroethylene (Chavant *et al.*, 2002) and polystyrene, under different nutrient conditions (Zeledon *et al.*, 2010). Its ability to form biofilms and colonise niches or harbourage sites within processing sites is believed to account for its persistence within processing environments (Carpentier & Cerf, 2011).

The ability of *L. monocytogenes* to form a biofilm is dependent on the stain and serotype, origin, temperature, nutrient level, previous exposure to sublethal stressors (acid) and topography and composition of the surface (Barbosa *et al.*, 2013; Kadam *et al.*, 2013; Mosquera-Fernandez *et al.*, 2014). The presence of other bacterial species can enhance its ability to form biofilms and survive within food processing environments (Bremer *et al.*, 2001) and can increase its resistance to sanitisers (Bremer *et al.*, 2002; Lourenco *et al.*, 2011).

Detection

Considering the severity of listeriosis outbreaks in high-risk populations, there is zero-tolerance for this organism in ready-to-eat foods in many countries. For example, European regulators have placed limits of <100 CFU/g or absence of the organism in 25 grams of the sample (Anon., 2014a).

Molecular methods, such as quantitative PCR, have been ideal for tracing contaminations on dairy equipment surfaces but cannot be applied to swab/culture-based enumeration. Molecular methods such as PFGE and MLST have been useful in determining *L. monocytogenes* serotypes and clonal diversities in food processing and the food chain (Schoder *et al.*, 2014). Similar molecular forensics were used to trace the origin of a multistate outbreak of listeriosis in the United States to six *L. monocytogenes* strains from Italian ricotta cheese produced in a dairy manufacturing plant in southern Italy (Chiara *et al.*, 2014).

MALDI-TOF MS provides a rapid, accurate and cost-effective proteomics-based approach to detecting *L. monocytogenes* directly from selective enrichment broths. Suspect milk samples are incubated in a broth for 24 hours, then put through secondary enrichment for a further 6 hours. The method is sensitive enough to detect 1 CFU/ml of *L. monocytogenes* (Jadhav *et al.*, 2014). Metabolomics-based identification of *Listeria* metabolites has recently been investigated in spiked milk samples (Beale *et al.*, 2014). This method is rapid and may hold promise if the *Listeria* metabolome is distinct from that of other contaminants.

Control

Cheeses are a major cause of outbreaks of listeriosis (Almeida *et al.*, 2013). HHP processing was trialled for post-packaging treatment of a Queso Fresco cheese spiked with a mixture of five strains of *L. monocytogenes*, but had limited success (Tomasula *et al.*, 2014). Some combinations of pressure, temperature and time decreased the *L. monocytogenes* count to below the detection level. The cells developed baroresistance with an increase in hold time, and grew during cold storage. Even in fluid models of contamination, it appears that the Gram-positive *Listeria* species are more resistant to HHP inactivation than the Gram-negative *E. coli* (Cavender & Kerr, 2011), indicating a need for secondary preservation techniques. Natural alternatives include bacterially produced peptides that inhibit multiple other species of bacteria. A combination of nisin and bovicin HC5 was trialled with fresh cheese artificially contaminated with 10^4 cells/g of *L. monocytogenes* Scott A and *S. aureus* ATCC 6538, and stored at 4 °C. After 15 days, there was complete inhibition of the former and over a 3 log decrease in the latter, as compared with the no-bacteriocin control cheese (Pimentel-Filho *et al.*, 2014). Nisin was also found to be effective in combination with caprylic acid in controlling the growth of a multispecies cocktail added to fresh cheese after 20 days of storage (Gadotti *et al.*, 2014). Grape polyphenols were also found to selectively inhibit laboratory cultures of *Listeria*, including *L. monocytogenes* (Rhodes *et al.*, 2006).

Other technologies that may reduce the *Listeria* burden in the food chain include UV irradiation of milk (Pereira *et al.*, 2014) and photohydroionisation, which relies on oxidation technology to decrease bacterial counts on abiotic surfaces (stainless steel) and biotic surfaces (infected chicken and turkey) (Saini *et al.*, 2014).

Recently developed mathematical models for *Listeria* contamination in raw milk and semisoft pasteurised cheese (Tiwari *et al.*, 2014) and soft cheese (Tenenhaus-Aziza *et al.*, 2014) identify and manage *Listeria* contamination during manufacture of soft cheese.

9.2.7 Staphylococcus

Staphylococcus is a ubiquitous genus, comprising Gram-positive aerobic cocci, and includes a large number of species. These species grow at temperatures between 7.0 and 47.8 °C (optimum 35 °C), at pH between 4.5 and 9.3 (optimum 7.0–7.5) and at low water activity. They are tolerant to the presence of salt and sugars and outcompete many other environmental organisms (Bennett *et al.*, 2013).

The species most associated with dairy product food poisoning is *S. aureus*, which produces heat-stable staphylococcal enterotoxins that can survive gastrointestinal digestion (Bennett *et al.*, 2013) and induce emesis (Hu & Nakane, 2014). Milk and dairy products account for 5% of known staphylococcal food poisoning in the European market (Bianchi *et al.*, 2014). Bovine mastitis may be a significant cause of the spread of these microorganisms if the microbial quality of milk products is not adequately controlled (Bardiau *et al.*, 2013).

The staphylococcal enterotoxin genes are varied, and a single microorganism is capable of carrying multiple genes on stable regions of its chromosome and on mobile genetic elements, resulting in multiple pathogenic toxin profiles (Bianchi *et al.*, 2014).

Mode of contamination/persistence

Staphylococci are found in the environment and on the skin of humans and animals. They can also be found attached to abiotic surfaces of processing equipment and in the food processing environment. They can grow rapidly in food that is frequently handled and is stored at an inappropriate temperature (Bennett *et al.*, 2013). Many of the genes that influence the virulence of *S. aureus* may also play a role in the persistence, via increased attachment, of these bacteria on both abiotic (stainless steel, glass) and biotic (teats and udders, nasal cavity, etc.) surfaces (Cucarella *et al.*, 2004; Kot *et al.*, 2013).

Most dairy isolates (82%) are capable of forming biofilms (Abdul-Ratha & Yarmorad, 2013), and this is recognised as their major mode of persistence on processing surfaces in the dairy manufacturing plant. The components in milk and the milk pH may influence bacterial adhesion. The level of inorganic compounds such as dipotassium hydrogen phosphate (Atulya *et al.*, 2014), the milk fat content and treatments such as ultra-high-temperature (UHT) temperature may influence biofilm development (Hamadi *et al.*, 2014). Older biofilms have more of the exopolysaccharide matrix, and this increases the resistance of the cells to sanitisers (Abdallah *et al.*, 2014).

Control

Use of sanitisers at specified concentrations and durations, such as sodium hypochlorite (NaOCl, 150 ppm for 10 minutes), has been known to reduce staphylococcal counts (Melo *et al.*, 2014). Natural antimicrobial compounds have been explored for use in controlling the growth and colonisation of surfaces by *S. aureus*. Such compounds may be incorporated into dairy products to control microbial counts and increase shelf life, and they may also prove to be a useful alternative to sanitisers. For example, tannic acid, a common tea polyphenol, was found to prevent bacterial colonisation of nasal epithelia by *S. aureus* (Payne *et al.*, 2013). Lysosyme has also been investigated for its ability to control biofilm formation by *S. aureus*, although it does not always demonstrate antimicrobial activity. For example, of 25 dairy strains of *S. aureus* studied, 6 survived and were able to form biofilms. The bacteriocin nisin was effective at preventing biofilm formation by *S. aureus* when applied at 25 μg/ml, although it was not effective at lower levels (Sudagidan & Yemenicioglu, 2012).

9.3 Yeasts and moulds

Yeasts and moulds are a relatively minor issue in dairy manufacturing plants. They can contaminate dairy products such as curd and cheese. Some such microorganisms are potentially pathogenic, but they do not survive pasteurisation; hence, their presence in dairy products indicates recontamination from environmental sources, such as walls and shelves, air, water and equipment (Torkar & Teger, 2006). In a dairy manufacturing plant in Serbia, both yeasts and moulds were isolated from 60% of cheese samples. They had average concentrations of 4.7 and 4.3 $\log_{10}$ CFU/g, respectively, with the genera *Geotrichum* (91.9%), *Moniliella* (5.4%) and *Aspergillus* (2.7%) being the most frequently isolated (Torkar & Teger, 2006). The *Aspergillus* isolates were not *Aspergillus flavus* or *Aspergillus parasiticus*, which are known to produce aflatoxins. The yeast and mould count increased tenfold during initial cheese-making stages, such as coagulation and whey drainage, and further increased to the maximum number in 2-week-old cheeses. Commonly, yeast and mould counts are found to be in the order of 1–3 $\log_{10}$ CFU/g, with a maximum of 6 $\log_{10}$ CFU/g (Brooks *et al.*, 2012; Mikulec *et al.*, 2012). The presence of moulds also appeared to be seasonal. However, *Penicillium* species were detected in fresh soft cheeses in Serbian dairy farms during all four seasons, and at different relative abundances than other mould species in the cheeses (Mikulec *et al.*, 2012).

Even moulds that are used to contribute to flavour and aroma during cheese ripening can have undesirable effects. For example, *Penicillium camemberti* and *Penicillium roqueforti* are known to metabolise the lactic acid produced by starter LAB during cheese making. The growth of these moulds can enhance the growth and survival of STEC O157:H7 in cheese, likely due to the stabilisation of or an increase in the pH of the cheese (Lee *et al.*, 2012).

9.4 Preventing contamination of dairy products by pathogenic microorganisms

9.4.1 *Pathogenic bacteria in raw milk*

Pathogenic bacteria originate from the general farm environment, including the feed and soil, the pasture and the faecal material that contaminate the cow's hide and udder, which subsequently contaminates the milk. Pathogens can also enter the milk through clinically infected udders (mastitis). Most pathogens are destroyed by thermal treatments during processing, but growth of pathogens in raw milk, before it is processed, can lead to the production of heat-resistant toxins and spores that survive into the final product. This, together with the desire by some sectors of the population to consume raw milk and raw milk products, increases the food safety risk associated with dairy products.

Raw milk may be contaminated by a range of different pathogenic bacteria, although, when present, these are typically at very low levels. For example, in the United States, Grade A raw silo milk may contain a total microbial load of 3×10^5 CFU/ml (Jackson *et al.*, 2012). In a recent survey of silo samples, 88.57% of the bulk silo tanks passed the Pasteurised Milk Ordinance, yet some samples had a small microbial load of pathogenic bacteria. Pathogens detected included *B. cereus* (8.91% of samples, with counts of 3.0–93.0 CFU/ml), *E. coli* O157:H7 (3.79–9.05% of samples, at 0.0055–1.10 CFU/ml, depending on the assay utilised), *Salmonella* species (21.96–57.94% of samples, at 0.0055–60.0 CFU/ml) and *L. monocytogenes* (50% of samples, at 0.0055–30.0 CFU/ml).

The contamination of raw milk by specific dairy pathogens at the dairy farm and during transportation is discussed in Chapter 5, as are practices used to reduce such contamination.

9.4.2 *Prevention of contamination at the dairy manufacturing plant*

As summarised in Figure 9.1, various biotic and abiotic factors in the immediate environment of the dairy manufacturing plant determine the survival of microbial contaminants and their potential to contaminate the final product.

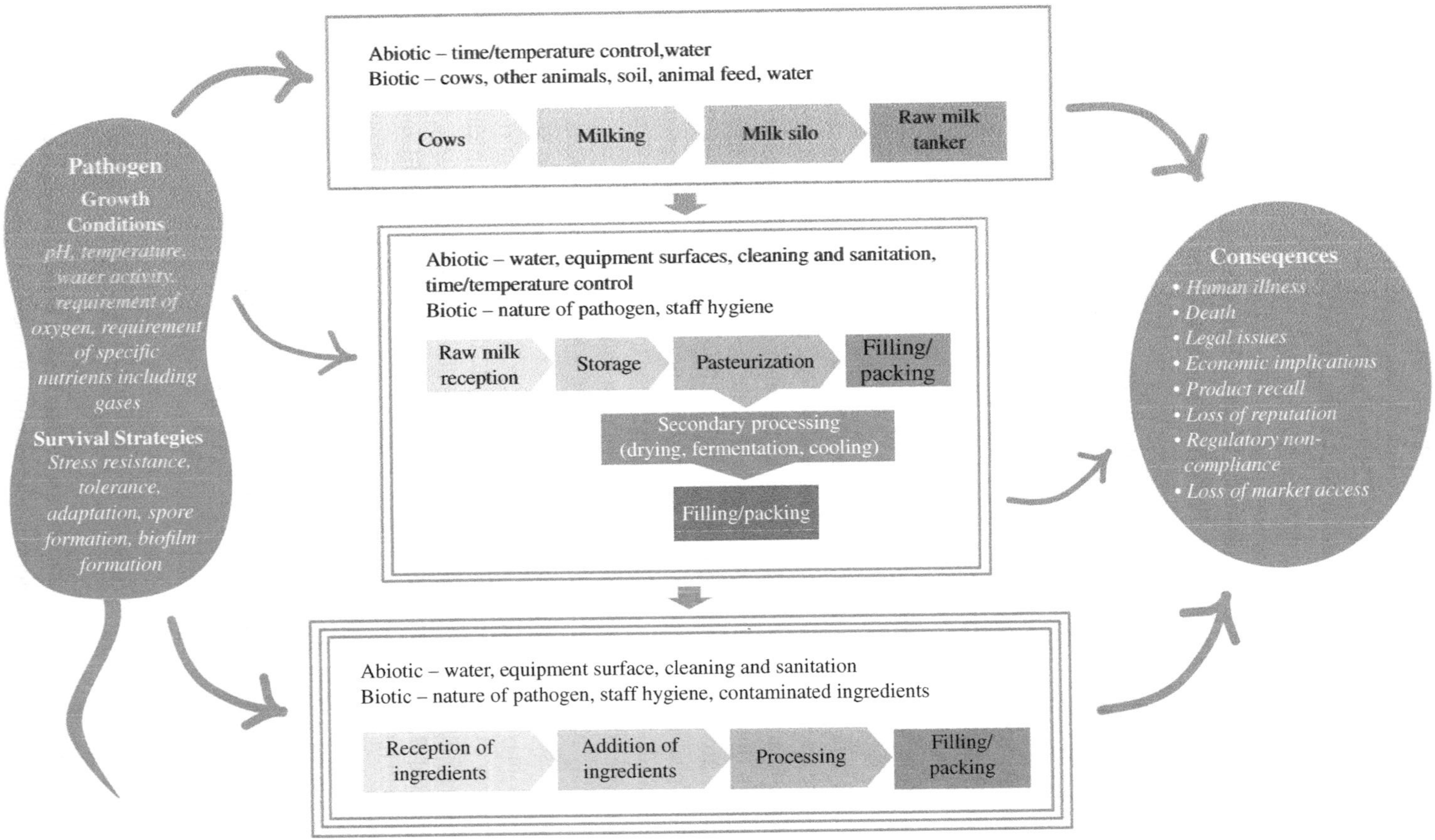

Figure 9.1 Mechanisms of pathogen contamination in the dairy processing environment and their consequences. Characteristics of the pathogenic organisms and biotic and abiotic factors in the processing environment influence the bacterial prevalence and survival in the dairy manufacturing plant. Boxes denote the primary (single outline), secondary (double outline) and tertiary environments (triple outline); exemplar workflows involved in these environments are included.

Design and operation of a dairy manufacturing plant

Dairy manufacturers must ensure that the storage and processing of raw materials, including raw milk, which are potentially contaminated with pathogenic bacteria is performed in locations that are separated from those where dairy products are packaged and stored.

Many practices and procedures (e.g. hazard analysis & critical control points (HACCP) and prerequisite programmes) are also implemented as part of food safety management programmes to prevent the entry of pathogenic bacteria into dairy manufacturing plants from a range of potential sources, including raw milk.

Processing

The fact that low-level contamination of raw milk by pathogenic bacteria occurs means that dairy food manufacturers must apply processing treatments that eliminate these microorganisms during manufacture. The prevention of pathogens surviving in dairy products is primarily achieved through application of a thermal processing treatment. The minimum heat treatment applied to milk in Australia and New Zealand is 72 °C for 15 seconds. Other manufacturing processes may employ more severe thermal treatments, but they are all designed to achieve the objective of inactivating pathogenic bacteria that may be present in raw milk.

Pathogenic bacteria in the processing environment

Foodborne pathogens can enter the food chain through the primary production environment (i.e. at the dairy farm) or through food processing environments (Larsen *et al.*, 2014). It is essential that dairy manufacturers understand the factors that influence the prevalence of pathogenic microorganisms in the processing environment. As summarised in Figure 9.1, growth of the pathogen is promoted by specific growth conditions (pH, temperature, water activity), but these may differ from the minimal conditions that are required just for survival.

Once pathogenic bacteria enter the processing environment, it can be difficult to remove them, due to their ability to persist within biofilms (Brooks & Flint, 2008). There are many factors that influence the adhesion of the microbes, including the cell surface characteristics of the bacteria (flagella, fimbriae) and the substrate to which they adhere (Sala *et al.*, 2012), the components in the milk (fat, proteins, inorganics) and the growth medium (Pagedar *et al.*, 2010). Biofilms provide a polysaccharide matrix that shelters the bacteria against physical and chemical methods of cleaning. Hard-to-access crevices in the equipment walls provide another reason for the reseeding of a factory environment following cleaning and sanitation (Cleto *et al.*, 2012). There are many factors that decide the strength of adhesion of the bacteria to substrates and the recalcitrance of biofilm to sanitisers. In one study, using stainless steel coupons of biofilms, alcohol-based sanitisers were found to be more effective than chlorine-based ones (Bae *et al.*, 2012).

Environmental pathogen monitoring programmes

Bacteria are known to persist in biofilms on surfaces within the dairy processing environment. Such biofilms may harbour potential foodborne pathogens, which may contaminate dairy products. Therefore, it is important to undertake routine pathogen testing programmes for both the dairy processing environment and the dairy products themselves.

Within the dairy industry, the most common environmental pathogen monitoring programmes, as a result of the previous history of outbreaks associated with dairy products, are for *L. monocytogenes* and for *Salmonella* species. In recent years, concern over *Cronobacter* species in PIF as a cause of life-threatening infections of neonates has led to the widespread implementation of environmental monitoring programmes for such species by manufacturers of milk powder and PIF.

End-product testing

The presence of foodborne pathogens in dairy products is of concern to the dairy industry, as most dairy products are ready-to-eat and many are prepared for consumption by consumers at greater risk of food poisoning than the general population (e.g. infants). Therefore, dairy products are routinely tested for foodborne pathogens before distribution.

Most dairy products are routinely tested for a range of common foodborne pathogens, including *Campylobacter*, *Salmonella*, *L. monocytogenes*, *S. aureus* and *E. coli*. Dairy products that are intended for infant consumption, such as infant formula, have stricter controls. Infant products are examined for *C. sakazakii,* which is associated with meningitis, necrotising enterocolitis and fatality in infected infants (Bowen and Braden, 2006; Caubilla-Barron *et al.*, 2007).

Most of the methods currently used in pathogen testing programmes are based on conventional microbiology methods that are often relatively cheap and simple to use, but typically require 3–5 days before obtaining a result. Pathogen routine testing programmes are usually based on detection in a certain quantity of a product, although enumeration procedures are also available for some pathogens. Conventional methods usually involve four steps: sample preparation, enrichment, selective plating and confirmation.

The samples are inoculated into a preenrichment medium to encourage the growth of the specific pathogens being tested for and to repair cells damaged during processing (Jantzen *et al.*, 2006; Wu, 2008). This may then be followed by a secondary or selective enrichment step. Preenrichment and selective enrichment steps are important in the detection of low numbers of pathogens (Feng, 2007). The enrichment media are always incubated at the optimum growth conditions of the pathogens being tested for.

The enrichment media are usually plated on selective media, such as mCCD agar for the detection of *Campylobacter*, to encourage the growth of specific pathogens. However, some dairy samples may be plated directly on to selective agar plates, without enrichment steps. This is the case in the detection of *S. aureus*, where the prepared samples are plated on to Baird–Parkar plates. Baird–Parkar plates are used extensively for the detection and enumeration of coagulase-positive staphylococci in dairy products. Furthermore, some selective agar plates contain chromogenic substrates permitting the pathogen to produce coloured colonies. For example, *Listeria* spp. may produce turquoise colonies on a selective chromogenic agar due to the β-D-glucosidase activities on the chromogenic substrate 5-bromo-4-chloro-3-indoxyl-β-D-glucopyranoside (Reissbrodt, 2004). The use of chromogenic substrates in selective agar plates may ease the detection of pathogens, although these media are generally more expensive than the traditional agars.

Some pathogens may need further identification and confirmation once they have been detected in a sample. Many of the more recent developments in pathogen testing have focused on the confirmation step, and in particular on reducing the time required to achieve a result.

Possibly the most important outcome of the performance of conventional testing for pathogens is when an isolate of the pathogen is obtained. This can be very important for public health and epidemiological purposes. Isolates can be characterised and typed using a variety of schemes (serotyping, PFGE, MLST) and can be used to develop a profile of strain types associated with food products and food industries. Strains isolated from food products may also be compared with outbreak strains to identify potential sources of foodborne outbreaks.

Microbial typing

We have discussed molecular typing for each pathogen in the preceeding sections. Molecular typing using methods such as PFGE provides a rapid and accurate method of detecting pathogenic Enterobacteriaceae that persist at different points in the dairy manufacturing environment and in dairy products (Popp *et al.*, 2010). PFGE typing has been used to trace the route of contamination by typing the clonal populations as they are transferred between different areas of a factory, such as tanker bays, evaporator rooms, an employee's shoes or external roofs (Craven *et al.*, 2010).

New methods

The use of novel, rapid detection methods and refinement of existing routine pathogen testing programmes may reduce the laborious process involved in conventional methods of detection. Some of the common molecular techniques include PCR and microarray. Several ISO standards have been established to provide guidelines for the detection of foodborne pathogens by PCR: ISO 22174:2004, ISO/Ts 20836:2005, ISO 20837:2006 and ISO 20838:2006 (Postollec *et al.*, 2011).

PCR has been used extensively in the research community as a diagnostic tool for food microbiology and other fields. It has also been used to detect and confirm the identities of pathogens (Daum *et al.*, 2002). PCR tests tend to be rapid, sensitive and capable of detecting subdominant foodborne pathogens from a variety of origins without the use of enrichment media (Malorny *et al.*, 2003; Postollec *et al.*, 2011). PCR can specifically identify dairy pathogens, including *L. monocytogenes*, *S. enterica*, *S. aureus* and *E. coli* O157:H7, and is more sensitive and rapid than culture-based methods (McLean *et al.*, 2010).

A combination of culture-based enrichment of enteric pathogens and PCR-based quantification of target bacteria has been explored for enteric pathogens. A simultaneous enrichment broth was used for *Salmonella* species, *E. coli* O157:H7, *Vibrio parahaemolyticus*, *S. aureus*, *B. cereus* and *L. monocytogenes* (Kobayashi *et al.*, 2009); this shortened the diagnostic protocol to 6–24 hours, as compared to 4–6 days for culture-only methods.

PCR has been shown to detect 1 CFU/ml in food products, and it can obtain results within 12 hours (Ellingson *et al.*, 2004). PCR using specific primers for the *mapA* gene of *C. jejuni* was found to be more sensitive in the detection of *C. jejuni* in spiked faecal and food samples (96% each) than were culture-based methods, which confirmed the pathogen in 87 and 80% of samples, respectively (Harkanwaldeep *et al.*, 2011).

L. monocytogenes and *Salmonella* species have been distinguished based on the distinct melt peaks of amplified products (Singh *et al.*, 2012). The detection sensitivity of the assay in reconstituted non-fat dried milk (NFDM, 11%) spiked with the target pathogens at different levels was 3 $\log_{10}$ CFU/ml of each pathogen. However, this was improved to 1 $\log_{10}$ CFU/ml by including a preenrichment step of 6 hours. On application of the assay to 60 market samples, one sample each of raw milk and ice cream was detected positive for *L. monocytogenes* and *Salmonella* species. The assay was quite specific, as no crossreactivity with non-target cultures could be observed. It was found to be useful in monitoring dairy products for the presence of *L. monocytogenes* and *Salmonella* species to ensure their microbiological quality and safety (Singh *et al.*, 2012).

RT-PCR has the potential to be used as a quantification tool in the enumeration of foodborne pathogens (Malorny *et al.*, 2003). The quantification of foodborne pathogens is estimated through the numbers of gene copies being detected (Postollec *et al.*, 2011), and RT-PCR has become the method of choice for the quantification of genes (Nolan *et al.*, 2006).

There are some limitations on the use of PCR. Dairy products have a complex matrix that can contain natural PCR inhibitors such as fats, proteinases and high concentrations of calcium ions, which may interfere with the PCR assay (Wilson, 1997). These inhibitors can be removed during sampling and preparation of the dairy samples prior to PCR (Cremonesi *et al.*, 2006). Furthermore, PCR will detect nucleic acids from both dead and live cells. This can be overcome by the use of ethidium monazide, a cross-linking agent which permeates the membranes of dead cells and forms irreversibly cross-linked DNA that can't be amplified in the subsequent PCR assay (Nogva *et al.*, 2003; Soejima *et al.*, 2008).

Microarray technology enables the detection and characterisation of multiple pathogens and genes in a single-array assay (Rasooly & Herold, 2008). Microarray technology offers similar advantages to PCR over conventional methods of foodborne pathogen detection, where it is emerging as a cost-effective, broad-spectrum platform for the detection of pathogens in food products. The microarray technology utilises biosensor chips incorporating single-stranded oligonucleotide probes that canbe hybridised with target DNA isolated from food sample (Uttamchandani *et al.*, 2009; Bai *et al.*, 2010). The sensitivity of microbial diagnostic microarrays can be categorised as either absolute or relative (Kostic *et al.*, 2010). Absolute sensitivity is defined as the least amount of nucleic acid needed for successful detection, while relative sensitivity is defined as the least detectable abundance of targeted microorganisms in a non-targeted background (Kostic *et al.*, 2010).

The use of molecular techniques in routine pathogen testing programmes may be rapid and accurate, but other factors need to be considered too, such as the operational cost and the technical skills required for their conduct. Furthermore, a standardised benchmark method is required to ensure that molecular techniques are used with confidence and to provide quality assurance to dairy products. A final limitation of these new techniques is that an isolate is not obtained from the sample. However, it is possible to employ a rapid technique to obtain a result quickly and then use this as an indication of whether or not to proceed with conventional testing in order to obtain an isolate of the pathogen.

References

Abdallah, M., Chataigne, G., Ferreira-Theret, P., Benoliel, C., Drider, D., Dhulster, P. & Chihib, N. E. 2014. Effect of growth temperature, surface type and incubation time on the resistance of *Staphylococcus aureus* biofilms to disinfectants. *Applied Microbiology and Biotechnology*, **98**, 2597–607.

Abdul-Ratha, H. A. & Yarmorad, M. A. 2013. Detection of biofilm producer *Staphylococcus aureus* and its susceptibility against antibiotics. *International Journal of Current Microbiology and Applied Sciences*, **2**, 493–500.

Adekunte, A., Valdramidis, V. P., Tiwari, B. K., Slone, N., Cullen, P. J., Donnell, C. P. & Scannell, A. 2010. Resistance of *Cronobacter sakazakii* in reconstituted powdered infant formula during ultrasound at controlled temperatures: a quantitative approach on microbial responses. *International Journal of Food Microbiology*, **142**, 53–9.

Almeida, G., Magalhaes, R., Carneiro, L., Santos, I., Silva, J., Ferreira, V., Hogg, T. & Teixeira, P. 2013. Foci of contamination of *Listeria monocytogenes* in different cheese processing plants. *International Journal of Food Microbiology*, **167**, 303–9.

Alwathnani, H. A. & Hessain, A. M. 2013. Molecular detection and characterization of *Escherichia coli* O157:H7 and O111 associated with milk samples by multiplex PCR. *Journal of Pure and Applied Microbiology*, **7**, 447–51.

Amer, A. A., Aiad, A. S. & Abd-Allah, M. A. 2010. Effect of yoghurt processing and ice cream manufacture on viability of some foodborne bacteria. *Assiut Veterinary Medical Journal*, **56**, 108–19.

Amin, W. F. & Shaker, E. M. 2011. Incidence and antibiotic resistance of psychrotolerant *Bacillus cereus* group in ice cream samples, with special reference to *Bacillus thuringiensis*. *Assiut Veterinary Medical Journal*, **57**, 157–68.

Anon. 2001. Microbiology of food and animal feeding stuffs. Horizontal method for the detection of *Escherichia coli* O157. European Standard EN ISO 16654:2001.

Anon. 2009. *Campylobacter jejuni* infection associated with unpasteurised milk and cheese – Kansas, 2007. Available from: http://www.cdc.gov/mmwr/preview/mmwrhtml/mm5751a2.htm (last accessed 12 March 2015).

Anon. 2013. Vital signs: listeria illnesses, deaths, and outbreaks – United States, 2009–2011. *Morbidity and Mortality Weekly Report*, **62**, 448–52.

Anon. 2014a. Commission Regulation (EC) No 2073/2005 of 15 November 2005 on microbiological criteria for foodstuff. Available from: http://eur-lex.europa.eu/legal-content/EN/TXT/PDF/?uri=CELEX:32005R2073&from=EN 2/11/2014 (last accessed 12 March 2015).

Anon. 2014b. Multistate outbreak of listeriosis linked to roos foods dairy products. Available from: http://www.cdc.gov/listeria/outbreaks/cheese-02-14/index.html (last accessed 12 March 2015).

Anon. 2014c. Proposed options for the sale of raw milk to consumers. Available from: http://www.foodsafety.govt.nz/elibrary/industry/raw-milk-sales-2014/ (last accessed 12 March 2015).

Atulya, M., Mathew, A. J., Rao, J. V. & Rao, C. M. 2014. Influence of milk components in establishing biofilm mediated bacterial mastitis infections in cattle: a fractional factorial approach. *Research in Veterinary Science*, **96**, 25–7.

Bae, Y.-M., Baek, S.-Y. & Lee, S.-Y. 2012. Resistance of pathogenic bacteria on the surface of stainless steel depending on attachment form and efficacy of chemical sanitizers. *International Journal of Food Microbiology*, **153**, 465–73.

Bai, S., Zhao, J., Zhang, Y., Huang, W., Xu, S., Chen, H., Fan, L.-M., Chen, Y. & Deng, X. W. 2010. Rapid and reliable detection of 11 food-borne pathogens using thin-film biosensor chips. *Applied Microbiology and Biotechnology*, **86**, 983–90.

Barbosa, J., Magalhaes, R., Santos, I., Ferreira, V., Brandao, T. R. S., Silva, J., Almeida, G. & Teixeira, P. 2013. Evaluation of antibiotic resistance patterns of food and clinical *Listeria monocytogenes* isolates in Portugal. *Foodborne Pathogens and Disease*, **10**, 861–6.

Bardiau, M., Yamazaki, K., Duprez, J. N., Taminiau, B., Mainil, J. G. & Ote, I. 2013. Genotypic and phenotypic characterization of methicillin-resistant *Staphylococcus aureus* (MRSA) isolated from milk of bovine mastitis. *Letters in Applied Microbiology*, **57**, 181–6.

Baumgartner, A. & Niederhauser, I. 2010. Occurrence of *Cronobacter* spp. in raw milk. *Journal of Consumer Protection and Food Safety*, **5**, 253.

Beale, D. J., Morrison, P. D. & Palombo, E. A. 2014. Detection of *Listeria* in milk using non-targeted metabolic profiling of *Listeria monocytogenes*: a proof-of-concept application. *Food Control*, **42**, 343–6.

Bennett, R. W., Hait, J. M. & Tallent, S. M. 2013. *Staphylococcus aureus*. In: Labbé, R. G. & García, S. (eds) Guide to Foodborne Pathogens. John Wiley & Sons, Chichester.

Bhatt, V. D., Ahir, V. B., Koringa, P. G., Jakhesara, S. J., Rank, D. N., Nauriyal, D. S., Kunjadia, A. P. & Joshi, C. G. 2012. Milk microbiome signatures of subclinical mastitis-affected cattle analysed by shotgun sequencing. *Journal of Applied Microbiology*, **112**, 639–50.

Bianchi, D. M., Gallina, S., Bellio, A., Chiesa, F., Civera, T. & Decastelli, L. 2014. Enterotoxin gene profiles of *Staphylococcus aureus* isolated from milk and dairy products in Italy. *Letters in Applied Microbiology*, **58**, 190–6.

Blackman, I. C., Park, Y. W. & Harrison, M. A. 2005. Effects of oxidative compounds on thermotolerance in *Escherichia coli* O157:H7 strains EO139 and 380-94. *Journal of Food Protection*, **68**, 2443–6.

Blaser, M. J. & Newman, L. S. 1982. A review of human salmonellosis: I. Infective dose. *Reviews of Infectious Diseases*, **4**, 1096–106.

Bonsaglia, E. C. R., Silva, N. C. C., Fernades Junior, A., Araujo Junior, J. P., Tsunemi, M. H. & Rall, V. L. M. 2014. Production of biofilm by *Listeria monocytogenes* in different materials and temperatures. *Food Control*, **35**, 386–91.

Bowen, A. B. & Braden, C. R. 2006. Invasive *Enterobacter sakazakii* disease in infants. *Emerging Infectious Diseases*, **12**, 1185–9.

Braoudaki, M. & Hilton, A. 2004. Adaptive resistance to biocides in *Salmonella enterica* and *Escherichia coli* O157 and cross-resistance to antimicrobial agents. *Journal of Clinical Microbiology*, **42**, 73–8.

Bremer, P. J., Monk, I. & Osborne, C. M. 2001. Survival of *Listeria monocytogenes* attached to stainless steel surfaces in the presence or absence of *Flavobacterium* spp. *Journal of Food Protection*, **64**, 1369–76.

Bremer, P. J., Monk, I. & Butler, R. 2002. Inactivation of *Listeria monocytogenes/Flavobacterium* spp. biofilms using chlorine: impact of substrate, pH, time and concentration. *Letters in Applied Microbiology*, **35**, 321–5.

Brichta-Harhay, D. M., Arthur, T. M., Bosilevac, J. M., Kalchayanand, N., Shackelford, S. D., Wheeler, T. L. & Koohmaraie, M. 2011. Diversity of multidrug-resistant *Salmonella enterica* strains associated with cattle at harvest in the United States. *Applied and Environmental Microbiology*, **77**, 1783–96.

Brooks, J. D. & Flint, S. H. 2008. Biofilms in the food industry: problems and potential solutions. *International Journal of Food Science and Technology*, **43**, 2163–76.

Brooks, J. C., Martinez, B., Stratton, J., Bianchini, A., Krokstrom, R. & Hutkins, R. 2012. Survey of raw milk cheeses for microbiological quality and prevalence of foodborne pathogens. *Food Microbiology*, **31**, 154–8.

Brouard, C., Espié, E., Weill, F-X., Kérouanton, A., Brisabois, A., Forgue, A-M., Vaillant, V. & De Valk, H. 2007. Two consecutive large outbreaks of *Salmonella enterica* serotype Agona infections in infants linked to the consumption of powdered infant formula. *The Pediatric Infectious Disease Journal*, **26**, 148–52.

Buchanan, R. L. & Edelson, S. G. 1999. Effect of pH-dependent, stationary phase acid resistance on the thermal tolerance of *Escherichia coli* O157:H7. *Food Microbiology*, **16**, 447–58.

Carlin, F., Brillard, J., Broussolle, V., Clavel, T., Duport, C., Jobin, M., Guinebretiere, M. H., Auger, S., Sorokine, A. & Nguyen-The, C. 2010. Adaptation of *Bacillus cereus*, an ubiquitous worldwide-distributed foodborne pathogen, to a changing environment. *Food Research International*, **43**, 1885–94.

Carpentier, B. & Cerf, O. 2011. Review – persistence of *Listeria monocytogenes* in food industry equipment and premises. *International Journal of Food Microbiology*, **145**, 1–8.

Carter, P. E., Mctavish, S. M., Brooks, H. J. L., Campbell, D., Collins-Emerson, J. M., Midwinter, A. C, & French, N. P. 2009. Novel clonal complexes with an unknown animal reservoir dominate *Campylobacter jejuni* isolates from river water in New Zealand. *Applied and Environmental Microbiology*, **75**, 6038–46.

Castrodale, L. J., Gerlach, R. F., Xavier, C. M., Smith, B. J., Cooper, M. P. & Mclaughlin, J. B. 2013. Sharing milk but not messages: campylobacteriosis associated with consumption of raw milk from a cow-share program in Alaska, 2011. *Journal of Food Protection*, **76**, 744–7.

Caubilla-Barron, J., Hurrell, E., Townsend, S., Cheetham, P., Loc-Carrillo, C., Fayet, O., Prere, M. F. & Forsythe, S. J. 2007. Genotypic and phenotypic analysis of *Enterobacter sakazakii* strains from an outbreak resulting in fatalities in a neonatal intensive care unit in France. *Journal of Clinical Microbiology*, **45**, 3979–85.

Cavender, G. A. & Kerr, W. L. 2011. Inactivation of vegetative cells by continuous high-pressure processing: new insights on the contribution of thermal effects and release device. *Journal of Food Science*, **76**, E525–9.

Ceuppens, S., Boon, N. & Uyttendaele, M. 2013. Diversity of *Bacillus cereus* group strains is reflected in their broad range of pathogenicity and diverse ecological lifestyles. *FEMS Microbiology Ecology*, **84**, 433–50.

Chandra, S., Motina, E. & Sunil, K. 2010. Microbiological quality of raw milk and its public health significance. *Journal of Dairying, Foods and Home Sciences*, **29**, 15–18.

Chavant, P., Martinie B, Meylheuc T, Bellon-Fontaine M N, Hebraud M. 2002. *Listeria monocytogenes* LO28: surface physicochemical properties and ability to form biofilms at different temperatures and growth phases. *Applied and Environmental Microbiology*, **68**, 728–37.

Chen, Z., Diao, J., Dharmasena, M., Ionita, C., Jiang, X. & Rieck, J. 2013. Thermal inactivation of desiccation-adapted *Salmonella* spp. in aged chicken litter. *Applied and Environmental Microbiology*, **79**, 7013–20.

Chia, T., Goulter, R., Mcmeekin, T., Dykes, G., Fegan, N. 2009. Attachment of different *Salmonella* serovars to materials commonly used in a poultry processing plant. *Food Microbiology*, **26**, 853–9.

Chiara, M., D'Erchia, A. M., Manzari, C., Minotto, A., Montagna, C., Addante, N., Santagada, G., Latorre, L., Pesole, G., Horner, D. S. & Parisi, A. 2014. Draft genome sequences of six *Listeria monocytogenes* strains isolated from dairy products from a processing plant in Southern Italy. *Genome Announcements*, **2**, e00282-14.

Ciotau, C. & Sindilar, E. V. 2010. Research on the frequency of *Bacillus cereus* in milk as raw material. *Lucrari Stiintifice – Medicina Veterinara, Universitatea de Stiinte Agricole si Medicina Veterinara 'Ion Ionescu de la Brad' Iasi*, **53**, 959–64.

Cleto, S., Matos, S., Kluskens, L. & Vieira, M. J. 2012. Characterization of contaminants from a sanitized milk processing plant. *PLoS One*, **7**, e40189.

Craven, H. M., Mcauley, C. M., Duffy, L. L. & Fegan, N. 2010. Distribution, prevalence and persistence of *Cronobacter* (*Enterobacter sakazakii*) in the nonprocessing and processing environments of five milk powder factories. *Journal of Applied Microbiology*, **109**, 1044–52.

Cremonesi, P., Castiglioni, B., Malferrari, G., Biunno, I., Vimercati, C., Moroni, P., Morandi, S. & Luzzana, M. 2006. Improved method for rapid DNA extraction of mastitis pathogens directly from milk. *Journal of Dairy Science*, **89**, 163–9.

Cucarella, C., Tormo, M. A., Ubeda, C., Trotonda, M. P., Monzon, M., Peris, C., Amorena, B., Lasa, I. & Penades, J. R. 2004. Role of biofilm-associated protein bap in the pathogenesis of bovine *Staphylococcus aureus*. *Infection and Immunity*, **72**, 2177–85.

Dancer, G., Mah, J. H. & Kang, D. H. 2009. Influences of milk components on biofilm formation of *Cronobacter* spp.(*Enterobacter sakazakii*). *Letters in Applied Microbiology*, **48**, 718–25.

Daum, L. T., Barnes, W. J., Mcavin, J. C., Neidert, M. S., Cooper, L. A., Huff, W. B., Gaul, L., Riggins, W. S., Morris, S., Salmen, A. & Lohman, K. L. 2002. Real-time PCR detection of *Salmonella* in suspect foods from a gastroenteritis outbreak in Kerr County, Texas. *Journal of Clinical Microbiology*, **40**, 3050–2.

De Jonghe, V., Coorevits, A., De Block, J., Van Coillie, E., Grijspeerdt, K., Herman, L., De Vos, P. & Heyndrickx, M. 2010. Toxinogenic and spoilage potential of aerobic spore-formers isolated from raw milk. *International Journal of Food Microbiology*, **136**, 318–25.

Desai, S. V. & Varadaraj, M. C. 2013. Behavior of vegetative cells and spores of *Bacillus cereus* in two food systems. *Journal of Food Safety*, **33**, 291–301.

Donnelly, C. W. & Diez-Gonzalez, F. 2013. *Listeria monocytogenes*. In: Labbé, R. G. & García, S. (eds) Guide to Foodborne Pathogens. John Wiley & Sons, Chichester.

Doyle, M. E., Hartmann, F. A. & Lee Wong, A. C. 2012. Methicillin-resistant staphylococci: implications for our food supply? *Animal Health Research Reviews/Conference of Research Workers in Animal Diseases*, **13**, 157–80.

Duynhoven, Y. T. H. P. V., Isken, L. D., Borgen, K., Besselse, M., Soethoudt, K., Haitsma, O., Mulder, B., Notermans, D. W., Jonge, R. D., Kock, P., Pelt, W. V., Stenvers, O. & Steenbergen, J. V. 2009. A prolonged outbreak of *Salmonella* Typhimurium infection related to an uncommon vehicle: hard cheese made from raw milk. *Epidemiology and Infection*, **137**, 1548–57.

Dzieciol, M., Fricker, M., Wagner, M., Hein, I. & Ehling-Schulz, M. 2013. A novel diagnostic real-time PCR assay for quantification and differentiation of emetic and non-emetic *Bacillus cereus*. *Food Control*, **32**, 176–85.

Eddy, D. W., Haynes, I. N., Veitch, M. G. K., Yi, L. H. & Hogg, G. G. 2010. Historical analysis of *Salmonella* in powdered dairy products in Victoria. *Australian Journal of Dairy Technology*, **65**, 58–62.

El-Gazzar, F. E. & Marth, E. H. 1992. Salmonellae, salmonellosis, and dairy foods: a review. *Journal of Dairy Science*, **75**, 2327–43.

Elango, A., Dhanalakshmi, B., Pugazhenthi, T. R., Jayalalitha, V., Kumar, C. N. & Doraisamy, K. A. 2010. Seasonal prevalence of *Campylobacter jejuni* isolated from raw milk samples in Chennai. *Egyptian Journal of Dairy Science*, **38**, 19–24.

Elhadidy, M. & Mohammed, M. A. 2013. Shiga toxin-producing *Escherichia coli* from raw milk cheese in Egypt: prevalence, molecular characterization and survival to stress conditions. *Letters in Applied Microbiology*, **56**, 120–7.

Ellingson, J. L. E., Anderson, J. L., Carlson, S. A. & Sharma, V. K. 2004. Twelve hour real-time PCR technique for the sensitive and specific detection of *Salmonella* in raw and ready-to-eat meat products. *Molecular and Cellular Probes*, **18**, 51–7.

Faille, C., Benezech, T., Blel, W., Ronse, A., Ronse, G., Clarisse, M. & Slomianny, C. 2013. Role of mechanical vs. chemical action in the removal of adherent *Bacillus* spores during CIP procedures. *Food Microbiology*, **33**, 149–57.

Faille, C., Benezech, T., Midelet-Bourdin, G., Lequette, Y., Clarisse, M., Ronse, G., Ronse, A. & Slomianny, C. 2014. Sporulation of *Bacillus* spp. within biofilms: a potential source of contamination in food processing environments. *Food Microbiology*, **40**, 64–74.

Fakruddin, M., Rahaman, M. M., Ahmed, M. M. & Hoque, M. M. 2013. *Cronobacter sakazakii* (*Enterobacter sakazakii*): an emerging food borne pathogen. *International Journal of Biomedical and Advance Research*, **4**, 349–59.

Farakos, S. M. S., Frank, J. F. & Schaffner, D. W. 2013. Modeling the influence of temperature, water activity and water mobility on the persistence of *Salmonella* in low-moisture foods. *International Journal of Food Microbiology*, **166**, 280–93.

Farrokh, C., Jordan, K., Auvray, F., Glass, K., Oppegaard, H., Raynaud, S., Thevenot, D., Condron, R., De Reu, K., Govaris, A., Heggum, K., Heyndrickx, M., Hummerjohann, J., Lindsay, D., Miszczycha, S., Moussiegt, S., Verstraete, K. & Cerf, O. 2013. Review of shiga-toxin-producing *Escherichia coli* (STEC) and their significance in dairy production. *International Journal of Food Microbiology*, **162**, 190–212.

Fegan, N. & Desmarchelier, P. 2010. Pathogenic *E. coli* in the dairy industry: implications for Australia. *Australian Journal of Dairy Technology*, **65**, 68–73.

Feng, P. 2007. Rapid methods for the detection of foodborne pathogens: current and next-generation technologies. In: Doyle, M. P. & Beuchat, L. R. (eds) Food Microbiology, Fundamentals and Frontiers, 3rd edn. ASM Press, Washington, DC, pp. 911–34.

Flint, S. H., Bremer, P. J. & Brooks, J. D. 1997. Biofilms in dairy manufacturing plant – description, current concerns and methods of control. *Biofouling*, **11**, 81–97.

Gadotti, C., Nelson, L. & Diez-Gonzalez, F. 2014. Inhibitory effect of combinations of caprylic acid and nisin on *Listeria monocytogenes* in queso fresco. *Food Microbiology*, **39**, 1–6.

Garcia, S. & Heredia, N. L. 2013. *Campylobacter*. In: Labbé, R. G. & García, S. (eds) Guide to Foodborne Pathogens. John Wiley & Sons, Chichester.

Gharst, G., Bark, D. H., Newkirk, R., Guillen, L., Wang, Q. & Abeyta, C., Jr. 2013. Evaluation and single-laboratory verification of a proposed modification to the us food and drug administration method for detection and identification of *Campylobacter jejuni* or *Campylobacter coli* from raw silo milk. *Journal of AOAC International*, **96**, 1336–42.

Greig, J. D. & Ravel, A. 2009. Analysis of foodborne outbreak data reported internationally for source attribution. *International Journal of Food Microbiology*, **130**, 77–87.

Guevremont, E., Lamoureux, L., Loubier, C. B., Villeneuve, S. & Dubuc, J. 2014. Detection and characterization of *Campylobacter* spp. from 40 dairy cattle herds in Quebec, Canada. *Foodborne Pathogens and Disease*, **11**, 388–94.

Hai, D. N. N. & Yuk, H. G. 2013. Changes in resistance of *Salmonella* Typhimurium biofilms formed under various conditions to industrial sanitizers. *Food Control*, **29**, 236–40.

Haldar, L. B. & Kuila, R. K. 2011. Psychrotrophic *Bacillus* in pasteurized market milk: a threat. *Environment and Ecology*, **29**, 202–6.

Haley, B. J., Luo, Y., Wang, C., Pettengill, J., Allard, M., Brown, E., Karns, J. S. & Van Kessel, J. A. 2014. Genome sequences of eight *Salmonella enterica* subsp. *enterica* serovars isolated from a single dairy farm. *Genome Announcements*, **2**, pii: e00082-14.

Hall, J. M., Herlihy, R. K., Dimond, M. P. S., Holbrook, J., Smith, L. H., Wagner, J. M., Clark, R. W. & Lanier, W. A. 2010. *Salmonella* Newport infections associated with consumption of unpasteurized milk – Utah, April–June 2010. *Morbidity and Mortality Weekly Report*, **59**, 817–18.

Hamadi, F., Asserne, F., Elabed, S., Bensouda, S., Mabrouk, M. & Latrache, H. 2014. Adhesion of *Staphylococcus aureus* on stainless steel treated with three types of milk. *Food Control*, **38**, 104–8.

Hanamant, P. S. & Bansilal, G. M. 2012. Proteolytic psychrotrophic *Bacillus cereus* from milk and fermented milk products. *Journal of Environmental Research and Development*, **6**, 660–6.

Harkanwaldeep, S., Rathore, R. S., Satparkash, S. & Singh Cheema, P. 2011. Comparative analysis of cultural isolation and PCR based assay for detection of *Campylobacter jejuni* in food and faecal samples. *Brazilian Journal of Microbiology*, **42**, 181–6.

Hartmann, I., Carranza, P., Lehner, A., Stephan, R., Eberl, L. & Riedel, K. 2010. Genes involved in *Cronobacter sakazakii* biofilm formation. *Applied and Environmental Microbiology*, **76**, 2251–61.

Haughton, P., Garvey, M. & Rowan, N. J. 2010. Emergence of *Bacillus cereus* as a dominant organism in Irish retailed powdered infant formulae (PIF) when reconstituted and stored under abuse conditions. *Journal of Food Safety*, **30**, 814–31.

Hauri, A., Just, M., Mcfarland, S., Schweigmann, A., Schlez, K. & Krahn, J. 2013. [Campylobacteriosis outbreaks in the state of Hesse, Germany, 2005–2011: raw milk yet again]. *Deutsche medizinische Wochenschrift (1946)*, **138**, 357–61.

Heyndrickx, M. 2011. The importance of endospore-forming bacteria originating from soil for contamination of industrial food processing. *Applied and Environmental Soil Science*, Article ID: 561975.

Hill, B., Smythe, B., Lindsay, D. & Shepherd, J. 2012. Microbiology of raw milk in New Zealand. *International Journal of Food Microbiology*, **157**, 305–8.

Honjoh, K., Hashimoto, Y., Shimotsu, S., Wen, H. M., Kiriki, M., Naito, K., Tokugawa, M., Satake, E., Kobayashi, H. & Miyamoto, T. 2009. Construction of several deletion mutants for genes involved in biofilm formation and recovery of heat-injured *Salmonella*: delta agfa and delta bcsa mutants of *Salmonella Enteritidis*; delta ahpc, delta ahpf, and delta katg mutants of *S.* Typhimurium; and delta rpoe, delta rpoh, and delta rpos mutants of *S.* Enteritidis and *S.* Typhimurium. *Journal of the Faculty of Agriculture Kyushu University*, **54**, 421–31.

Hornstra, L. M., De Leeuw, P. L. A., Moezelaar, R., Wolbert, E. J., De Vries, Y. P., De Vos, W. M. & Abee, T. 2007. Germination of *Bacillus cereus* spores adhered to stainless steel. *International Journal of Food Microbiology*, **116**, 367–71.

Hu, D. L. & Nakane, A. 2014. Mechanisms of staphylococcal enterotoxin-induced emesis. *European Journal of Pharmacology*, **722**, 95–107.

Hunter, C. J. & Bean, J. F. 2013. *Cronobacter*: an emerging opportunistic pathogen associated with neonatal meningitis, sepsis and necrotizing enterocolitis. *Journal of Perinatology*, **33**, 581–5.

International Commission on Microbiological Specification for Foods. 2002. Microbiological Testing in Food Safety Management, Vol. **7**. Academic/Plenum Publisher, New York, NY.

Jackson, E. E., Erten, E. S., Maddi, N., Graham, T. E., Larkin, J. W., Blodgett, R. J., Schlesser, J. E. & Reddy, R. M. 2012. Detection and enumeration of four foodborne pathogens in raw commingled silo milk in the United States. *Journal of Food Protection*, **75**, 1382–93.

Jadhav, S., Sevior, D., Bhave, M. & Palombo, E. A. 2014. Detection of *Listeria monocytogenes* from selective enrichment broth using MALDI-TOF mass spectrometry. *Journal of Proteomics*, **97**, 100–6.

Jantzen, M. M., Navas, J., De Paz, M., Rodriguez, B., Da Silva, W. P., Nunez, M. & Martinez-Suarez, J. V. 2006. Evaluation of ALOA plating medium for its suitability to recover high pressure-injured *Listeria monocytogenes* from ground chicken meat. *Letters in Applied Microbiology*, **43**, 313–17.

Jay-Russell, M. T., Mandrell, R. E., Yuan, J., Bates, A., Manalac, R., Mohle-Boetani, J., Kimura, A., Lidgard, J. & Miller, W. G. 2013. Using major outer membrane protein typing as an epidemiological tool to investigate outbreaks caused by milk-borne *Campylobacter jejuni* isolates in California. *Journal of Clinical Microbiology*, **51**, 195–201.

Jayarao, B. M., Donaldson, S. C., Straley, B. A., Sawant, A. A., Hegde, N. V. & Brown, J. 2006. A survey of foodborne pathogens in bulk tank milk and raw milk consumption among farm families in Pennsylvania. *Journal of Dairy Science*, **89**, 2451–8.

Jesudhasan, P. R., Cepeda, M. L., Widmer, K., Dowd, S. E., Soni, K. A., Hume, M. E., Zhu, J. & Pillai, S. D. 2010. Transcriptome analysis of genes controlled by luxS/autoinducer-2 in *Salmonella enterica* Serovar Typhimurium. *Foodborne Pathogens and Disease*, **7**, 399–410.

Jo, S.-H., Baek, S.-B., Ha, J.-H. & Ha, S.-D. 2010. Maturation and survival of *Cronobacter* biofilms on silicone, polycarbonate, and stainless steel after uv light and ethanol immersion treatments. *Journal of Food Protection*, **73**, 952–6.

Kadam, S. R., Den Besten, H. M. W., Van Der Veen, S., Zwietering, M. H., Moezelaar, R. & Abee, T. 2013. Diversity assessment of *Listeria monocytogenes* biofilm formation: impact of growth condition, serotype and strain origin. *International Journal of Food Microbiology*, **165**, 259–64.

Kandhai, M. C., Reij, M. W., Gorris, L. G. M., Guillaume-Gentil, O. & Van Schothorst, M. 2004. Occurrence of *Enterobacter sakazakii* in food production environments and households. *Lancet*, **363**, 39–40.

Kim, J.-M., Forghani, F., Kim, J.-B., Park, Y.-B., Park, M.-S., Wang, J., Park, J. H. & Oh, D.-H. 2012. Improved multiplex PCR assay for simultaneous detection of *Bacillus cereus* emetic and enterotoxic strains. *Food Science and Biotechnology*, **21**, 1439–44.

Klein, D., Alispahic, M., Sofka, D., Iwersen, M., Drillich, M. & Hilbert, F. 2013. Prevalence and risk factors for shedding of thermophilic *Campylobacter* in calves with and without diarrhea in Austrian dairy herds. *Journal of Dairy Science*, **96**, 1203–10.

Knochel, S. 2010. *Listeria monocytogenes* – not only bad but also persistent. *Australian Journal of Dairy Technology*, **65**, 50–7.

Knowles, J., Roller, S., Murray, D. & Naidu, A. 2005. Antimicrobial action of carvacrol at different stages of dual-species biofilm development by *Staphylococcus aureus* and *Salmonella enterica* serovar Typhimurium. *Applied and Environmental Microbiology*, **71**, 797–803.

Kobayashi, H., Kubota, J., Fujihara, K., Honjoh, K., Ito, M., Fujiki, N., Nakabe, M., Oda, S., Satoyama, T., Takasu, K., Nakanishi, H. & Miyamoto, T. 2009. Simultaneous enrichment of *Salmonella* spp, *Escherichia coli* O157:H7, *Vibrio parahaemolyticus*, *Staphylococcus aureus*, *Bacillus cereus*, and *Listeria monocytogenes* by single broth and screening of the pathogens by multiplex real-time PCR. *Food Science and Technology Research*, **15**, 427–38.

Koch, J., Schrauder, A., Alpers, K., Werber, D., Frank, C., Prager, R., Rabsch, W., Broll, S., Feil, F., Roggentin, P., Bockemuhl, J., Tschape, H., Ammon, A. & Stark, K. 2005. *Salmonella* Agona outbreak from contaminated aniseed, Germany. *Emerging Infectious Diseases*, **11**, 1124–7.

Korber, D., Choi, A., Wolfaardt, G., Ingham, S., & Caldwell, D 1997. Substratum topography influences susceptibility of *Salmonella enteritidis* biofilms to trisodium phosphate. *Applied and Environmental Microbiology*, **63**, 3352–8.

Kostic, T., Stessl, B., Wagner, M., Sessitsch, A. & Bodrossy, L. 2010. Microbial diagnostic microarray for food- and water-borne pathogens. *Microbial Biotechnology*, **3**, 444–54.

Kot, B., Binek, T., Piechota, M., Wolska, K. M., Zdunek, E. & Platkowska, K. 2013. Virulence factors and ability of staphylococci from bovine milk and the cowshed environment to biofilm formation. *Polish Journal of Veterinary Sciences*, **16**, 639–45.

Kucerova, E., Joseph, S. & Forsythe, S. 2011. The *Cronobacter* genus: ubiquity and diversity. *Quality Assurance and Safety of Crops & Foods*, **3**, 104–22.

Kumar, C. & Anand, S. 1998. Significance of microbial biofilms in food industry: a review. *International Journal of Food Microbiology*, **42**, 9–27.

Kumari, S. & Sarkar, P. K. 2014. *In vitro* model study for biofilm formation by *Bacillus cereus* in dairy chilling tanks and optimization of clean-in-place (CIP) regimes using response surface methodology. *Food Control*, **36**, 153–8.

Kuo, L.-S., Wang, B.-J., He, Y.-S. & Weng, Y.-M. 2013. The effects of ultraviolet light irradiation and drying treatments on the survival of *Cronobacter* spp. (*Enterobacter sakazakii)* on the surfaces of stainless steel, teflon and glass. *Food Control*, **30**, 106–10.

Larsen, M. H., Dalmasso, M., Ingmer, H., Langsrud, S., Malakauskas, M., Mader, A., Møretrø, T., Smole Možina, S., Rychli, K. & Wagner, M. 2014. Persistence of foodborne pathogens and their control in primary and secondary food production chains. *Food Control*, **44**, 92–109.

Lee, D. J. 2014. *Salmonella*. Available from: http://www.foodsafety.govt.nz/industry/general/foodborne-illness/salmonella/ (last accessed 12 March 2015).

Lee, K. A., Moon, S. H., Kim, K. T., Nah, S. Y. & Paik, H. D. 2011. Antimicrobial effect of kaempferol on psychrotrophic *Bacillus cereus* strains outbreakable in dairy products. *Korean Journal for Food Science of Animal Resources*, **31**, 311–15.

Lee, K.-I., Watanabe, M., Sugita-Konishi, Y., Hara-Kudo, Y. & Kumagai, S. 2012. *Penicillium camemberti* and *Penicillium roqueforti* enhance the growth and survival of Shiga toxin-producing *Escherichia coli* O157 under mild acidic conditions. *Journal of Food Science*, **77**, M102–7.

Lehner, A. 2010. *Cronobacter* (former *Enterobacter sakazakii*) spp. – its effect in dairy product safety, quality and trade. *Australian Journal of Dairy Technology*, **65**, 63–7.

Lehner, A. & Stephan, R. 2004. Microbiological, epidemiological, and food safety aspects of *Enterobacter sakazakii*. *Journal of Food Protection*, **67**, 2850–7.

Lehner, A., Fricker-Feer, C., Gschwend, K. & Stephan, R. 2010. [Identification of Enterobacteriaceae and *Cronobacter* spp. in raw milk, milk concentrate and milk powder: prevalence and genotyping]. *Archiv für Lebensmittelhygiene*, **5**, 22–6.

Liu, Q., Lu, X., Swanson, B. G., Rasco, B. A. & Kang, D.-H. 2012. Monitoring ultraviolet (UV) radiation inactivation of *Cronobacter sakazakii* in dry infant formula using Fourier transform infrared spectroscopy. *Journal of Food Science*, **77**, M86–93.

Longenberger, A. H., Palumbo, A. J., Chu, A. K., Moll, M. E., Weltman, A. & Ostroff, S. M. 2013. *Campylobacter jejuni* infections associated with unpasteurized milk-multiple states, 2012. *Clinical Infectious Diseases*, **57**, 263–6.

Lourenco, A., Machado, H. & Brito, L. 2011. Biofilms of *Listeria monocytogenes* produced at 12 degrees C either in pure culture or in co-culture with *Pseudomonas aeruginosa* showed reduced susceptibility to sanitizers. *Journal of Food Science*, **76**, M143–8.

Lovdal, I. S., Hovda, M. B., Granum, P. E. & Rosnes, J. T. 2011. Promoting *Bacillus cereus* spore germination for subsequent inactivation by mild heat treatment. *Journal of Food Protection*, **74**, 2079–89.

Lu, Y., Chen, Y., Lu, X. A., Lv, J., Man, C. X., Chai, Y. L. & Jiang, Y. J. 2014. Comparison of methods for the microbiological identification and typing of *Cronobacter* species in infant formula. *Journal of Dairy Science*, **97**, 632–41.

Lucking, G., Stoeckel, M., Atamer, Z., Hinrichs, J. & Ehling-Schulz, M. 2013. Characterization of aerobic spore-forming bacteria associated with industrial dairy processing environments and product spoilage. *International Journal of Food Microbiology*, **166**, 270–9.

Malorny, B., Tassios, P. T., Radstrom, P., Cook, N., Wagner, M. & Hoorfar, J. 2003. Standardization of diagnostic PCR for the detection of foodborne pathogens. *International Journal of Food Microbiology*, **83**, 39–48.

Martinez-Rodriguez, Y., Acosta-Muniz, C., Olivas, G. I., Guerrero-Beltran, J., Rodrigo-Aliaga, D. & Sepulveda, D. R. 2012. High hydrostatic pressure processing of cheese. *Comprehensive Reviews in Food Science and Food Safety*, **11**, 399–416.

Mataragas, M., Greppi, A., Rantsiou, K. & Cocolin, L. 2014. Gene transcription patterns of pH- and salt-stressed *Listeria monocytogenes* cells in simulated gastric and pancreatic conditions. *Journal of Food Protection*, **77**, 254–61.

McLean, S. K., Dunn, L. A. & Palombo, E. A. 2010. Applications of polymerase chain reaction in the dairy industry. *Australian Journal of Dairy Technology*, **65**, 81–5.

Melo, J., Andrew, P. W. & Faleiro, M. L. 2013. Different assembly of acid and salt tolerance response in two dairy *Listeria monocytogenes* wild strains. *Archives of Microbiology*, **195**, 339–48.

Melo, P. D. C., Sousa, C., Botelho, C., Oliveira, R. & Nader Filho, A. 2014. NaOCl effect on biofilm produced by *Staphylococcus aureus* isolated from the milking environment and mastitis infected cows. *Pesquisa Veterinaria Brasileira*, **34**, 109–13.

Mikulec, D. P., Tambur, Z., Radakovic, S., Mileusnic, I. & Jevremovic, D. 2012. Rate of potentially pathogenic bacteria in fresh soft cheeses made on small dairy farms in Serbia. *International Journal of Research and Reviews in Applied Sciences*, **12**, 247–51.

Miszczycha, S. D., Perrin, F., Ganet, S., Jamet, E., Tenenhaus-Aziza, F., Montel, M.-C. & Thevenot-Sergentet, D. 2013. Behavior of different Shiga toxin-producing *Escherichia coli* serotypes in various experimentally contaminated raw-milk cheeses. *Applied and Environmental Microbiology*, **79**, 150–8.

Mosquera-Fernandez, M., Rodriguez-Lopez, P., Cabo, M. L. & Balsa-Canto, E. 2014. Numerical spatio-temporal characterization of *Listeria monocytogenes* biofilms. *International Journal of Food Microbiology*, **182**, 26–36.

Nakano, M., Itoh, Y., Yamada, Y., Nakamura, K., Sumitomo, M. & Nitta, M. 2012. Enhancement of enterohemorrhagic *Escherichia coli* O157:H7 stress tolerance via pre-heating. *Molecular Medicine Reports*, **5**, 715–20.

Nassib, T. A., El-Din, M. Z. & El-Sharoud, W. M. 2003. Viability of *Salmonella enterica* subsp *enterica* during the preparation and cold storage of Egyptian soft cheeses and ice-cream. *International Journal of Dairy Technology*, **56**, 30–4.

Nazarowec-White, M., Mckellar, R. C. & Piyasena, P. 1999. Predictive modelling of *Enterobacter sakazakii* inactivation in bovine milk during high-temperature short-time pasteurization. *Food Research International*, **32**, 375–9.

Neaves, P. 2013. Hygienic quality of UK raw milk: is it 'horrendous' or 'the best milk supply in the world'? *World Food Regulation Review*, **22**, 28–9.

Nicolay, N., Thornton, L., Cotter, S., Garvey, P., Bannon, O., Mckeown, P., Cormican, M., Fisher, I., Little, C., Boxall, N., De Pinna, E., Peters, T., Cowden, J., Salmon, R., Mason, B., Irvine, N., Rooney, P. & O'Flanagan, D. 2011. *Salmonella enterica* serovar Agona European outbreak associated with a food company. *Epidemiology and Infection.*, **139**, 1272–80.

Nogva, H. K., Dromtorp, S. M., Nissen, H. & Rudi, K. 2003. Ethidium monoazide for DNA-based differentiation of viable and dead bacteria by 5′-nuclease PCR. *Biotechniques*, **34**, 804–13.

Nolan, T., Hands, R. E. & Bustin, S. A. 2006. Quantification of mRNA using real-time RT-PCR. *Nature Protocols*, **1**, 1559–82.

Norberg, S., Stanton, C., Ross, R. P., Hill, C., Fitzgerald, G. F. & Cotter, P. D. 2012. *Cronobacter* spp. in powdered infant formula. *Journal of Food Protection*, **75**, 607–20.

Oermeci, E. & Oezdemir, H. 2007. Prevalence of thermophilic *Campylobacter* spp. in raw milk. *Milk Science International*, **62**, 264–5.

Ohnishi, M., Okatani, A. T., Esaki, H., Harada, K., Sawada, T., Murakami, M., Marumo, K., Kato, Y., Sato, R., Shimura, K., Hatanaka, N. & Takahashi, T. 2013. Herd prevalence of Enterobacteriaceae producing CTX-M-type and CMY-2-lactamases among Japanese dairy farms. *Journal of Applied Microbiology*, **115**, 282–9.

Oliwa-Stasiak, K., Kolaj-Robin, O. & Adley, C. C. 2011. Development of real-time PCR assays for detection and quantification of *Bacillus cereus* group species: differentiation of *B. weihenstephanensis* and Rhizoid *B. pseudomycoides* isolates from milk. *Applied and Environmental Microbiology*, **77**, 80–8.

Opiyo, B. A., Wangoh, J. & Njage, P. M. K. 2013. Microbiological performance of dairy processing plants is influenced by scale of production and the implemented food safety management system: a case study. *Journal of Food Protection*, **76**, 975–83.

Osaili, T. & Forsythe, S. 2009. Desiccation resistance and persistence of *Cronobacter* species in infant formula. *International Journal of Food Microbiology*, **136**, 214–20.

Pagedar, A., Singh, J. & Batish, V. K. 2010. Surface hydrophobicity, nutritional contents affect *Staphylococcus aureus* biofilms and temperature influences its survival in preformed biofilms. *Journal of Basic Microbiology*, **50**, S98–106.

Pagotto, F. J. & Farber, J. M. 2009. *Cronobacter* spp. (*Enterobacter sakazakii*): advice, policy and research in Canada. *International Journal of Food Microbiology*, **136**, 238–45.

Payne, D. E., Martin, N. R., Parzych, K. R., Rickard, A. H., Underwood, A. & Boles, B. R. 2013. Tannic acid inhibits *Staphylococcus aureus* surface colonization in an isaA-dependent manner. *Infection and Immunity*, **81**, 496–504.

Pena, W. E. L., De Andrade, N. J., Soares, N. F. F., Alvarenga, V. O., Rodrigues, S., Granato, D., Zuniga, A. D. G. & St'ana, A. D. 2014. Modelling *Bacillus cereus* adhesion on stainless steel surface as affected by temperature, pH and time. *International Dairy Journal*, **34**, 153–8.

Peng, S., Stephan, R., Hummerjohann, J., Blanco, J. & Zweifel, C. 2012. *In vitro* characterization of Shiga toxin-producing and generic *Escherichia coli* in respect of cheese production-relevant stresses. *Archiv Fur Lebensmittelhygiene*, **63**, 136–41.

Pereira, R. V., Bicalho, M. L., Machado, V. S., Lima, S., Teixeira, A. G., Warnick, L. D. & Bicalho, R. C. 2014. Evaluation of the effects of ultraviolet light on bacterial contaminants inoculated into whole milk and colostrum, and on colostrum immunoglobulin G. *Journal of Dairy Science*, **97**, 2866–75.

Pexara, A. & Govaris, A. 2010. *Bacillus cereus*: an important foodborne pathogen. *Journal of the Hellenic Veterinary Medical Society*, **61**, 127–33.

Pimentel-Filho, N. D. J., Mantovani, H. C., De Carvalho, A. F., Dias, R. S. & Vanetti, M. C. D. 2014. Efficacy of bovicin HC5 and nisin combination against *Listeria monocytogenes* and *Staphylococcus aureus* in fresh cheese. *International Journal of Food Science & Technology*, **49**, 416–22.

Pina-Perez, M. C., Rodrigo, D. & Martinez, A. 2012. *Cronobacter sakazakii* growth prediction in reconstituted powdered infant formula milk. *Archiv Fur Lebensmittelhygiene*, **63**, 168–74.

Pinto, A. D., Bonerba, E., Bozzo, G., Ceci, E., Terio, V. & Tantillo, G. 2013. Occurence of potentially enterotoxigenic *Bacillus cereus* in infant milk powder. *European Food Research and Technology*, **237**, 275–9.

Popp, A., Fricker-Feer, C., Gschwend, K. & Stephan, R. 2010. PFGE-typing of *Enterobacter cloacae*, *Klebsiella pneumoniae* and *Leclercia adecarboxylata* isolates from an infant formula processing plant. *Archiv Fur Lebensmittelhygiene*, **61**, 128–31.

Postollec, F., Falentin, H., Pavan, S., Combrisson, J. & Sohier, D. 2011. Recent advances in quantitative PCR (qPCR) applications in food microbiology. *Food Microbiology*, **28**, 848–61.

Pui, C. F., Wong, W. C., Chai, L. C., Tunung, R., Jeyaletchumi, P., Noor Hidayah, M. S., Ubong, A., Farinazlenn, M. G., Cheah, Y. K. & Son, R. 2011. *Salmonella*: a foodborne pathogen. *International Food Research Journal*, **18**, 465–73.

Quigley, L., O'Sullivan, O., Stanton, C., Beresford, T. P., Ross, R. P., Fitzgerald, G. F. & Cotter, P. D. 2013. The complex microbiota of raw milk. *FEMS Microbiology Reviews*, **37**, 664–98.

Rasooly, A. & Herold, K. E. 2008. Food microbial pathogen detection and analysis using DNA microarray technologies. *Foodborne Pathogens and Disease*, **5**, 531–50.

Reich, F., Koenig, R., Von Wiese, W. & Klein, G. 2010. Prevalence of *Cronobacter* spp. in a powdered infant formula processing environment. *International Journal of Food Microbiology*, **140**, 214–17.

Reissbrodt, R. 2004. New chromogenic plating media for detection and enumeration of pathogenic *Listeria* spp. – an overview. *International Journal of Food Microbiology*, **95**, 1–9.

Renter, D. G., Morris, J. G., Sargeant, J. M., Hungerford, L. L., Berezowski, J., Ngo, T., Williams, K. & Acheson, D. W. K. 2005. Prevalence, risk factors, O serogroups, and virulence profiles of Shiga toxin-producing bacteria from cattle production environments. *Journal of Food Protection*, **68**, 1556–65.

Rhodes, P. L., Mitchell, J. W., Wilson, M. W. & Melton, L. D. 2006. Antilisterial activity of grape juice and grape extracts derived from *Vitis vinifera* variety Ribier. *International Journal of Food Microbiology*, **107**, 281–6.

Ricke, S. C., Ok-Kyung, K., Foley, S. & Nayak, R. 2013. *Salmonella*. In: Labbé, R. G. & García, S. (eds) Guide to Foodborne Pathogens. John Wiley & Sons, Chichester.

Ronner A. B. & Wong, A. C. L. 1993. Biofilm development and sanitizer inactivation of *Listeria monocytogenes* on stainless steel and Buna-n-rubber. *Journal of Food Protection*, **56**, 750–8.

Russo, E. T., Biggerstaff, G., Hoekstra, M. R., M. Eyer, S., Patel, N., Miller, B., Quick, R. 2013. A recurrent, multistate outbreak of *Salmonella* serotype Agona infections associated with dry, unsweetened cereal consumption, United States, 2008. . *Journal of Food Protection*, **76**, 227–30.

Ryu, J. H. & Beuchat, L. R. 2005. Biofilm formation and sporulation by *Bacillus cereus* on a stainless steel surface and subsequent resistance of vegetative cells and spores to chlorine, chlorine dioxide, and a peroxyacetic acid-based sanitizer. *Journal of Food Protection*, **68**, 2614–22.

Sadek, Z. I., Fathi, F. A. & Salem, M. M. E. 2006. Incidence, survival and biocontrol of psychrotrophic *Bacillus cereus* and its potential for toxin production in milk and Tallaga cheese. *Polish Journal of Food and Nutrition Sciences*, **15**, 419–25.

Saini, J. K., Marsden, J. L., Getty, K. J. K. & Fung, D. Y. C. 2014. Advanced oxidation technology with photohydroionization as a surface treatment for controlling *Listeria monocytogenes* on stainless steel surfaces and ready-to-eat cheese and turkey. *Foodborne Pathogens and Disease*, **11**, 295–300.

Sala, C., Morvay, A. A., Imre, K., Nichita, I., Cerna, D. D. & Morar, A. 2012. Study of bacterial biofilm stage formation on stainless steel surfaces with varying degrees of roughness. *Lucrari Stiintifice – Universitatea de Stiinte Agricole a Banatului Timisoara, Medicina Veterinara*, **45**, 141–7.

Salustiano, V. C., Andrade, N. J., Ribeiro Junior, J. I., Fernandes, P. E., Lopes, J. P., Bernardes, P. C. & Portugal, J. G. 2010. Controlling *Bacillus cereus* adherence to stainless steel with different cleaning and sanitizing procedures used in dairy plants. *Arquivo Brasileiro De Medicina Veterinaria E Zootecnia*, **62**, 1478–83.

Schmidt, V. S. J., Kaufmann, V., Kulozik, U., Scherer, S. & Wenning, M. 2012. Microbial biodiversity, quality and shelf life of microfiltered and pasteurized extended shelf life (ESL) milk from Germany, Austria and Switzerland. *International Journal of Food Microbiology*, **154**, 1–9.

Schoder, D., Stessl, B., Szakmary-Braendle, K., Rossmanith, P. & Wagner, M. 2014. Population diversity of *Listeria monocytogenes* in quargel (acid curd cheese) lots recalled during the multinational listeriosis outbreak 2009/2010. *Food Microbiology*, **39**, 68–73.

See, K. C., Than, H. A. & Tang, T. 2007. *Enterobacter sakazakii* bacteraemia with multiple splenic abscesses in a 75-year-old woman: a case report. *Age and Ageing*, **36**, 595–6.

Shaheen, R., Svensson, B., Andersson, M. A., Christiansson, A. & Salkinoja-Salonen, M. 2010. Persistence strategies of *Bacillus cereus* spores isolated from dairy silo tanks. *Food Microbiology*, **27**, 347–55.

Sharma, M., Ryu, J. H. & Beuchat, L. R. 2005. Inactivation of *Escherichia coli* O157:H7 in biofilm on stainless steel by treatment with an alkaline cleaner and a bacteriophage. *Journal of Applied Microbiology*, **99**, 449–59.

Shrestha, S., Grieder, J. A., Mcmahon, D. J. & Nummer, B. A. 2011. Survival of *Listeria monocytogenes* introduced as a post-aging contaminant during storage of low-salt Cheddar cheese at 4, 10, and 21 degrees C. *Journal of Dairy Science*, **94**, 4329–35.

Silva, R., Cruz, A. G., Faria, J. A. F., Moura, M. M. L., Carvalho, L. M. J., Water, E. H. M. & Sant'ana, A. S. 2010. Pasteurized milk: efficiency of pasteurization and its microbiological conditions in Brazil. *Foodborne Pathogens and Disease*, **7**, 217–19.

Singh, J., Batish, V. K. & Grover, S. 2012. Simultaneous detection of *Listeria monocytogenes* and *Salmonella* spp. in dairy products using real time PCR-melt curve analysis. *Journal of Food Science and Technology-Mysore*, **49**, 234–9.

Soejima, T., Iida, K. I., Qin, T., Taniai, H., Seki, M. & Yoshida, S. I. 2008. Method to detect only live bacteria during PCR amplification. *Journal of Clinical Microbiology*, **46**, 2305–13.

Solheim, H., Sekse, C., Urdahl, A., Wasteson, Y. & Nesse, L. 2013. Biofilm as an environment for dissemination of stx genes by transduction. *Applied and Environmental Microbiology*, **79**, 896–900.

Sonbol, H., Joseph, S., Mcauley, C. M., Craven, H. M. & Forsythe, S. J. 2013. Multilocus sequence typing of *Cronobacter* spp. from powdered infant formula and milk powder production factories. *International Dairy Journal*, **30**, 1–7.

Songzhe, F., Jianxin, G., Ying, L. & Haiying, C. 2011. Isolation of *Cronobacter* spp. isolates from infant formulas and their survival in the production process of infant formula. *Czech Journal of Food Sciences*, **29**, 391–9.

Spencer, S. E. F., Marshall, J., Pirie, R., Campbell, D., Baker, M. G. & French, N. P. 2012. The spatial and temporal determinants of campylobacteriosis notifications in New Zealand, 2001–2007. *Epidemiology and Infection*, **140**, 1663–77.

Sudagidan, M. & Yemenicioglu, A. 2012. Effects of nisin and lysozyme on growth inhibition and biofilm formation capacity of *Staphylococcus aureus* strains isolated from raw milk and cheese samples. *Journal of Food Protection*, **75**, 1627–33.

Sundberg, M., Christiansson, A., Lindahl, C., Wahlund, L. & Birgersson, C. 2011. Cleaning effectiveness of chlorine-free detergents for use on dairy farms. *Journal of Dairy Research*, **78**, 105–10.

Tabak, M., Scher, K., Hartog, E., Romling, U., Matthews, K. R., Chikindas, M. L. & Yaron, S. 2007. Effect of triclosan on *Salmonella typhimurium* at different growth stages and in biofilms. *FEMS Microbiology Letters*, **267**, 200–6.

Tenenhaus-Aziza, F., Daudin, J.-J., Maffre, A. & Sanaa, M. 2014. Risk-based approach for microbiological food safety management in the dairy industry: the case of *Listeria monocytogenes* in soft cheese made from pasteurized milk. *Risk Analysis*, **34**, 56–74.

Tiwari, U., Walsh, D., Rivas, L., Jordan, K. & Duffy, G. 2014. Modelling the interaction of storage temperature, pH, and water activity on the growth behaviour of *Listeria monocytogenes* in raw and pasteurised semi-soft rind washed milk cheese during storage following ripening. *Food Control*, **42**, 248–56.

Todd, E. C. D. 2011. The international risk governance council framework and its application to Listeria monocytogenes in soft cheese made from unpasteurised milk. *Food Control*, **22**, 1513–24.

Tomasula, P. M., Renye, J. A., Van Hekken, D. L., Tunick, M. H., Kwoczak, R., Toht, M., Leggett, L. N., Luchansky, J. B., Porto-Fett, A. C. S. & Phillips, J. G. 2014. Effect of high-pressure processing on reduction of *Listeria monocytogenes* in packaged Queso Fresco. *Journal of Dairy Science*, **97**, 1281–95.

Torkar, K. G. & Teger, S. G. 2006. The presence of some pathogenic microorganisms, yeasts and moulds in cheese samples produced in small dairy-processing plants. *Acta Agriculturae Slovenica*, **88**, 37–51.

Trachoo, N. & Brooks, J. D. 2005. Attachment and heat resistance of *Campylobacter jejuni* on *Enterococcus faecium* biofilm. *Pakistan Journal of Biological Sciences*, **8**, 599–605.

Trevisani, M., Mancusi, R., Riu, R. & Serraino, A. 2013. Quantitative detection of Shiga toxin-producing and enteropathogenic *Escherichia coli* Serotypes O157 and O26 in bulk raw milk. *Food Analytical Methods*, **6**, 1750–8.

Unicomb, L. E., Fullerton, K. E., Kirk, M. D. & Stafford, R. J. 2009. Outbreaks of Campylobacteriosis in Australia, 2001 to 2006. *Foodborne Pathogens and Disease*, **6**, 1241–50.

Uttamchandani, M., Neo, J. L., Ong, B. N. Z. & Moochhala, S. 2009. Applications of microarrays in pathogen detection and biodefence. *Trends in Biotechnology*, **27**, 53–61.

Vencia, W., Nogarol, C., Bianchi, D. M., Gallina, S., Zuccon, F., Adriano, D., Gramaglia, M. & Decastelli, L. 2014. Validation according to ISO 16140:2003 of a commercial real-time PCR-based method for detecting *Campylobacter jejuni*, *C. coli*, and *C. lari* in foods. *International Journal of Food Microbiology*, **177**, 78–80.

Verhoeff-Bakkenes, L. 2012. *Campylobacter jejuni*: exposure assessment and hazard characterization: growth, survival and infectivity of *Campylobacter jejuni*. PhD thesis, Wageningen University, Netherlands.

Vestby, L., Moretro, T., Langsrud, S., Heir, E., Nesse, L. 2009. Biofilm forming abilities of *Salmonella* are correlated with persistence in fish meal and feed factories. *BMC Veterinary Research*, **5**, 20.

Walsh, D., Molloy, C., Iversen, C., Carroll, J., Cagney, C., Fanning, S. & Duffy, G. 2011. Survival characteristics of environmental and clinically derived strains of *Cronobacter sakazakii* in infant milk formula (IMF) and ingredients. *Journal of Applied Microbiology*, **110**, 697–703.

Wang, R., Bono, J. L., Kalchayanand, N., Shackelford, S. & Harhay, D. M. 2012. Biofilm formation by Shiga toxin-producing *Escherichia coli* O157:H7 and non-O157 strains and their tolerance to sanitizers commonly used in the food processing environment. *Journal of Food Protection*, **75**, 1418–28.

Wijman, J. G. E., De Leeuw, P. P. L. A., Moezelaar, R., Zwietering, M. H. & Abee, T. 2007. Air–liquid interface biofilms of *Bacillus cereus*: formation, sporulation, and dispersion. *Applied and Environmental Microbiology*, **73**, 1481–8.

Wilson, I. G. 1997. Inhibition and facilitation of nucleic acid amplification. *Applied and Environmental Microbiology*, **63**, 3741–51.

Wu, V. C. H. 2008. A review of microbial injury and recovery methods in food. *Food Microbiology*, **25**, 735–44.

Wysok, B. & Uradzinski, J. 2009. *Campylobacter* spp. – a significant microbiological hazard in food. I. Characteristics of *Campylobacter* species, infection source, epidemiology. *Polish Journal of Veterinary Sciences*, **12**, 141–8.

Yemis, G. P., Pagotto, F., Bach, S. & Delaquis, P. 2012. Thermal tolerance and survival of *Cronobacter sakazakii* in powdered infant formula supplemented with vanillin, ethyl vanillin, and vanillic acid. *Journal of Food Science*, **77**, M523–7.

Zeledon, G. C., Solano, M. R. & Echandi, M. L. A. 2010. Biofilm formation capacity of *Listeria monocytogens* strains isolated from soft cheese from Costa Rica. *Archivos Latinoamericanos De Nutricion*, **60**, 175–8.

Zogaj, X., Bokranz, W., Nimtz, M. & Römling, U. 2003. Production of cellulose and curli fimbriae by members of the family Enterobacteriaceae isolated from the human gastrointestinal tract. *Infection and Immunity*, **71**, 4151–8.

10 Biofilm Issues in Dairy Waste Effluents

Michael Dixon[1], Steve Flint[1], Koon Hoong Teh[1] and Kieran Mellow[2]

[1]Institute of Food, Nutrition and Human Health, Massey University, Palmerston North, New Zealand
[2]Institute of Fundamental Sciences, Massey University, Palmerston North, New Zealand

10.1 Introduction

The dairy industry has seen large growth throughout the world. The disposal of effluent produced by this industry is a challenge. Dairy effluent is considered one of the most polluting types of effluent, not only because of the amount generated, but also because of its composition, which has been shown to have high chemical oxygen demand (COD) and biological oxygen demand (BOD), and its content of nutrients, organic and inorganic (Kushwaha *et al.*, 2011). The actual composition of this effluent depends on what the plant is processing; as the composition changes, so do the bacterial species that develop in the effluent (Kushwaha *et al.*, 2011). Therefore, if this effluent is discharged without proper treatment, pollution of the environment occurs, resulting in eutrophication of the waterways.

Biofilm formation is a commonly known problem throughout the dairy industry, with studies in different areas showing the presence of varying bacterial biofilms. Both thermophilic and mesophilic biofilms have been found. Formation occurs on many surfaces that are exposed to product, including membranes, stainless steel and plastics. Regular cleaning schedules and well-developed cleaning-in-place (CIP) processes leave these biofilms little time to develop and become mature. In waste treatment, however, there are no regular CIP processes, so the biofilms have time to mature fully. As the effluent is generated by cleaning the plant, any and all bacteria that are present at any stage of the process will also be present in the effluent. Therefore, any biofilm that is already known to occur in the dairy industry could also be a problem in the effluent treatment systems.

As all of the wastewater is destined for a natural water body, the load of these pollutants must be reduced considerably. Fortunately, biofilms are one of the most economical, satisfactory and energy-efficient hazardous pollutants to deal with in wastewater produced from agricultural, municipal, industrial and mining sources. In the United States, 240 liters of wastewater per capita is produced daily, containing 240 mg/l biosolids (Hammer & Hammer, 2001). According

Biofilms in the Dairy Industry, First Edition.
Edited by Koon Hoong Teh, Steve Flint, John Brooks and Geoff Knight.

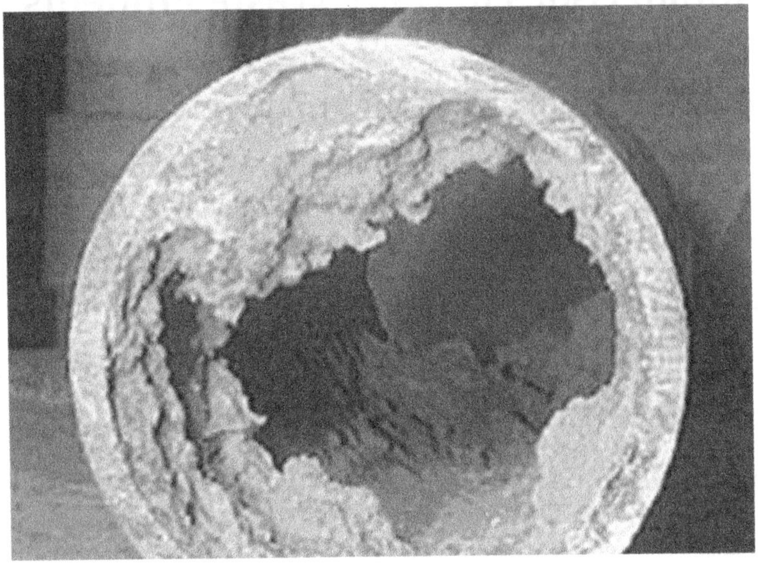

Figure 10.1 Dairy wastewater pipe, showing corrosion caused by growth of a biofilm.

to one study, more than 8×10^6 dry tons of biosolids are disposed of per annum in the United States (Environmental Protection Agency, 2006), and 2.39×10^6 in Europe (Chang *et al.*, 2002).

One treatment method that is very effective and that has been used for years are irrigation systems. This is used in areas where discharge into waterways is not possible or is not allowed by the district or regional governing bodies. However, in the last couple of years, our laboratory has seen corrosion and reduced flow to complete blockage of the pipes used in dairy wastewater systems (Figure 10.1). The processes that can be used to treat effluent or clean pipework are limited because the effluent is discharged on to fields that are used to grow crops. For example, common methods such as acid washes cannot be used because the plants would die if acid were sprayed on to the fields.

10.2 Overview of dairy effluent treatment

Dairy effluent treatment practices change according to where a plant is located. Dairy plants can use a range of different treatments, producing highly processed to minimally processed effluents. These varying treatment practices may be a reason for the slime build-up that is sometimes seen. Some regulatory bodies require a large amount of treatment before a company is allowed to discharge effluent. Other sites are allowed to employ minimal treatment.

A few common methods used to treat dairy effluent include clarification, oil–water separation, grease trapping and solids separation. Dairy effluent is also usually treated by biological means, as all components of dairy effluent are biodegradable; however, fats and proteins are not as easily degraded (Kushwaha *et al.*, 2011). Both aerobic and anaerobic treatment processes are available, with anaerobic processes being more widely used in industry.

Aerobic treatments include processes such as activated sludge, trickling filters and aeration ponds, or a combination of these. As fat has an inhibitory effect on these processes, it is often removed in an earlier step, such as a dissolved air flotation (DAF) tank. Kushwaha *et al.* (2011) state that among the various aerobic processes, sequential batch reactors seem promising. Effluent is added to activated sludge contained in these reactors, then treated and discharged. Equalisation, aeration and clarification of the effluent can all take place in a DAF tank. A membrane filtration system can also be attached to increase the treatment, resulting in a suspended solids (SS)-free effluent.

However, although studies show the advantages of aerobic treatment, there are also drawbacks. Aerobic treatment often has high energy requirements and a high area demand (aeration ponds). Upflow anaerobic sludge blanket (UASB) reactors are therefore commonly used for effluent treatment in the dairy industry. These reactors have the effluent enter from the bottom and travel up through the sludge while gas and solid separation take place. However, due to the inhibitory effect of fat in anaerobic treatment, fast and efficient treatment is not possible. Enzymatic pretreatment might provide one way of removing this fat.

Table 10.1, developed from Kushwaha *et al.* (2011), compares aerobic and anaerobic treatment of dairy effluent.

A highly efficient aerobic wastewater treatment system is the ICI Deep Shaft System (Walker & Wilkinson, 2006). This evolved from the basic aerobic fermentation technology used in the production of single-cell protein from methanol. The technology uses a novel pressure-cycle fermenter, in which the air provides oxygen for microbial fermentation and mixing of the liquid, giving an oxygen transfer efficiency of approximately 50%. This is excessive for waste treatment, so modifications have been made to give a longer bubble contact time by increasing the height of the fermenter. Such systems have been used to

Table 10.1 Comparison of aerobic and anaerobic processing. Adapted from Kushwaha *et al.* (2011).

Factors	Aerobic process	Anaerobic process
Reactor type	Aerated lagoons, ponds, trickling filters, biological disks, rotating biological contactor	UASB, packed bed reactor, CSTR, fixed film reactor, buoyant filter bioreactor
Reactor size	Large area generally required	Smaller reactor size
Effluent quality	Excellent	COD removal fair, nutrient removal low, further treatment required
Energy	High energy input required	Can produce energy, i.e. methane
Biomass yield	6–8 times greater biomass produced	Lower biomass produced
Loading rate	9000 g COD/m^3 max. reported	Very large – up to 3.5 times greater than anaerobic
Oil /grease removal	Do not cause serious problems	Inhibitory action during treatment
Shock loading	Excellent performance	Showed bad response to shock loading
Alkalinity addition	N/A	Needed to maintain pH, due to acid production from lactose

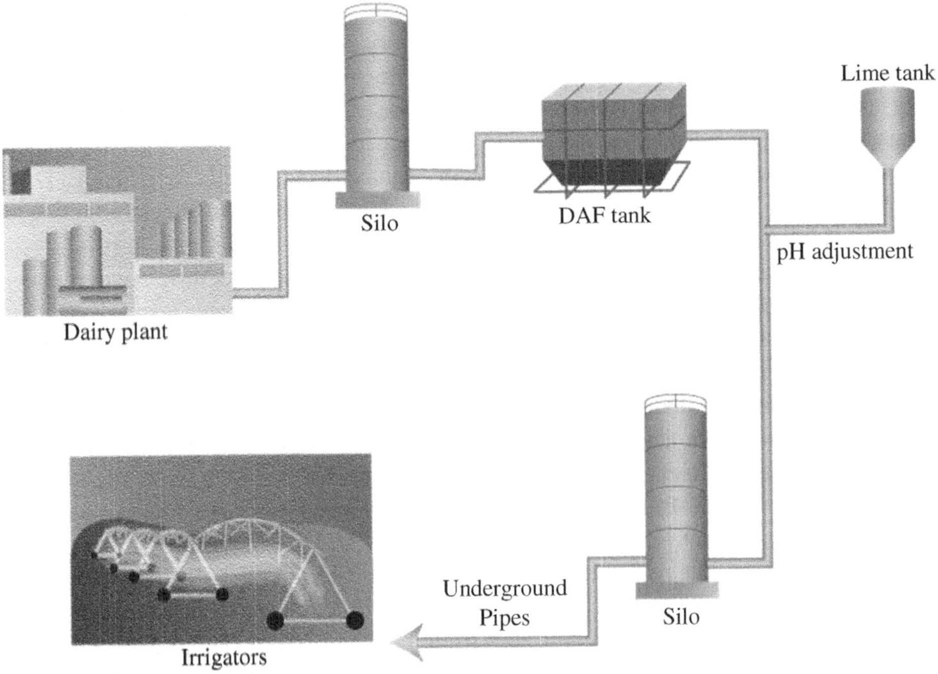

Figure 10.2 Basic dairy effluent system.

treat effluent from potato starch plants, breweries and various chemical manufacturing plants, but not, to our knowledge, in the dairy industry.

Figure 10.2 shows a basic treatment process for dairy effluent. This process is for a dairy powder plant that requires minimal treatment processing, involving only the removal of fat.

10.3 Dairy farm waste treatment

Effluent stored in a dairy farm waste effluent pond on-farm mainly consists of animal excreta and wastewater generated from farm management practices. The effluent stored on-farm may account for 25% of total on-farm emissions of methane to the atmosphere (Mosier *et al.*, 1998; MFE, 2010). Methane emissions generated by a 1000 m^2 dairy farm waste effluent pond for a dairy herd of 450 cattle are estimated at 26 m^3/day (Craggs *et al.*, 2008). Currently, it is not cost-effective for the average dairy farm to harness the methane production as a renewable source of energy (Pratt *et al.*, 2012). However, an anaerobic digester has been used to treat wastewater and to produce biogas. The microbial communities in the anaerobic digester will aggregate to form flocs or sludge granules that may facilitate the production of methane.

Methane emission is seasonal, with higher levels during summer than winter, perhaps due to the methane influx rate, temperature and the possibility of a shift in the microbial population within a dairy farm waste effluent pond (Pratt *et al.*, 2012). However, it may also be explained by there being less waste produced during winter, as milking is reduced compared with the summer. Methanogenesis is dependent on the concentration of dairy farm waste, which provides the energy source for methanogens: a group of archaea responsible for methane production as a byproduct of their metabolic pathway.

Several factors influence methanogensis, including hydrolysis, acidogenesis and acetogenesis, which are performed by specific microbial communities that are dependent on each other for their metabolisms (Weiland, 2010). Therefore, methanogenesis may be enhanced in a complex microbial community within a biofilm because of the enclosed structure of the biofilm and the availability of substrates. The physicochemical properties and structure of a biofilm can provide an anoxic environment for growth of methanogens. It has been suggested that microbial diversity within biofilms predominates in nature and that they may function at a level that is similar to that of multicellular organisms. Enzymatic activities within biofilms tend to be greater than those in planktonic cultures (Frølund *et al.*, 1995; Teh *et al.*, 2014).

The profile of methanogens found in an anaerobic manure digester has been suggested to be influenced by management practice and by years of operation (St-Pierre & Wright, 2013). *Methanosarcina thermophila* was found to be dominant in two digesters, while another comprised four phylogenetic groups: Methanomirobiales, Methanosarcinales, Methanoplasmatales and Methanobacteriales (St-Pierre & Wright, 2013). Furthermore, the microbial community composition in a dairy farm waste effluent pond consists primarily of *Bacillus*, *Clostridium*, *Mycoplasma*, *Eubacterium* and *Proteobacteria* species that originate from the gastrointestinal tracts of ruminants (Ibekwe *et al.*, 2003). Certain strains of the methanogens and ruminant microorganisms have been sequenced by the Hungate 1000 Project (www.hungate1000.org.nz), a catalogue of reference genomes from the rumen microbiome (Kelly *et al.*, 2013). The reference genomes can be used to support international efforts to develop methane mitigation and rumen adaptation technologies. The diversity of microbial communities in the dairy farm effluent pond may influence the rate of methanogenesis.

10.4 Composition of biofilms

The composition and quantity of microorganisms in a biofilm are highly influenced by physiochemical factors provided for the development and sustenance of biofilms. In aerobic, anoxic and anaerobic environments, different microorganisms can survive and perform according to their genetic potentials. Biofilm composition is decidedly manipulated by the composition of wastewater received (Lydmark *et al.*, 2007).

The composition and morphology of biofilms have been intensively studied in recent years. Different molecular biology techniques have been applied to explore the nature and function of microorganisms and related components of biofilms (Ivnitsky *et al.*, 2007; Sanz & Kochling, 2007). Advancement of confocal laser scanning microscopy and other techniques in biophysics has helped us a lot in deepening our knowledge of the structure of biofilms and the anatomical interaction of microorganisms (McLean *et al.*, 2008).

Techniques such as DNA extraction, polymerase chain reaction (PCR), denaturing gradient gel electrophoresis polymerase chain reaction (DGGE-PCR), gene sequencing and fluorescent in situ hybridisation (FISH) have been employed to identify and differentiate the microorganisms structuring biofilms and developing active consortia. In one study, 14 different experiments were conducted to find the composition of a biofilm. Of a total of 22 sequences obtained, 19 were Gram-negative. In all scenarios, *Proteobacteria* was seen to be a ubiquitous group (16 of 22 sequences), with the predominant member the b-subdivision (8 sequences), followed by the g-subdivision (5 sequences). *Pseudomonas/Burkholderia*, *Ralstonia*, Bacteroidetes and *Sphingomonas* were the dominant groups found in most cases (Ivnitsky *et al.*, 2007).

Analyses of a dairy wastewater system in our laboratory have shown the Enterobacteriaciae family to be the dominant bacteria present. The ability of these bacteria to form biofilms and their role in preventing biofilms in wastewater treatment are not known and should be studied further.

Aerobic granules have been extensively studied by different researchers. They are made up of different layers of microorganisms, especially bacteria, with each layer made of characteristic species and involved in a different activity, such as nitrification, denitrification, ammonia oxidation or simple respiration (Tsuneda *et al.*, 2004). Weber *et al.* (2007) have studied the importance of ciliated protozoa in the formation of biofilms with a granular appearance. Swarming protozoa provide support for the attachment of bacteria, which subsequently develop layers on the stalks of protozoa; these initially developed central masses are further joined by swarming protozoa with colonising bacteria.

Sometimes, fungi can provide some support, and this process keeps on going, forming larger sludge granules. With the help of FISH and Gram and Neisser stains, it has been shown that most of the filaments are related to the genus *Thiothrix* or to *Sphaerotilus natans*. Synthetic wastewater granules have been found to be dominated by cocci tetrads and a great quantity of extracellular polymeric substances (EPS) (Weber *et al.*, 2007).

EPS are the predominant component of a biofilm, making up 50–90% of the solid phase of a sludge. If methods for the removal of EPS are evolved, a great amount of sludge can be removed (Tian, 2008). Various studies on the types, characteristics, synthesis, functioning and significance of EPS in an activated sludge have been reviewed by Raszka *et al.* (2006) and Tian (2008). EPS provide stability and strength to a biofilm against shearing forces, and a very small quantity can work as a glue to keep the flocs joined together (Sheng *et al.*, 2006). EPS are important to the structure and stability of the granule. It has been found that extracted EPS consist of 45–55% proteins, 30–33% humic substances and 10% carbohydrates. The protein component of EPS is involved in foam formation in sludge (Nakajima & Mishima, 2005).

Phospholipid-derived fatty acids (PLFAs) can be used to determine the composition and microbial mass of a biofilm, due to the unique fatty acids that many bacteria produce. Analysis using methods such as gas chromatography–mass spectroscopy (GC-MS) provides information on the identity and quantity of each PLFA present. Results in one study showed varying total amounts of PLFA in the emitter heads of irrigation systems, but the diversity of PLFAs was low. Emitters that experienced higher clogging rates had greater amounts of PLFA. Protein and polysaccharide concentrations in the biofilm were arguably related to the stress experienced in each emitter head (Yan *et al.*, 2009). The composition of the EPS matrix was found to be very heterogenous.

EPS are involved in biosorption of heavy metals like copper, lead and cadmium. At first it was thought that bacterial cell surfaces were the sites of biosorption of the metals, but later EPS was found to perform that duty. Comte *et al.* (2008) have observed that a change in pH changes the ability of EPS to bind metals; increasing pH from 4 to 8 increased the binding.

The mineral content ranges from 10 to 90% of the total dry weight of the granules taken from sludge, depending on the wastewater composition and so forth. The main components of the ash are calcium, potassium, sodium and phosphorous (Schmidt & Ahring, 1997).

10.5 Application of biofilms in dairy wastewater treatment

While it is true that biofilms were initially recognised and studied because of the negative impact they had on food industries and human health, they can be utilised beneficially in some circumstances, as in the case of wastewater treatment strategies. Prokaryotes are the most common microorganisms used in biological wastewater treatment processes (Bitton, 2005). The aim of biological wastewater treatment is to facilitate the removal of organic and/or inorganic compounds from wastewater by utilising the metabolic and respiratory processes that key bacteria are able to deploy (Andersson, 2009). Bacteria are able to absorb dissolved nutrients and utilise them for growth and survival, removing material from water and converting it into biomass in the bacteria themselves, or else metabolising these nutrients into compounds in a gaseous phase (Andersson, 2009). Prokaryotes belonging to the classes α-, β- and γ-proteobacteria, Bacteroidetes and Actinobacteria are commonly found in biological wastewater treatment systems (Wagner & Loy, 2002). Excess nutrients in wastewater, particularly nitrogen and phosphorus, and a lack of oxygen content are of great concern for the environment. Eutrophication is the excessive growth of algae, leading to abnormal changes in a habitat; this process is facilitated by runoff wastewater resulting from human activities (Kloc, 2012). While the environment is capable of removing wastewater material on its own, the increase in wastewater runoff has prompted the development of vigorous wastewater treatment research, including research into biological wastewater treatment, in particular the use of biofilms and nitrogen/ammonia removal.

Effluent containing high concentrations of COD, BOD, nitrogen, minerals, complex chemicals and ammonia can be harmful for aquatic life in a water body into which it is discharged. Standards have therefore been set to ensure that effluent received by any water body contains these components at an acceptable level. To achieve this criterion, all wastewater needs to be treated before passing into a body of water. Chemical and physical treatment can increase cost to an unbearable level. However, major quantities of certain components can be removed by microbial degradation very efficiently and economically.

There are two different types of degradation – aerobic and anaerobic – and different kinds of bioreactor have been developed using biofilms operating in either of these modes.

Membrane-aerated biofilm reactors are used in some waste treatment systems. They can deliver oxygen at high rates and transfer efficiencies, leading to increased biofilm activity. In a thick membrane, oxygen supply is the reaction limiting factor. Decreased oxygen supply caused by a high organic load rate, low hydraulic retention time and high temperature can further be augmented by thick membrane structures. A novel method has been proposed in which biofilm is developed on oxygen-permeable membranes provided with a supply of pure oxygen (Syron & Casey, 2008).

Simultaneous nitrification and denitrification (SND) was achieved by developing a single compact suspended carrier biofilm reactor (SCBR) and the nutrient removal capacity of the reactor was studied. The response of the microbial community structure to different ratios of carbon to nitrogen (C/N) was determined using denaturing gel gradient electrophoresis (DGGE), profiles of the 16S rDNA V3 region and *amoA* gene amplifications. Population growth curves of ammonia-oxidising bacteria (AOB) and nitrite-oxidising bacteria (NOB) were estimated by FISH with 16S rDNA-targeted oligonucleotide probes. This study showed that the two-in-one SCBR was efficient for the treatment of municipal wastewater (Xia *et al.*, 2008).

Nitrogen removal from water is achieved using bacteria capable of converting aqueous ammonia into gaseous nitrogen (Madigan, 2005). The first step in nitrogen removal is facilitated by AOB, such as *Nitrosomonas*, that convert ammonia into nitrate:

$$15CO_2 + 13NH_4^+ \rightarrow 10NO_2^- + 3C_5H_7NO_2 + 23H^+ + 4H_2O \tag{10.1}$$

The nitrite produced via the oxidation of ammonia is then converted into nitrate by nitrite oxidising bacteria:

$$5CO_2 + NH_4^+ + 10NO_2^- + 2H_2O \rightarrow 10NO_3^- + C_5H_7NO_2 + H^+ \tag{10.2}$$

Under anaerobic conditions, denitrification results in the utilisation of the nitrate or nitrite for respiration, producing nitrogen gas and ultimately removing previously aqueous nitrogen compounds from the water (Henze, 2002). Denitrification is widespread throughout heterotrophic bacteria and researchers are still attempting to determine the organisms and conditions most efficient at carrying out *in situ* denitrification (Andersson, 2009). The overall process of denitrification with acetate as an electron donor is as follows:

$$5CH_3COOH + 8NO_3 \rightarrow 8HCO_3 + 2CO_2 + 6H_2O + 4N_2 \tag{10.3}$$

Removal of phosphorus from wastewater involves the use of polyphosphate-accumulating organisms. In contrast to nitrogen removal, phosphorus is not removed via a phase change, but instead by bacterial uptake of phosphorus for cell growth and development. Thus, phosphorus in wastewater is utilised for biomass. In order to achieve sufficient uptake of phosphorus, alternating aerobic and anaerobic conditions are required – an undesirable requirement in this process, because a significant energy input is needed. However, bacteria capable of enhanced phosphorus uptake in purely aerobic conditions have been discovered. Bacteria are also able to remove other compounds from water, including dissolved metals such as iron (Andersson, 2009).

10.6 Irrigation systems

A number of different irrigation schemes exist in the treatment of wastewater and dairy effluent. Common irrigation systems include macroirrigators, in which the manifold spans the field and droppers hang from it as shown in Figure 10.3. Another is the drip irrigator: the application of water or effluent though a line source with emitters at or below the surface of the soil, using low operating pressures and small discharge rates. A similar approach is the microirrigator, in which water is applied both by emitters at or below the surface and by sprayers above the soil (Yan *et al.*, 2009). Drip irrigation is a plausible solution to the treatment of effluent where other

Figure 10.3 Macroirrigator, with organic material (noodle-like material) blocking the nozzles.

methods are inappropriate due to factors such as a high water table, space constraints and so on (Yan *et al.*, 2009). Treated effluent is also increasingly being used in crop irrigation to supplement the demand for fresh water in the industrial and domestic sectors, especially in areas where there is a fresh water shortage (Liu & Huang, 2009).

A major concern when implementing the treatment of wastewater with an irrigation system is partial or complete blockage of the emitter heads. Due to the small size of the emitter heads, they are vulnerable to a number of possible obstructions. These can be caused by physical blockages, such as sand or rust build-up, chemical contamination, such as precipitated salts, or biological contamination, such as the formation of biofilms or the growth of algae.

If the biomass of this biofilm reaches a sufficiently high level, EPS can detach from the surface of the irrigator pipes where it formed and cause blockages in the irrigation system. This is especially the case in drip irrigator nozzles, due to the low flow rates and small size of the emitter heads. Yan *et al.* (2009) say that 'More than 90% of the clogging composition included biological species and the clogging process was usually initiated by bacterial biofilms'. The emitters that experienced the greatest amount of clogging had the highest content of EPS present in the emitter head, including the highest amount of polysaccharides, proteins and PLFAs. However, in many cases it is not the detached EPS that causes the blockage of the irrigators, but rather the reactions, both physical and chemical, that take place in or around them (Adin & Slacks, 1991). The major problem encountered was particles becoming entrapped in the biofilm EPS and forming sediment, reducing flow. This reduced flow could then help increase the formation of biofilms in the emitter heads, as the amounts of protein and polysaccharides were related to the stress imposed by the fluid flow. The bacterial growth could cause the precipitation of ions present in the water and effluent or the EPS could act like an adhesive, causing the fine particles in the solution (clay and sand from sources such as open silos and rust) to agglutinate and cause clogging of the irrigation system.

Adin and Slacks (1991) state that 'The clogging rate is more affected by particle size than by particle number density'. This is because the EPS entraps suspended particles and forms a three-dimensional structure, with the larger particles forming a larger structure for potential clogging.

In effluent treatment, emitter clogging varies with effluent quality, filtering methods, environmental conditions, flow rate and the size of the emitter heads. Some preventative methods can reduce clogging or clean blocked lines, such as the use of chlorination, acid injection, antagonistic bacteria or line flushing systems (Ravina *et al.*, 1997; Dosti *et al.*, 2005; Sahin *et al.*, 2005; Liu and Huang, 2009).

10.7 Controlling biofilms in waste treatment systems

Turbulent flow through the emitters in a drip irrigator system may be able to reduce the amount of clogging that occurs by causing the larger particles entrapped by the EPS to be flushed out. However, emitters with low flow rates or low hydrodynamic forces, as seen in drip irrigation systems, are more prone to clogging (Oliver *et al.*, 2014). Liu and Huang (2009) agree that emitters with high flow rates experience less clogging, but also state that the tailing part of the irrigator laterals – those furthest from pump – no matter what the flow rate, will experience more clogging than those closer to the source. Another preventative method found effective by Oliver *et al.* (2014) is the use of weblike filters. These do not prevent the formation of biofilms but work as a 'trap' for larger solids passing through. This trap helps prevent the build-up of the three-dimensional structure found to cause clogging.

Sahin *et al.* (2005) comment on the use of antagonistic bacteria to eliminate clogging in drip irrigation systems. Three bacterial strains (*Bacillus* OUS-142, *Bacillus* ERZ and *Burkholdria* OUS-7) were determined to exhibit a strong antagonistic activity, although the mechanism of this activity was not reported. A system of two irrigator laterals (one for control) was used to determine whether antagonistic bacteria could be used. The irrigator lines were run for 8 hours per day for 30 days when all emitters were partially or totally blocked. The antagonistic bacteria were added to one irrigator line, while the other was flushed with sterile water. After 2 weeks, the maximum discharge rate for the lateral treated with antagonistic bacteria was observed to increase, while the lateral treated with sterile water showed no improvement.

Chemical addition to the effluent could also be used to reduce the amount of clogging encountered. Direct acid injection would be beneficial in chemical clogging of the emitters, while chlorination could be used to reduce the amount of bacterial build-up in the pipes (Sahin *et al.*, 2005; Liu & Huang, 2009). Dosti *et al.* (2005) found that an ozonation treatment consisting of 0.6 ppm for 10 minutes effectively reduced biofilm build-up on stainless steel coupons. However, a shorter period of exposure (1 minute) did not significantly reduce the bacterial population. Chlorination (100 ppm for 2 minutes) was also trialled under the same conditions and significantly reduced the bacterial populations. The results showed that there was no difference between the 10-minute ozonation and chlorination methods, except in one of the biofilms tested (*Pseudomonas putida*). However, injection of chemicals into effluent that is to be sprayed on fields might produce a negative impact. In cases where direct acid injection can be used, the environmental impact should be taken into account. Spraying of an acidic effluent could damage crops and cause plant death (Oliver *et al.*, 2014).

Due to the nature of biofilm growth in waste treatment systems, growth often becomes a problem only when physical blockage of pipelines occurs. Physical means can be used to reduce the amount of biofilm build-up, but this will not destroy the biofilm, removing only the build-up present in the piping. Physical methods might include flushing with high-pressure water or the use of other equipment, such as cleaning balls. Cleaning balls are usually flexible and are sized 1–3 mm larger than the inner diameter of the pipe. They are forced through by the pressure of the fluid and rub the walls, keeping them clean (Al-Bakeri & El Hares, 1993). This is a purely mechanical method of removing biofilm build-up and does not tackle the problem of bacterial growth in the effluent.

Ultraviolet (UV) irradiation is one of the most common disinfection methods used in the treatment of wastewater. However, secondary use of this UV-treated effluent, such as in irrigation systems, can be potentially hazardous due to the possible photo and dark repair of reversibly damaged bacteria (Haaken *et al.*, 2014). Haaken *et al.* (2014) state that UV irradiation works to reduce the number of bacteria present (*Escherichia coli*). However, at high total SS the process was less effective, because the UV radiation was absorbed by the particles present besides the bacteria, allowing their survival. The effect of UV radiation was also limited by the formation of biofouling and scaling on the quartz sleeve of the UV lamps. However, the combination of UV irradiation and electrolysis was found to yield a reliable bacterial reduction and prevent reactivation. The only limits encountered to this combination treatment was in wastewaters containing very high total SS, which in practice were present only in poorly functioning treatment systems (Haaken *et al.*, 2014). Therefore, for dairy effluent, a UV/electrolysed treatment step could help reduce the problem of biofilms in the wastewater. Removal of SS would need to take place first.

Recent studies show that quorum sensing and cell-to-cell signalling between bacteria in the biofilm can also negate the use of cleaning chemicals. This signalling allows bacteria to monitor the environment around the EPS matrix and then alter their gene expression, allowing for further resistance to chemical cleaning. Interference with this quorum sensing can be used as a different approach to the control of biofilm growth, such as by controlling the production of EPS. Following the application of quenchers that prevent this quorum sensing, biofilms have been found to be more readily removed with bactericidal chemicals (Anand *et al.*, 2014).

10.8 Conclusion

Biofilms in dairy wastewater treatment can have both positive and negative effects on the treatment systems. Biofilms can help to remove both organic and inorganic substances. Biofilm reactors can be used to remove nutrients such as nitrogen and ammonia, as well as the heavy metals that are often found in wastewater systems. However, the environment provided by the wastewater also allows for the rapid growth of biofilms that have a negative impact on treatment. Biofilm formation in pipes can slow down, and in some cases prevent, distribution of the wastewater. The EPS matrix is responsible for forming slimy structures, entrapping inorganic particles such as clay and causing blockages.

More studies should be conducted to explore the genetic potential of microorganisms involved in wastewater treatment and knowledge must be applied to handle newly emerging issues of industrial and agricultural byproducts and waste.

References

Adin, A. & Slacks, M. 1991. Dripper-clogging factors in wasterwater irrigation. *Journal of Irrigation and Drainage Engineering*, **117**, 813–26.

Al-Bakeri, F. & El Hares, H. 1993. Experimental optimization of sponge ball cleaning system operaiton in Umm Al Nar MSF desalination plants. *Desalination*, **94**, 133–50.

Anand, S., Singh, D., Avadhanula, M. & Marka, S. 2014. Development and control of bacterial biofilms on dairy processing membranes. *Comprehensive Reviews in Food Science and Food Safety*, **13**, 18–33.

Andersson, S. 2009. Characterization of Bacterial Biofilms for Wastewater Treatment. Royal Institute of Technology, Stockholm.

Bitton, G. 2005. Wastewater Microbiology. John Wiley & Sons, Chichester.

Chang, A. C., Pan, G., Page, A. L. & Asano, T. 2002. Developing Human Health-Related Chemical Guidelines for Reclaimed Water and Sewage Sludge Applications in Agriculture. World Health Organization, Geneva.

Comte, S., Guibaud, G. & Baudu, M. 2008. Biosorption properties of extracellular polymeric substances (EPS) towards Cd, Cu and Pb for different pH values. *Journal of Hazardous Materials*, **151**, 185–93.

Craggs, R., Park, J. & Heubeck, S. 2008. Methane emissions from anaerobic ponds on a piggery and a dairy farm in New Zealand. *Australian Journal of Experimental Agriculture*, **48**, 142–6.

Dosti, B., Guzel-Seydim, Z. & Greene, A. K. 2005. Effectiveness of ozone, heat and chlorine for destroying common food spoilage bacteria in synthetic media and biofilms. *International Journal of Dairy Technology*, **58**, 19–24.

Environmental Protection Agency. 2006. Emerging Technologies for Biosolids Managment. Environmental Protection Agency, Washington, DC.

Frølund, B., Griebe, T. & Nielsen, P. H. 1995. Enzymatic activity in the activated sludge floc matrix. *Applied Microbiology and Biotechnology*, **43**, 755–61.

Haaken, D., Dittmar, T., Schmalz, V. & Worch, E. 2014. Disinfection of biologically treated wastewater and prevention of biofouling by UV/electrolysis hybrid technology: influence factors and limits for domestic wastewater reuse. *Water Research*, **52**, 20–8.

Hammer, M. J. & Hammer, M. J. J. 2001. Water and Wastewater Technology. Prentice Hall, Upper Saddle River, NJ.

Henze, M. 2002. Wastewater Treatment: Biological and Chemical Processes. Springer, New York, NY.

Ibekwe, A. M., Grieve, C. M. & Lyon, S. R. 2003. Characterization of microbial communities and composition in constructed dairy wetland wastewater effluent. *Applied and Environmental Microbiology*, **69**, 5060–9.

Ivnitsky, H., Katz, I., Minz, D., Volvovic, G., Shimoni, E., Kesselman, E., Semiat, R. & Dosoretz, C. G. 2007. Bacterial community composition and structure of biofilms developing on nanofiltration membranes applied to wastewater treatment. *Water Research*, **41**, 3924–35.

Kelly, W. J., Attwood, G. T., Cookson, A. L., Henderson, G., Lambie, S. C., Perry, R. & Leahy, S. C. 2013. The Hungate 1000: a catalogue of reference genomes from the rumen microbiome. *Advances in Animal Biosciences*, **4**, 406.

Kloc, J. 2012. The Study of Biological Wastewater Treatment through Biofilm Development on Synthetic Material vs. Membranes. Worcester Polytechnic Institute, Worcester, MA.

Kushwaha, J. P., Srivastava, V. C. & Mall, I. D. 2011. An overview of various technologies for the treatment of dairy wastewaters. *Critical Reviews in Food Science and Nutrition*, **51**, 442–52.

Liu, H. & Huang, G. 2009. Laboratory experiment on drip emitter clogging with fresh water and treated sewage effluent. *Agricultural Water Management*, **96**, 745–56.

Lydmark, P., Almstrand, R., Samuelsson, K., Mattsson, A., Sorensson, F., Lindgren, P. E. & Hermansson, M. 2007. Effects of environmental conditions on the nitrifying population dynamics in a pilot wastewater treatment plant. *Environmental Microbiology*, **9**, 2220–33.

Madigan, M. T. 2005. Brock Biology of Microorganisms, 11th edn. Prentice Hall, Upper Saddle River, NJ.

McLean, J. S., Ona, O. N. & Majors, P. D. 2008. Correlated biofilm imaging, transport and metabolism measurements via combined nuclear magnetic resonance and confocal microscopy. *Isme Journal*, **2**, 121–31.

MFE. 2010. New Zealand's Greenhouse Gas Inventory 1990–2008. Ministry for the Environment, New Zealand.

Mosier, A. R., Duxbury, J. M., Freney, J. R., Heinemeyer, O., Minami, K. & Johnson, D. E. 1998. Mitigating agricultural emissions of methane. *Climatic Change*, **40**, 39–80.

Nakajima, J. & Mishima, I. 2005. Measurement of foam quality of activated sludge in MBR process. *Acta Hydrochimica et Hydrobiologica*, **33**, 232–9.

Oliver, M. M. H., Hewa, G. A. & Pezzaniti, D. 2014. Bio-fouling of subsurface type drip emitters applying reclaimed water under medium soil thermal variation. *Agricultural Water Management*, **133**, 12–23.

Pratt, C., Walcroft, A. S., Tate, K. R., Ross, D. J., Roy, R., Reid, M. H. & Veiga, P. W. 2012. Biofiltration of methane emissions from a dairy farm effluent pond. *Agriculture Ecosystems & Environment*, **152**, 33–9.

Raszka, A., Chorvatova, M. & Wanner, J. 2006. The role and significance of extracellular polymers in activated sludge. Part I: Literature review. *Acta Hydrochemica et Hydrobiologica*, **34**, 411–24.

Ravina, I., Paz, E., Sofer, Z., Marm, A., Schischa, A., Sagi, G., Yechialy, Z. & Lev, Y. 1997. Control of clogging in drip irrigation with stored treated municipal sewage effluent. *Agricultural Water Management*, **32**, 127–37.

Sahin, U., Anapali, O., Donmez, M. F. & Sahin, F. 2005. Biological treatment of clogged emitters in drip irrigation systems. *Journal of Environmental Management*, **76**, 338–41.

Sanz, J. L. & Kochling, T. 2007. Molecular biology techniques used in wastewater treatment: an overview. *Process Biochemistry*, **42**, 119–33.

Schmidt, J. E. & Ahring, B. K. 1997. Granular sludge formation in upflow anaerobic sludge blanket (UASB) reactors. *Biotechnology and Bioengineering*, **53**, 442.

Sheng, G. P., Yu, H. Q. & Li, X. Y. 2006. Stability of sludge flocs under shear conditions: roles of extracellular polymeric substances (EPS). *Biotechnology and Bioengineering*, **93**, 1095–102.

St-Pierre, B. & Wright, A.-D. G. 2013. Metagenomic analysis of methanogen populations in three full-scale mesophilic anaerobic manure digesters operated on dairy farms in Vermont, USA. *Bioresource Technology*, **138**, 277–84.

Syron, E. & Casey, E. 2008. Membrane-aerated biofilms for high rate biotreatment: performance appraisal, engineering principles, scale-up, and development requirements. *Environmental Science and Technology*, **42**, 1833–44.

Teh, K. H., Flint, S., Palmer, J., Andrewes, P., Bremer, P. & Lindsay, D. 2014. Biofilm – an unrecognised source of spoilage enzymes in dairy products? *International Dairy Journal*, **34**, 32–40.

Tian, Y. 2008. Behaviour of bacterial extracellular polymeric substances from activated sludge: a review. *International Journal of Environment and Pollution*, **32**, 78–89.

Tsuneda, S., Ejiri, Y., Nagano, T. & Hirata, A. 2004. Formation mechanism of nitrifying granules observed in an aerobic upflow fluidized bed (AUFB) reactor. *Water Science and Technology*, **49**, 27–34.

Walker, J. & Wilkinson, G. W. 2006. The treatment of industrial effluents using the deep shaft process. *Annals of the New York Academy of Sciences*, **326**, 181–91.

Wagner, M. & Loy, A. 2002. Bacterial community composition and function in sewage treatment systems. *Current Opinion in Biotechnology*, **13**, 218–27.

Weber, S. D., Ludwig, W., Schleifer, K. H. & Fried, J. 2007. Microbial composition and structure of aerobic granular sewage biofilms. *Applied and Environmental Microbiology*, **73**, 6233–40.

Weiland, P. 2010. Biogas production: current state and perspectives. *Applied Microbiology and Biotechnology*, **85**, 849–60.

Xia, S., Li, J. & Wang, R. 2008. Nitrogen removal performance and microbial community structure dynamics response to carbon nitrogen ratio in a compact suspended carrier biofilm reactor. *Ecological Engineering*, **32**, 256–62.

Yan, D., Bai, Z., Rowan, M., Gu, L., Ren, S. & Yang, P. 2009. Biofilm structure and its influence on clogging in drip irrigation emitters distributing reclaimed wastewater. *Journal of Environmental Science*, **21**, 834–41.

11 Biofilm Modelling

John Brooks[1], Isabel Li[1] and Robin Hankin[2]

[1]School of Applied Sciences, Faculty of Health and Environmental Sciences, Auckland University of Technology, Auckland, New Zealand

[2]School of Computer and Mathematical Sciences, Auckland University of Technology, Auckland, New Zealand

11.1 Introduction

As we have seen elsewhere in this book, microbial biofilms cause a number of problems in the dairy industry. Understanding the interactions of bacteria with the environment – the physical properties of the substratum, the characteristics of the fluid (such as pH, temperature and availability of nutrients) and the intrinsic properties of bacteria, yeasts and moulds – is the first step in predicting the development of biofilms in a dairy manufacturing plant. This information can be used to build mathematical models of bacterial attachment to surfaces, biofilm development and the behaviour of biofilms in response to processing variables.

This chapter introduces different types of models and their features, then presents a case study of mathematical modelling of typical thermophilic dairy biofilms grown in a laboratory reactor.

11.2 What is a model?

Modern food processing plants are often very complex, being constructed of many different functional components. Each of these components carries out a unit operation on the material being processed. A unit operation is a basic step that involves some sort of physical change or chemical transformation. The process can therefore be broken down into a series of unit operations that are common to many different plants, such as separation, evaporation, crystallisation and drying. By studying the individual unit operations, we can build up an understanding of the whole process.

A model is simply a representation of the system. This may take the form of a physical, small-scale representation of the components of the process as an aid during the design

Biofilms in the Dairy Industry, First Edition.
Edited by Koon Hoong Teh, Steve Flint, John Brooks and Geoff Knight.

phase in laying out the equipment in the plant, or it may be a mathematical representation of the various unit operations, which can be combined to describe the whole process. The mathematical model is the subject of this chapter.

Mathematical models may be empirical correlations representing one aspect of the biofilm, such as substrate utilisation with respect to mass transport into the biofilm, or they may be very complex mechanistic models, attempting to relate many factors to the nonuniform three-dimensional development of the biofilm. Single-species biofilms are obviously far less complicated than multispecies biofilms, although the latter are more usual in nature. The biofilm wastewater treatment system is an example of an extremely complex environment, in which multispecies biofilms grow on multisubstrate feeds and the biofilm is very thick, resulting in very different conditions at the biofilm–liquid surface and at the biofilm–substratum boundary.

In all models, particularly less complex ones, many assumptions and simplifications have to be made. The more detailed models dealing with two- and three-dimensional biofilm systems attempt to reduce the number of assumptions by including terms to describe such things as rates of diffusion, nutrient concentration in the bulk phase and concentration gradients in the biofilms. Unfortunately, as models become more detailed, the computing power required to run them also increases.

There is a significant body of literature on biofilm modelling. Biofilms have been used in water treatment for over a century, but only since the 1980s have biofilm processes been studied (Beg & Chaudhry, 1999). The development of mathematical models to describe biofilm processes has followed a parallel path.

11.3 Why construct a model?

If we can represent bacterial growth and destruction with mathematical expressions, we have a tool to optimise processing conditions in order to minimise the impact of biofilms on the manufacturing process and the finished product. But more than this, we have a tool to test our understanding of the way that biofilms develop. These two objectives require rather different approaches.

The selection of the particular model depends upon the application – it may be that the objective is to study the relationship between temperature and nutrient concentration in the growth of the biofilm, or the objective may be highly applied, such as predicting the available runtime of a milk evaporator before the limit for thermophilic spores in the finished product is reached.

The International Water Association (IWA) Task Group on Biofilm Modelling has produced a very useful and comprehensive monograph on the subject (Wanner *et al.*, 2006). Wanner *et al.* (2006) proposed five potential goals for biofilm modelling:

- to understand fundamental mechanisms;
- to link different types of mechanisms;
- to premodel experimental designs;
- to create novel process designs;
- to improve the performance of a process.

In many cases, the modelling can reduce the cost of experiments by permitting the performance of initial experiments *in silico* to identify the most promising conditions for physical experiments. However, it is important to understand that models are only as good as our understanding of the system. If this understanding is flawed, our models may predict the behaviour of the system incorrectly.

11.4 Types of model available

Most food scientists and microbiologists will be familiar with mathematical techniques to predict spoilage or development of unsafe food. These models fall broadly into four categories: probabilistic, kinetic, analytical and numerical.

11.4.1 *Probabilistic models*

Probabilistic models provide a quantitative estimate of the likelihood of a particular microbiological outcome occurring in a given time, such as the probability of toxin formation by *Clostridium botulinum* in a particular food within a given time. This probability is often described by a regression equation incorporating a number of terms representing factors that might alter the probability of toxin production. This equation can be used to produce a response surface that allows visualisation of the combined effects of factors on the event being modelled. This ability might be very useful in predicting the development of a biofilm in a piece of equipment within a particular runtime. Unfortunately, probabilistic models provide little information on *rates* of change, although the response surface can indicate whether the factors interact or are independent.

11.4.2 *Kinetic models*

In kinetic models, a mathematical function is fitted to the response variable; for example, an equation that describes the growth curve is fitted to the experimentally derived growth rate. This is a 'logistic equation' (Vandermeer, 2010; Weisstein, 2013), a term apparently first used by Verhulst (1845). Examples of kinetic models are the Gompertz equation (Zwietering *et al.*, 1990), the parameters of which can be applied to the lag time and growth rate of microbial populations; the Arrhenius equation (Del Mundo *et al.*, 2014) and its variants, such as the Ratkowsky square root model (Ratkowsky *et al.*, 1983); and the more general Bělehrádek power function (Robertson, 1998).

Kinetic models are capable of modelling the various phases of microbial growth, such as the lag phase, the exponential phase and so on, which is important because the response of the population in individual phases of microbial growth may vary depending on the conditions. Once the important factors have been determined, response surface methodology can be used to evaluate several factors simultaneously, using factorial designed experiments, reducing the number of experiments required and yielding information on interaction of factors.

Many models have mathematical parameters (*a*, *b*, *c* etc.) with no biological meaning, and it can be difficult to interpret them. The equations include experimentally derived constants, such as those found in the Gompertz equation:

$$y = a\,exp\left[-exp\left(b - ct\right)\right] \tag{11.1}$$

where *y* is relative bacterial concentration, $\frac{lnN}{lnNo}$, *a*, *b* and *c* are constants and *t* is time.

Zwietering *et al.* (1990) showed how such equations can be reparameterised by substituting the mathematical parameters with lag time (λ), maximum specific growth rate (μ_m) and the asymptote (A). Zwietering *et al.* (1991) demonstrated how this and various other equations could be used to predict experimentally determined numbers of microorganisms as a function of temperature and time.

Wanner *et al.* (2006) have provided a summary of the various types of model that can be applied to biofilm systems. It is not appropriate to go into detail here: the reader is referred to their monograph.

The conservation of mass is one of the most important principles of any quantitative system. The mass balance can be expressed as:

Net rate of accumulation of mass of component in the system = Rate of mass influx of component to the system + Net rate of generation of the component in the system − Rate of mass efflux of component from the system (11.2)

Most of the following models are derived from the principal mass balance. The differences between these models are in the number of assumptions and computations and in their flexibility.

11.4.3 *Analytical models*

Analytical models are widely used to model general biofilm systems. They do not require a high level of mathematical knowledge and they can be solved using mathematical derivation, omitting the need for numerical techniques. Each term and its effects can be analysed directly and separately. The analytical model can be used only for a simple biofilm system with simple conditions, such as a biofilm that is homogeneous and has only one rate-limiting substrate. Zero-order kinetics is assumed where the concentration of the limiting substrate in the bulk fluid is higher than the half saturation concentration (K_s), whereas first-order kinetics is used where the substrate concentration is below K_s (Wanner *et al.*, 2006).

Pseudoanalytical models

A pseudoanalytical model is a less complex alternative to an analytical model, in which assumptions are made to simplify the conditions and maintain the robustness and predictive power of the model. A simplified form of pseudoanalytical model can be designed only for a highly specific system with many known parameters, such as biofilm thickness, whether the system is single-substrate limited and whether the kinetics is first-order or zero-order. As this model is simpler than an analytical model, the algebraic equations within it are solvable by hand or using a spreadsheet.

Such a model might be a mass balance on a reactor system using the concentration of the bulk as input and concentration in the outflow as output, based on the principle of conservation of mass (Rittmann & Sáez, 2004).

The basic pseudoanalytical model applies only to a specific system, such as a steady-state biofilm with one microbial species and one rate-limiting substrate. However, it can be adapted to fit a multispecies environment, which makes the multispecies system model more accessible to nonspecialist modellers. It illustrates the important interactions between different materials or biomasses in a multispecies biofilm system, such as the *Geobacillus*/*Anoxybacillus* biofilms in the dairy industry.

11.4.4 *Numerical models*

Numerical one-dimensional dynamic models

A dynamic model is often used if the prediction concerns how a biofilm forms and develops over time in one dimension perpendicular to the substratum. This model is normally applied to a more complex system, such as a multispecies biofilm in a multisubstrate environment, and can be used to study the biofilm formation process, the microbial composition of the system and the impact of detachment processes on the biofilm.

Because of the one-dimensional nature of the model, with gradients of variables perpendicular to the substratum, local prediction may not hold true when applied to the whole system. For example, if a microcolony of one species utilises the product of an adjacent second species, a local concentration gradient parallel to the substrate will result. Another limitation is that this model considers the bulk liquid to be a fully mixed homogeneous environment without any clusters of microbes, sediments, lumps or other particulates (such as protein).

Numerical one-dimensional steady-state models

Analytical and pseudoanalytical models involve expression of key processes and variables using linear algebraic expressions to approximate nonlinear equations. Using a computer to apply numerical methods, it is possible to solve nonlinear equations, at least approximately. Compared with the one-dimensional dynamic model, the numerical one-dimensional steady-state model excludes consideration of the dynamic development of the biofilm with time. Because of this, the assumptions need to be carefully selected.

Multidimensional numerical models

This type of model can simulate the heterogeneity of complex biofilm systems. Whether biofilm structure is heterogeneous or homogeneous depends on how it forms and the environment within it (Picioreanu *et al.*, 1998). A multidimensional numerical model can provide some insight into the details of this process, such as how bacteria interact with each other and with the substratum and how they form different biofilms (Picioreanu & Van Loosdrecht, 2002).

Some assumptions in the less complicated analytical models and one-dimensional models no longer hold true and cannot be applied in multidimensional models. Instead, in many cases, the biofilm is assumed to be a uniform structure with cells evenly distributed in the microcolony (Picioreanu *et al.*, 2000). This model is much more complicated and realistic than any of the previously mentioned ones, with far fewer idealised and simplified assumptions. In the real world, a biofilm is a three-dimensional structure. The multidimensional model allows us to simulate more realistic situations, such as flows into and out of the biofilm (Picioreanu *et al.*, 2000).

The crucial limitation of this type of model for application in industry is the complexity of the mathematical equations. Such models have much more freedom, but they are no longer solvable by hand or with simple computer skills. Models that attempt to describe, from first principles, all stages of microbial biofilm growth are exceedingly complex. Solving multidimensional models requires heavy computing power, and the useful information gained from them may still be limited.

Some aspects of biofilm growth, such as the attachment of bacteria to surfaces, have been studied extensively, and research on the modelling and prediction of biofilm structure has been conducted by Picioreanu *et al.* (2000).

11.5 Modelling dairy biofilms

According to the IWA Task Group, there are six steps in designing and using a mathematical model for biofilms (Wanner *et al.*, 2006):

1. Identification of important processes and variables existing in the system.
2. Expression of the identified processes in mathematical terms.
3. Implementation of mass, energy or momentum balances to combine the mathematical expressions.
4. Assignation of appropriate values (such as values according to literature or experiments) to the mathematical terms in the modelling equations.
5. Solution of the mathematical equations using suitable techniques, ranging from simple spreadsheets to numerical methods.
6. Description of the properties of the system, represented by the model's variables.

To this list might be added the most important consideration: 'What is the purpose of this model? How will it be used?'

After identifying variables and processes, the appropriate mass balance needs to be determined, together with expressions for each variable within it. There are two types of parameter: system specific and universal. System-specific parameters, such as biofilm thickness and density, are dependent on the targeted biofilm system and may vary when the system varies. Universal parameters, such as kinetic parameters for microbial reactions, are obtained from the literature or from other experiments that are independently conducted on the biofilm being modelled. Universal parameters do not change with the system.

In its simplest form, a biofilm can be described in terms of compartments, as shown in Figure 11.1. This schematic diagram of the process contains two compartments: the bulk

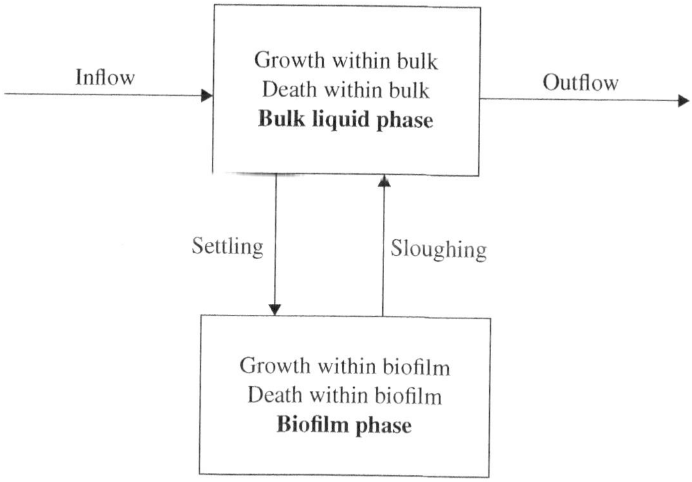

Figure 11.1 Schematic diagram of the processes and compartments of a biofilm in a reactor system.

liquid flowing past the biofilm and the biofilm itself. Strictly, there is another compartment: the boundary liquid layer, which is a very thin layer of liquid formed between the bulk fluid and the biofilm. Particles within the boundary layer have less momentum, due to the reduced flow. In a simplification of this system, the boundary layer is ignored. There is bacterial interchange between the bulk liquid phase and the biofilm phase. This is often called the 'biotransfer potential'. 'Settling' is the movement of bacteria towards and attachment to a substrate or an existing biofilm, while 'sloughing' is detachment from the biofilm – as clumps of cells or pieces of biofilm – and reentry into the bulk phase.

Further simplification can be achieved by assuming that there is no growth within the bulk liquid phase. This is appropriate for dairy systems, as the mean residence time in processing equipment is usually too short to allow an increase in bacterial numbers. There is no death in either the bulk liquid phase or the biofilm phase. The sloughing event is random and is the only process that decreases the bacterial population in the biofilm.

11.6 Example of biofilm modelling

There is a remarkable paucity of published models of biofilm formation in dairy manufacturing plants. Most mathematical models refer to very specific aspects, such as initial attachment or mass transfer through the boundary layer. A notable exception is the work of de Jong *et al* (2002), who developed a mathematical model that describes the contamination of food as a result of attachment, growth and release of bacteria in processing equipment. They studied the effects of fluid flow on the adherence of *Streptococcus thermophilus* and estimated model parameters in a plate heat exchanger. Their model was validated during whey processing in a full-scale cheese plant, predicting the growth of *S. thermophilus*. There was some lack of agreement between measured and predicted outflow levels, but this was thought to be the result of an underestimation of the wall coverage in the plate heat exchanger.

11.6.1 *Model laboratory system*

In our laboratory, we built a small laboratory system to model biofilm formation by *Geobacillus stearothermophilus* in heated dairy processing equipment (Figure 11.2). The system consisted of a reservoir of milk held at 4 °C, a preheating section to bring the milk up to the operating temperature and a small flat hexagonal reactor, in which biofilm growth was modelled. Two preheating pipes and reactors provided duplication to obtain an estimate of reproducibility in the experiments. The hexagonal reactors were placed side by side on a modified thermocycler, which enabled control of the local temperature. The milk was drawn through each reactor by a peristaltic pump and discharged to waste.

In this example, the objective was to model biofilm development by *G. stearothermophilus* in a reactor at a constant temperature. While conducting investigations using this system, it became apparent that biofilm development also occurred in the preheating section. We therefore also modelled biofilm development in the preheating pipe. The compartmental model for the system, which separates the preheating pipe and hexagonal reactor, is shown in Figure 11.3.

11.6.2 *Pipe model*

The preheating section consisted of a narrow-bore silicone tube (hereafter refered to as 'the pipe') submerged in a water bath held at 55 °C. The feed milk contained *G. stearothermophilus* at approximately 10^3 CFU/ml – higher than would be expected in good-quality milk. The bacteria

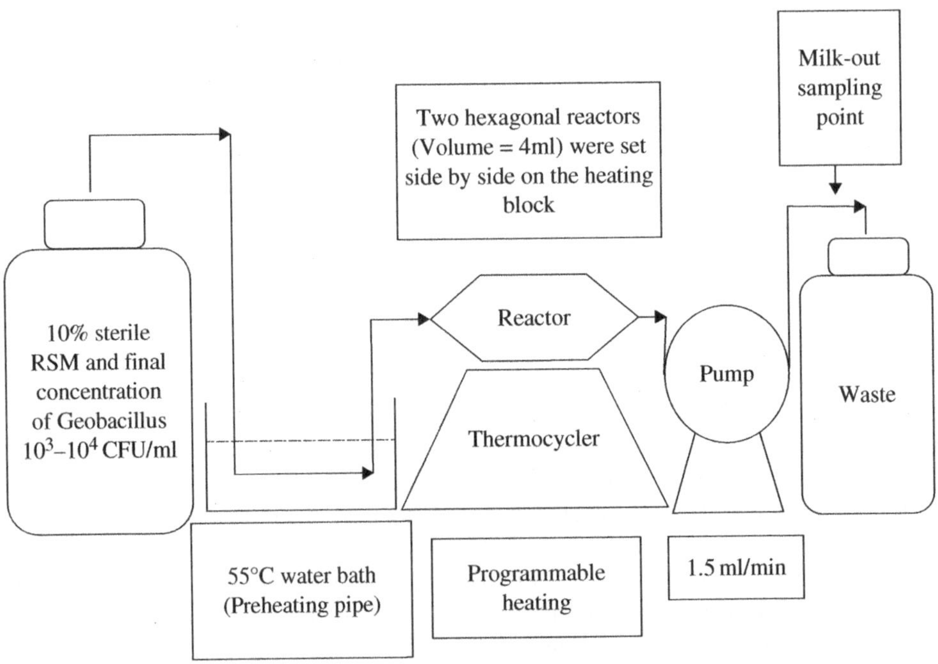

Figure 11.2 Laboratory system developed by Massey University to study biofilm development. Milk, inoculated with *G. stearothermophilus*, was held in a reservoir on ice, drawn through silicone tubing submerged in a 55 °C water bath (preheating pipe) into the two custom-made hexagonal reactors (set up in parallel and placed on top of a modified PCR thermocycler) and pumped into a waste container.

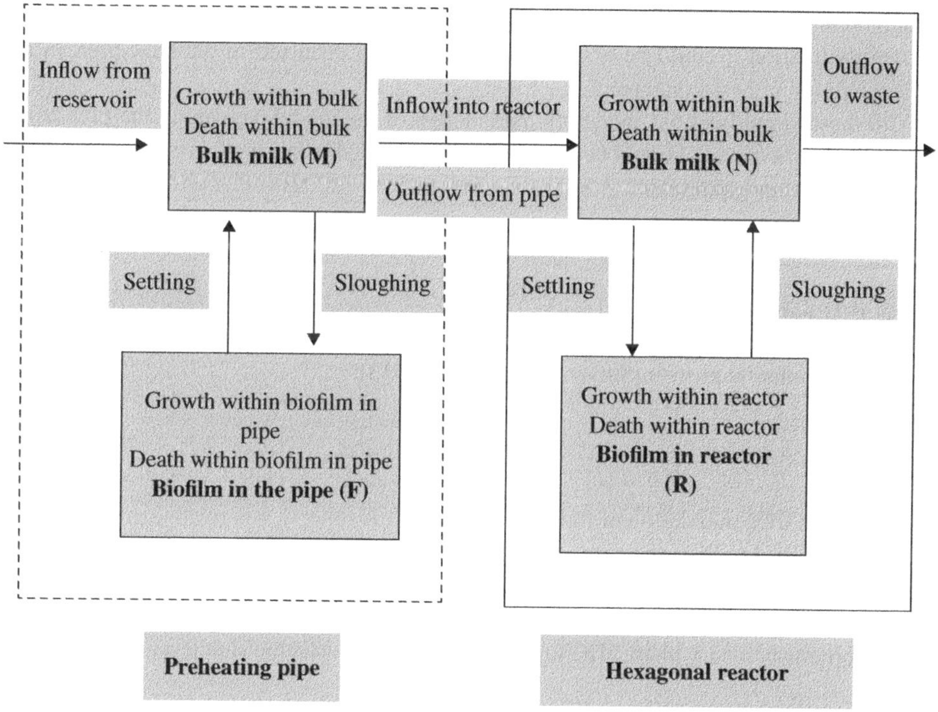

Figure 11.3 Schematic diagram of processes in the preheating section and hexagonal reactor. The dashed-line box is the preheating section and the solid-line box is the reactor section. M is the total number of bacteria in the bulk milk phase in the preheating section; F is the total number of bacteria in the biofilm phase in the preheating section; N is the total number of bacteria in the bulk milk phase in the hexagonal reactor; and R is the total number of bacteria in the biofilm phase in the hexagonal reactor section. All have units of CFU.

entering the preheating section could either remain within the bulk milk phase or settle on to the pipe surface. The settled bacteria could detach (slough off) or grow and form a biofilm, with progeny cells either extending the biofilm or detaching and entering the bulk milk phase. Cells within the bulk milk phase would be washed out of the system in the outflowing milk.

The system started with a known concentration of bacteria in the incoming milk and a sterile reactor. To a certain extent, this represents the situation in a dairy manufacturing plant; that is, the milk entering the manufacturing process will contain low numbers of bacteria and the plant surfaces can be assumed to be clean at the start of a manufacturing run. The measurable factors in the experiment were the concentration of bacteria in the inflowing and outflowing milk and the final counts on the reactor surfaces. A biofilm could develop in the preheating pipe and unattached or detached cells would enter the reactor.

Under normal circumstances in an actual dairy manufacturing plant, death of the bacteria may occur both in the bulk milk phase and in the biofilm. These cells no longer take any part in growth, although the nutrients released from the dead cells within the biofilm may permit some growth in the biofilm that would not otherwise occur. The model could be simplified by assuming that no death occurred in either the bulk or the biofilm phase. The death term in the biofilm compartment could be replaced by the sloughing term; sloughing is a random event. In addition, if the mean residence time in the preheating section and hexagonal reactor were set significantly shorter than the doubling time of the bacteria, to a first approximation it could be assumed that no growth occurred in the bulk milk in the preheating pipe or hexagonal reactor.

With these simplifications made, the parameters of the model describing the preheating section ('the pipe model') could be selected and expressed using mathematical terms, then later organised into equations.

Development of the pipe model

Law of population growth

The key theory behind this model is a modified logistic equation (Weisstein, 2013), which is the law of population growth (Stover & Weisstein, 2013):

$$\frac{dP}{dt} = rP\left(1 - \frac{P}{k}\right) \tag{11.3}$$

where P is the total population of the system (CFU), t is time(s), r is relative growth rate of the population (/s) and k is the carrying capacity of the system (CFU).

Equation 11.3 describes the sigmoidal nature of restricted population growth, in which dP/dt eventually approaches zero. Bacterial growth shows an exponential growth phase and an equilibrium stationary phase. The logistic growth equation is the simplest model to fit this situation:

$$\frac{dP}{dt} = rP - \frac{rP^2}{k} = \text{Growth rate} - \text{Death rate} \tag{11.4}$$

The term rP is the unrestricted growth rate; at some point, the rate of change of the population becomes zero, and the rate of death is equal to the rate of growth.

$$\frac{dP}{dt} = 0 \text{ so, } rP = \frac{rP^2}{k}$$

$$\text{Therefore, } P = k \tag{11.5}$$

In other words, at equilibrium, the population size is equal to the carrying capacity of the system. Therefore, the unit of the carrying capacity is the same as that of the population, which is CFU.

Expression of the law of population growth for biofilms

The only process that decreases the bacterial population of biofilms is sloughing. Bacteria can slough from the biofilm by detaching as single cells or as clumps of biomass. Sloughing is the average rate for both forms of detachment. Thus, the death term in Equation 11.4 is interpreted as the sloughing term. Using F as the population in the biofilm phase, Equation 11.3, describing the change in the biofilm population (rate of change in F), can be rewritten as:

$$\frac{dF}{dt} = \text{Growth within the biofilm} - \text{Sloughing} \tag{11.6}$$

The *e*-folding time is used to describe the exponential growth phenomenon in the logistic growth equation. The *e*-folding time is the time interval in which population increases by a factor of *e* ($e = 2.71$). This term is used as the exponential base analogue of the doubling time, t_D.

Therefore, the growth and sloughing terms can be expressed as:

$$\text{Growth rate} = rF\left(\text{CFU/s}\right) \tag{11.7}$$

$$\text{Sloughing rate} = \frac{rF^2}{k_{pipe}}\left(\text{CFU/s}\right) \tag{11.8}$$

where r is the growth rate within the biofilm based on *e*-folding time (/s) and k_{pipe} is the carrying capacity of the tube wall (CFU).

Parameters of the pipe model

In the preheating section, the response of the model used is the rate of change of the number of bacteria in each of the compartments minus the bulk milk phase or the biofilm phase (Table 11.1). Bacteria enter the system via inflowing milk and leave the system via the outflow. The main processes in the two compartments are growth, settling and sloughing. The variables from the main processes are listed in Table 11.2. There are some constants in the system, referred to as system parameters, which do not change within the experiments (Table 11.3).

Inflow and outflow terms

The milk reservoir remains cold (< 4 °C) for the duration of biofilm development experiments. This temperature is below the minimum growth temperature for *G. stearothermophilus* and, as a result, growth does not occur in the reservoir. Therefore, the inflow to the heating pipe (CFU/s) can be treated as a constant.

$$\mathit{Inflow}\left(\text{CFU/s}\right) = \text{Incoming milk concentration}\left(\text{CFU/ml}\right) \times \mathit{Flowrate}\left(\text{ml/s}\right) \tag{11.9}$$

The rate at which bacteria exit from the preheating pipe in the outflow (CFU/s) is given by:

$$\text{Outflow}\left(\text{CFU/s}\right) = \frac{M\left(\text{CFU}\right) \times \mathit{Flowrate}\left(\text{ml/s}\right)}{\mathit{Volume}_{pipe}\left(\text{ml}\right)} \tag{11.10}$$

Table 11.1 Response of the pipe model.

Notation	Definition	Unit
dF/dt	Rate of change of the total number of bacteria present in the biofilm on the wall of the preheating pipe	CFU/s
dM/dt	Rate of change of the total number of bacteria present in the bulk milk in the preheating pipe	CFU/s

Table 11.2 Notation used for process expressions and parameters in the pipe model.

Notation	Definition	Unit
α	Settling rate of the bacteria based on e – Fold settling time $= \frac{1}{t_{S,pipe}}$	/s
F	Total number of bacteria present in the biofilm on the preheating pipe	CFU
F_0	Initial number of bacteria present in the biofilm on the preheating pipe at time 0	CFU
$Growth_{pipe}$	Rate of change of the number of bacteria present in the biofilm on the preheating pipe	CFU/s
$Inflow_{pipe}$	Rate at which bacteria enter the preheating pipe in the inflow of milk from the reservoir	CFU/s
k	Carrying capacity of the system	CFU
k_{pipe}	Carrying capacity of the preheating pipe wall	CFU
M_t	Total number of bacteria present in the bulk milk in the preheating pipe at any particular time after the start of the run	CFU
M_0	Initial number of bacteria present in the bulk milk in the preheating pipe at time 0	CFU
$Outflow_{pipe}$	Rate at which bacteria exit from the preheating pipe in the outflow of the bulk milk	CFU/s
P	Total population of the system (surface + bulk fluid) in the logistic equation	CFU
r	Relative growth rate of the population in the logistic equation	/s
$Settling_{pipe}$	Rate at which bacteria settle from the bulk milk on to the surface in the preheating pipe	CFU/s
$Sloughing_{pipe}$	Rate at which bacteria slough from the surface into the bulk milk in the preheating pipe	CFU/s
$t_{S,pipe}$	Settling e-folding time of the preheating pipe	s
$t_{G,pipe}$	Growth e-folding time of the preheating pipe	s
t_D	Doubling time of the microorganism	min

Table 11.3 Notation used for system parameters in the pipe model.

Notation	Definition	Unit
$Volume_{pipe}$	Volume of the preheating pipe	cm^3
$Flowrate$	System flowrate	ml/min
$Milk\ conc_{income}$	Concentration bacteria in the incoming milk from the reservoir	CFU/ml

Process equation for bacteria in the bulk milk

For the bulk phase milk in the preheating pipe, the rate of change in the number of bacteria is affected only by the rates of the inflow, outflow, settling and sloughing processes, because there is no growth in the bulk milk. Bacteria enter the pipe via the inflow and exit via the outflow. Bacteria in the bulk milk may settle on the pipe wall and attach to form a biofilm. Bacteria from the biofilm may slough into the bulk milk stream. The rate of change of the total number of bacteria in the bulk milk phase (the rate of change of M) over time is expressed as:

$$\frac{dM}{dt} = \text{Inflow} - \text{Outflow} - \text{Settling} + \text{Sloughing} \tag{11.11}$$

Process equation for bacteria in the biofilm phase

For the biofilm phase, the rate of change of the total number of bacteria is influenced by the rate of the settling and sloughing processes, as well as by the rate of growth within the biofilm. Bacterial numbers in the biofilm increase with time. The rate of change of the total number of bacteria in the biofilm phase (the rate of change of F) over time (CFU/s) is expressed as:

$$\frac{dF}{dt} = \text{Settling} - \text{Sloughing} + \text{Growth within biofilm} \tag{11.12}$$

Growth within biofilm term

The relative growth rate, r, represents the proportional increase in the population in one unit of time. The constant r can be calculated from the growth e-folding time ($t_{G,pipe}$, s).

$$r = \text{Relative growth rate within the biofilm based on } e-\text{folding time} = \frac{1}{t_{G,pipe}} \tag{11.13}$$

Equation 11.14 is used to express mathematically the number of bacteria in a biofilm on the preheating pipe walls (F) at time t with respect to the initial population, F_0:

$$F_t = F_0 \times e^{rt} \tag{11.14}$$

Substituting for r, this gives:

$$F_t = F_0 \times e^{t/t_{G.pipe}} \tag{11.15}$$

After differentiation:

$$\frac{dF}{dt} = F_0 \times \frac{1}{t_{G.pipe}} \times e^{t/t_{G.pipe}} \tag{11.16}$$

After rearrangement:

$$\frac{dF}{dt} = \frac{1}{t_{G.pipe}} \times F_0 \times e^{t/t_{G.pipe}} \tag{11.17}$$

Substituting Equation 11.15 into Equation 11.17 gives:

$$\frac{dF}{dt} = \frac{F_t}{t_{G,pipe}} = \text{Growth within biofilm}\left(\text{CFU/s}\right) \tag{11.18}$$

Settling term

A similar approach can be used for the settling term. The settling e-folding time (t_S) is used. The settlement term describes the process of the decrease in the numbers of bacteria in the bulk milk as they leave to form a biofilm. Therefore, the settlement term in relation to the rate of change in M should be negative:

$$\text{Settlement rate} = -\alpha M \tag{11.19}$$

The *e*-folding settling rate and the *e*-folding settling time are used because:

$$M_t = M_0 \times e^{-\alpha t} \tag{11.20}$$

where M_0 is the initial population.

Substituting for α in Equation 11.20 gives:

$$M_t = M_0 \times e^{-t/t_{S.pipe}} \tag{11.21}$$

After differentiation:

$$\frac{dM}{dt} = M_0 \times \left(-\frac{1}{t_{S.pipe}}\right) \times e^{-t/t_{S.pipe}} \tag{11.22}$$

After rearrangement:

$$\frac{dM}{dt} = \left(-\frac{1}{t_{S.pipe}}\right) \times M_0 \times e^{-t/t_{S.pipe}} \tag{11.23}$$

Substituting Equation 11.21 into Equation 11.23 gives:

$$\frac{dM}{dt}(\text{CFU/s}) = -\frac{M_t}{t_{S,pipe}} \tag{11.24}$$

The rate of settling is negative in relation to the bulk milk phase because it describes bacteria leaving the bulk milk. However, this settlement term becomes positive when it describes the addition of bacteria to the biofilm phase:

$$\frac{dM}{dt}(\text{CFU/s}) = \frac{M_t}{t_{S,pipe}} \tag{11.25}$$

Sloughing term

The sloughing term is derived from Equation 11.8 and is modified by substituting in the *e*-folding growth rate (r) from Equation 11.13:

$$\text{Sloughing} = \frac{F^2}{k_{pipe} \times t_{G,pipe}}(\text{CFU/s}) \tag{11.26}$$

Returning to the two main process equations

Substituting for the inflow, outflow, growth in biofilms, settling and sloughing terms, the two main process equations (Equations 11.11 and 11.12) can be rewritten as:

$$\frac{dM}{dt} = \text{Inflow} - \text{Outflow} - \text{Settling} + \text{Sloughing} =$$

$$\left(Milk\ conc_{inflow} \times Flowrate\right) - \left(\frac{M_t \times Flowrate}{Volume_{pipe}}\right) - \left(\frac{M_t}{t_{S,pipe}}\right) + \left(\frac{F^2}{k_{pipe} \times t_{G,pipe}}\right) \tag{11.27}$$

$$\frac{dF}{dt} = \text{Settling} - \text{Sloughing} + \text{Growth} = \left(\frac{M_t}{t_{S,pipe}}\right) - \left(\frac{F^2}{k_{pipe} \times t_{G,pipe}}\right) + \left(\frac{F}{t_{G,pipe}}\right) \quad (11.28)$$

Optimisation of the pipe model

The model needs to be optimised in order to find values for the parameters ($t_{S,pipe}$, $t_{G,pipe}$ and k_{pipe}) that allow it to best fit the measured data. These values can then be used in the model for later prediction. We used the statistical software 'R' (version 2.15.3, Institute of Statistics and Mathematics of Wirtschaftsuniversität Wien, Vienna, Austria) in this part of the study.

Determination of best fit is achieved using the following process:

- The model (Equations 11.27 and 11.28) is rewritten into 'R'.
- After inserting different values for each of the parameters, the output is given as the estimated M and F values for each time interval.
- Data from experiments on the heating pipe are converted into M values using Equation 11.10.
- The differences between the converted M values and the estimated M values from the model are used to calculate 'badness' (lack of fit), according to Equation 11.29:

$$\text{Total badness} = \sum \frac{(\text{Estimation} - \text{Observation})^2}{\text{Estimation}} \quad (11.29)$$

- The badness value is minimised by inserting different values for the three unknown parameters, using an iteration process.

After the optimisation process is completed, the best values for these parameters are: $t_{S,pipe} = 4.853 \times 10^4$ seconds; $t_{G,pipe} = 6.473 \times 10^3$ seconds; and $k_{pipe} = 4.067 \times 10^8$ CFU.

Validation of the pipe model

Parameters generated by the model need to be validated through a comparison with observed data. An example of such a parameter is the specific growth rate. The measured specific growth rates for *G. stearothermophilus*, growing as biofilms on stainless steel coupons in a CDC reactor in 10% solids skim milk, are 1.47/h at 50 °C, 1.60/h at 55 ° C and 1.20/h at 60 °C.

The specific growth rate (/h) and the doubling time (t_D, h) can be calculated from the optimised value of $t_{G,pipe}$. The relationship between t_D and $t_{G,pipe}$ is shown in Equation 11.30. The relationship between the specific growth rate and t_D is shown in Equation 11.31.

$$t_D = t_{G,pipe} \times ln2 \quad (11.30)$$

Using the optimised value of $t_{G,pipe}$ and solving for t_D:

$$\text{Predicted } t_D = 6{,}473 \times ln2 \approx 4486 \text{ seconds} \approx 1.25 \text{ hours}$$

$$\text{Predicted specific growth rate} = \frac{ln2}{t_D} \approx 0.56 / h \quad (11.31)$$

The specific growth rate calculated from the estimated value for $t_{G,pipe}$ using Equations 11.30 and 11.31 is lower than the observed specific growth rate for *G. stearothermophilus* in biofilms measured in a CDC reactor. This lower apparent growth rate is likely to be the result of the temperature gradient along the preheating pipe, some of the pipe wall being below the minimum temperature for growth. Thus, the optimised value of $t_{G,pipe}$ is likely to be an overestimate, which can be refined if the temperature profile in the preheating pipe is known.

The bacterial numbers in the bulk milk phase estimated using the model and calculated from the observed data (the counts for milk at the outflow) for two trials are plotted together in Figure 11.4. The model predicts growth with lag, log and stationary phases of similar lengths to those observed. However, the reduction in bacterial numbers in the lag phase is not predicted, perhaps reflecting rapid initial attachment to the pipe. The observed data are noisy, possibly as a result of the randomness of detachment or sloughing events and the presence of cell clumps in the outflow. This plot demonstrates that the model can successfully estimate the levels of *G. stearothermophilus* exiting from the preheating section and can be used to calculate the inflow of *G. stearothermophilus* for the reactor model.

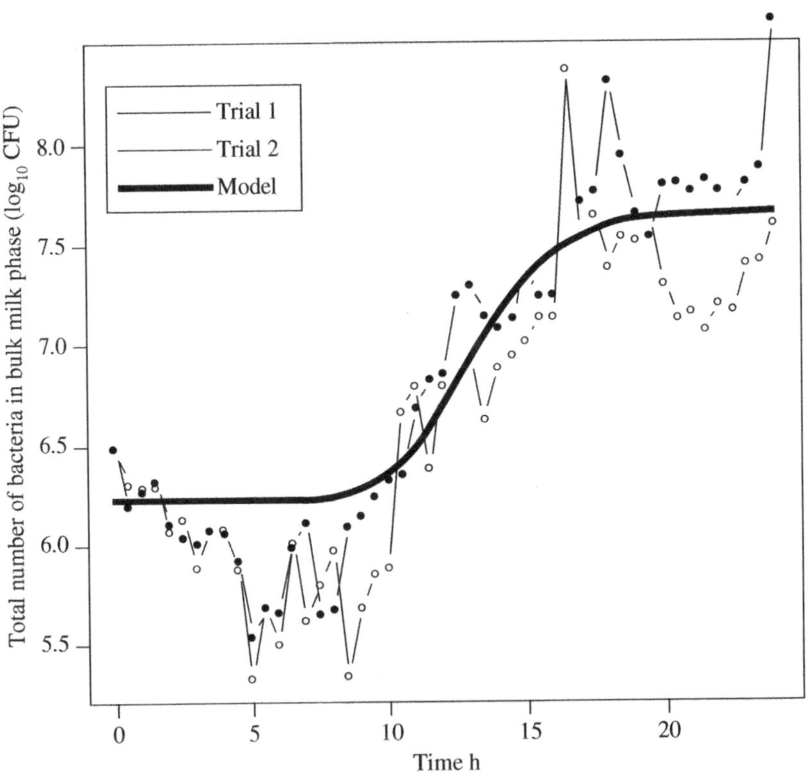

Figure 11.4 Bacterial numbers in the bulk milk phase calculated from the model using the optimised parameters (black line) and observed values for bacterial numbers in the bulk milk phase obtained from two trials (filled and open circles). Observed bacterial numbers in the bulk milk phase (CFU) were calculated from counts (CFU/ml) for milk samples taken at the outlet of the preheating pipe using the formula $\log_{10}$ (Observed bacterial numbers in bulk milk phase) = $\log_{10}$ (Bacterial count determined for bulk milk at outlet × Volume of pipe).

11.6.3 *Reactor model*

The pipe model can be used to estimate the number of *G. stearothermophilus* in the bulk milk phase in the preheating section. The output from the pipe model, the M values, are used to feed into the reactor model to take account of bacterial growth in the preheating pipe before the milk enters the reactor system.

Parameters for the reactor model

As in the preheating section, the model response used is the rate of change of the number of bacteria in each of the compartments: the bulk milk phase or the biofilm phase (Table 11.4). Bacteria enter the system via the inflowing milk from the preheater pipe and leave the system via the outflow. The main processes in the two compartments are growth, settling and sloughing. The variables from the main processes are listed in Table 11.5. There are some constants in the system, referred to as 'system parameters', which do not change within the experiments (Table 11.6).

Assumptions of the reactor model

The approach and parameters used for the reactor model are very similar to those for the pipe model. All assumptions in the pipe model still hold. However, there are more assumptions that need to be considered in the reactor system, such as the carrying capacity of stainless steel compared with silicone tubing and the velocity with which bacteria settle on to the reactor surface.

Bacterial attachment to the stainless steel of the hexagonal reactor and the silicone tubing of the preheating section are assumed to be similar. Therefore, the carrying capacity of the pipe, k_{pipe}, is used to calculate the carrying capacity of the hexagonal reactor ($k_{reactor}$) (Equations 11.32–11.34). It is also assumed that the carrying capacity does not change with temperature.

$$k_{pipe} = \text{Carrying capacity per unit area} \times Area_{pipe}\,(\text{CFU}) \tag{11.32}$$

$$k_{reactor} = \text{Carrying capacity per unit area} \times Area_{reactor}\,(\text{CFU}) \tag{11.33}$$

The carrying capacity of the reactor is rewritten using the known k_{pipe} term:

$$k_{reactor} = \frac{k_{pipe} \times Area_{reactor}}{Area_{pipe}}\,(\text{CFU}) \tag{11.34}$$

Table 11.4 Responses for the reactor model.

Notation	Definition	Unit
dN/dt	Rate of change of the total number of bacteria in the bulk milk in the hexagonal reactor	CFU/s
dR/dt	Rate of change of the total number of bacteria present in the biofilm on the walls of the hexagonal reactor	CFU/s

Table 11.5 Notation for process expressions and parameters in the reactor model.

Notation	Definition	Unit
x	Distance travelled by the bacteria before attachment	cm
t_{set}	Time taken for the bacteria to travel that distance	s
$Growth_{reactor}$	Rate of change of the number of bacteria present in the biofilm on the walls of the reactor	CFU/s
$Inflow_{reactor}$	Rate of change of the number of bacteria present in the inflow to the hexagonal reactor	CFU/s
$k_{reactor}$	Carrying capacity of the hexagonal reactor	CFU
N_t	Total number of bacteria present in the bulk milk in the hexagonal reactor at any particular time	CFU
N_0	Initial number of bacteria present in the bulk milk in the reactor at time 0	CFU
$Outflow_{reactor}$	Rate at which bacteria exit from the hexagonal reactor in the outflow of the bulk milk	CFU/s
R	Total number of bacteria present in the biofilm on the hexagonal reactor surface	CFU
r	Relative growth rate of the population in the logistic equation	/h
$Settling_{reactor}$	Rate at which bacteria settle from the bulk milk on to the surface in the hexagonal reactor	CFU/s
$Sloughing_{reactor}$	Rate at which bacteria slough from the surface into the bulk milk in the hexagonal reactor	CFU/s
$t_{S,reactor}$	Settling e-folding time of the hexagonal reactor	s
$t_{G,reactor}$	Growth e-folding time of the hexagonal reactor	s
t_D	Doubling time of the microorganism	min
$Velocity$	Specific settling velocity	cm/s

Table 11.6 System parameters.

Notation	Definition	Unit
$Area_{reactor}$	Surface area of the hexagonal reactor	cm^2
$Flowrate$	System flowrate	ml/min
$Volume_{pipe}$	Volume of the preheating tube	cm^3
$Volume_{reactor}$	Volume of the hexagonal reactor	cm^3

Settling term

It is assumed that the bacterial settlement and attachment process does not involve acceleration, as described in Figure 11.5. The distance that bacteria travel before reaching the surface is δx and time taken is δt. Both values are assumed to be very small. N represents the number of bacteria in the bulk phase in the reactor (CFU). The volume and the area are properties of the system; that is, of the preheating pipe or the reactor. The number of bacteria settling is proportional to the number of bacteria in the bulk phase.

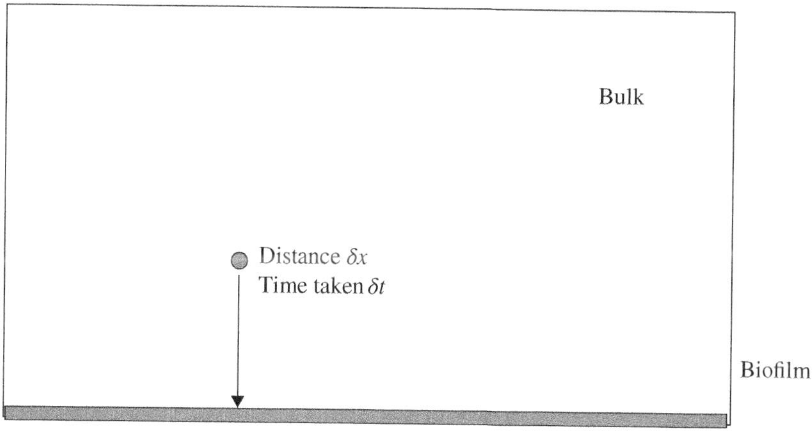

Figure 11.5 Schematic diagram of bacteria settling from the bulk phase on to the surface. The dot represents a single bacterium.

Settling decreases the number of bacteria in the bulk phase. From the bulk phase point of view, the settling term is negative:

$$\text{Bacteria settled} = -\frac{\delta x \times Area_{reactor} \times N_t}{Volume_{reactor}} \tag{11.35}$$

The settling velocity is defined by the distance and the time taken to travel that distance. Therefore, this relationship is written as:

$$\text{Distance } \delta x = Velocity \times \delta t_{Set} \tag{11.36}$$

After replacing the distance term in Equation 11.35:

$$\text{Bacteria settled} = -\frac{\delta t_{Set} \times Velocity \times Area_{reactor} \times N_t}{Volume_{reactor}} \tag{11.37}$$

Over the period δt_{Set}, which is very small, N_t can be regarded as constant.

After rearrangement:

$$\text{Bacteria settled} = -N_t \times \left(\frac{Area_{reactor}}{Volume_{reactor}}\right) \times Velocity \times \delta t_{Set} \tag{11.38}$$

After differentiation with respect to time:

$$\frac{dN}{dt} = -N_t \times \left(\frac{Area_{reactor}}{Volume_{reactor}}\right) \times Velocity \tag{11.39}$$

The term dN/dt is the rate of depletion of bacteria from the bulk milk. The settling term can also be described using the settling *e*-folding time t_S, as in the previous pipe model. Switching N for M in Equation 11.24, we get:

$$\frac{dN}{dt} = \left(-\frac{1}{t_S}\right) \times N_t \tag{11.40}$$

These two expressions (Equations 11.39 and 11.40) can be linked to establish the relationship between velocity and the settling *e*-folding time, t_S:

$$-N_t \times \left(\frac{Area}{Volume}\right) \times Velocity = \left(-\frac{1}{t_S}\right) \times N_t \tag{11.41}$$

After cancelling terms:

$$\left(\frac{Area}{Volume}\right) \times Velocity = \frac{1}{t_S} \tag{11.42}$$

After rearrangement:

$$Velocity = \frac{Volume}{Area \times t_S} (\mathrm{cm/s}) \tag{11.43}$$

$$t_S = \frac{Volume/Area}{Velocity} (s) \tag{11.44}$$

The volume and area parameters are specific for the system. Therefore:

$$t_{S,pipe} = \frac{Volume_{pipe}/Area_{pipe}}{Velocity} (\mathrm{s}) \tag{11.45}$$

$$t_{S,reactor} = \frac{Volume_{reactor}/Area_{reactor}}{Velocity} (\mathrm{s}) \tag{11.46}$$

Development of the reactor model

The basis for this model is to use a similar approach and a similar logistic equation as in the pipe model.

Process equation for bacteria in the bulk milk and biofilm phases

The rate of change of the total number of bacteria in the bulk milk phase (the rate of change of N) can be expressed as the net result of the rates of all processes in the schematic diagram in Figure 11.3, with units for all terms of CFU/s:

$$\frac{dN}{dt} = Inflow_{reactor} - Outflow_{reactor} - Settling_{reactor} + Sloughing_{reactor} \tag{11.47}$$

The rate of change of the total number of bacteria in the biofilm phase in the hexagonal reactor (the rate of change of R) can be expressed, with units for all terms of CFU/s, as:

$$\frac{dR}{dt} = Settling_{reactor} - Sloughing_{reactor} + Growth_{reactor} \tag{11.48}$$

Inflow and outflow terms

The inflow to the reactor is the outflow from the preheating section, which is determined using Equation 11.10 from the pipe model:

$$Inflow_{reactor}(\text{CFU/s}) = Outflow_{pipe}(\text{CFU/s}) = \frac{M_t \times Flowrate}{Volume_{pipe}} \tag{11.49}$$

The rate at which bacteria exit from the reactor in the outflow is related to the total number of bacteria present in the bulk milk phase, N_t, and the volume of the reactor:

$$Outflow_{reactor}(\text{CFU/s}) = \frac{N_t \times Flowrate}{Volume_{reactor}} \tag{11.50}$$

Settling term

From the assumptions for the pipe model, it is known that the process of settling on to surfaces is related to the settling e-folding time. Substituting N for M in Equation 11.24, we get:

$$Settling = \frac{dN}{dt} = -\frac{N_t}{t_{S,reactor}} = (\text{CFU/s}) \tag{11.51}$$

Combining Equations 11.44 and 11.51 gives:

$$Settling_{reactor} = -\frac{N_t}{t_{S,reactor}} = -\frac{Velocity \times Area_{reactor} \times N_t}{Volume_{reactor}}(\text{CFU/s}) \tag{11.52}$$

Sloughing term

The sloughing term from the pipe model (Equation 11.26), modified for the hexagonal reactor, becomes:

$$Sloughing_{reactor} = \frac{R^2}{k_{reactor} \times t_{G,reactor}}(\text{CFU/s}) \tag{11.53}$$

Using the relationship between $k_{reactor}$ and k_{pipe} in Equation 11.34, we get:

$$Sloughing_{reactor} = \frac{R^2 \times Area_{pipe}}{k_{pipe} \times Area_{reactor} \times t_{G,reactor}}(\text{CFU/s}) \tag{11.54}$$

Growth-within-biofilm term

The growth term for bacteria in the pipe model (Equation 11.18) can be modified for the reactor model and is expressed as:

$$Growth_{reactor}(\text{CFU/s}) = \frac{R}{t_{G,reactor}} \tag{11.55}$$

The value for $t_{G,reactor}$ in the growth term is calculated within 'R' using a function that describes the relationship between growth rate and temperature.

Returning to the two main process equations

Substituting for the inflow, outflow, growth-within-biofilm, settling and sloughing terms, the two main process equations (Equations 11.47 and 11.48) become:

$$\frac{dN}{dt} = Inflow_{reactor} - Outflow_{reactor} - Settling_{reactor} + Sloughing_{reactor}$$
$$= \left(\frac{MM_t \times Flowrate}{Volume_{pipe}}\right) - \left(\frac{NN_t \times Flowrate}{Volume_{reactor}}\right) - \left(\frac{Velocity \times Area_{reactor} \times NN_t}{Volume_{reactor}}\right) + \left(\frac{R^2 \times Area_{pipe}}{k_{pipe} \times Area_{reactor} \times t_{G,reactor}}\right) \quad (11.56)$$

$$\frac{dR}{dt} = Settling_{reactor} - Sloughing_{reactor} + Growth_{reactor}$$
$$= \left(\frac{Velocity \times Area_{reactor} \times N_t}{Volume_{reactor}}\right) - \left(\frac{R^2 \times Area_{pipe}}{k_{pipe} \times Area_{reactor} \times t_{G,reactor}}\right) + \left(\frac{R}{t_{G,reactor}}\right) \quad (11.57)$$

Application of the reactor model

Knight *et al.* (2004) were the first to report the effect of temperature cycling on the control of dairy biofilms by either interruption of exponential growth or prevention of cell attachment to surfaces. In our study, we used the thermal cycler to shift the temperature of the hexagonal reactor out of the temperature zone for growth on a regular basis to disrupt biofilm formation or exponential growth.

To construct the model for temperature cycling in the reactor, we need to know the relationship between specific growth rate and temperature. A series of experiments was conducted in a CDC biofilm reactor to determine the maximum specific growth rate of *G. stearothermophilus* on stainless steel coupons at various temperatures (Table 11.7).

For sine wave thermocycling, the temperature and time relationship is expressed as in Equation 11.58 to calculate the temperature at any specific given time point during the thermocycling experiments; T_{max} is the maximum temperature and T_{min} the minimum; time is in units of seconds and period is in units of minutes.

$$\text{Temperature at any given time} = \left(\frac{T_{min} + T_{max}}{2}\right) - \left(\frac{T_{max} - T_{min}}{2}\right) \times \cos\left(\frac{2\pi \times time}{period \times 60}\right) \quad (11.58)$$

For square wave thermospiking, the temperature and time relationship is constructed using the modulus of integer division. The total time since start of thermospiking is divided by the

Table 11.7 Maximum specific biofilm growth rate versus temperature for *G. stearothermophilus* in 10% reconstituted skim milk.

Temperature (°C)	Maximum specific biofilm growth rate (/h)
35	1×10^{-6} (no growth observed in 36 hours)[a]
40	0.8465
50	1.4661
60	1.1977
70	1.0520
80	1×10^{-6} (no growth observed in 24 hours)[a]

[a]A very small result (1×10^{-6}/h) was given to the no-growth situations to avoid 'divide by zero' errors in the model.

sum of times spent at high (t_{high}) and low (t_{low}) temperatures; that is, by the period of one cycle. If the remainder is larger than the time at high temperature, then at the given time point, the temperature is low. If the remainder is smaller than the time at high temperature, then at the given time point, the temperature is at the high setpoint. Therefore, using 'R', the preceeding statements can be expressed as true (1) or false (0):

$$\text{if Modulus of}\left\{\text{specific time}/\left(t_{high}+t_{low}\right)\right\}<t_{high},\text{it is true},1 \tag{11.59}$$

$$\text{if Modulus of}\left\{\text{specific time}/\left(t_{high}+t_{low}\right)\right\}>t_{high},\text{it is false},0 \tag{11.60}$$

Then, if the statement is true, the temperature at a specific time after the start of the run is equal to the high temperature (T_{max}, expressed as T_{min} + difference between T_{max} and T_{min}). If the statement is false, the temperature at a specific time is equal to the low temperature (T_{min}, expressed as T_{min} + 0). Thus, the temperature at a specific time is described as follows:

$$\text{Temperature at any given time}=T_{min}+\left(T_{max}-T_{min}\right)\times\left(1\,\text{or}\,0\right) \tag{11.61}$$

An important assumption implicit in this model is that the bacteria respond instantly to a change in temperature with a change in growth rate. This is unlikely to be true and the resulting predictions can be expected to underestimate the effects of thermal cycling.

The model was applied to a square wave temperature-cycled system (preheater pipe and hexagonal reactor) to compare the observed bacterial counts with the predicted bacterial counts in the outflowing bulk milk phase (Figure 11.6).

The logistic equation is used widely to predict populations in ecology and biology. This model provided a reasonable prediction of the observed biofilms growth in this system. It was also a reasonable first approximation for the temperature-cycled system. It predicted the general growth behaviour of the bacteria in the temperature-cycling experiments and the duration of each phase – the lag phase, log phase and stationary phase – of the bacteria in the outflow from the temperature-cycled reactors. It did not predict the initial reduction in bacterial counts in the outflow during the lag phase.

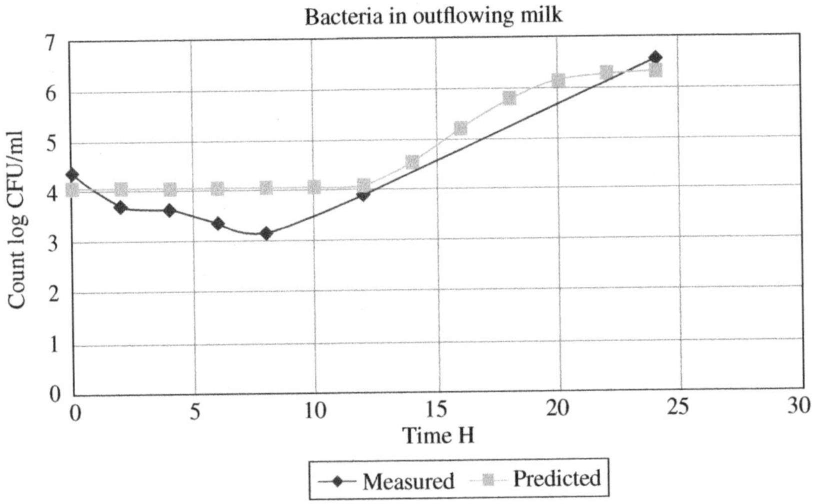

Figure 11.6 Comparison between the estimated bacterial concentration in the outflowing bulk phase and the actual outflowing bacterial concentration (CFU/ml) using the reactor model with square wave thermospiking at 55 °C/15 minutes, 35 °C/35 minutes in 10% RSM.

This model is very simple to use under laboratory conditions. Before it can be used in an actual dairy process, it must be validated using pilot plant trials. The outcome and the assumptions need to be tested to determine whether the predictions of this reactor model still hold.

The model provides a first approximation to modelling the biofilm behaviour of *G. stearothermophilus* in a flowing, heated system. As it was designed based on simulation of simplified and idealised processes, rather than on what actually occurs in the plant, there are some limitations in its application to real manufacturing situations and some validation work is needed. After validation in the pilot plant, the model may be used in manufacturing plants as a predictive tool by changing its parameters. For example, the area and volume values can be changed easily.

11.7 Conclusion

The example described in this chapter shows the principles involved in biofilm modelling and provides two mathematical models tested with experimentally derived data. The pipe model demonstrates the large changes that may arise in the feed to a process when biofilm formation is able to occur in equipment such as plate heat exchangers in preheaters. Logistic theory is the foundation for both the pipe model and the hexagonal reactor model. These two models can estimate biofilm growth with a known level of incoming thermophilic bacteria under constant or cycling temperatures. A rapid microbiological tool, such as flow cytometry, can be used to determine the incoming milk counts within a short period of time. These values can then be used in the model to estimate the potential plant runtime and the quality of the final product, thus providing plant management with a tool to maximise runtime while maintaining levels of *Geobacillus* within specification.

In milk powder manufacturing plants, there are two main sites for the development of biofilms of thermophilic bacteria that cause noticeable increases in the levels of thermophilic vegetative cells and spores in the outflowing product. These are the plate heat exchanger before the evaporators and the first two passes of the evaporators (Scott *et al.*, 2007). The preheating plate heat exchanger before the evaporators is a suitable site for implementation of mathematical modelling, as it is easy to set up and is unlikely to affect the operation of the evaporator. Implementing temperature cycling in evaporator passes 1 and 2 is less feasible because of the impact it would have on the subsequent evaporation process. Another site where temperature cycling may be introduced is the heat exchanger used to heat milk before separation. Modelling of thermophilic biofilm growth at this stage may permit optimisation of the separation of cream and skim milk to minimise the growth of thermophiles.

These models need to be scaled up in pilot plant-scale trials and validated before they can be implemented in manufacturing plants. They can be modified for *Anoxybacillus* species or mixed species to assist in the control of thermophilic bacteria in dairy manufacturing plants.

References

Beg, S. A. & Chaudhry, M. A. S. 1999. A review of mathematical modelling of biofilm pro-cesses: advances in modelling of selected biofilm processes. *International Journal of Environmental Studies*, **56**, 285–313.

De Jong, P., Te Giffel, M. C. & Kiezebrink, E. A. 2002. Prediction of the adherence, growth and release of microorganisms in production chains. *International Journal of Food Microbiology*, **74**, 13–25.

Del Mundo, G., Moussa, K., Chacha, P., Odufalu, D-F., Mudda, G. & Kan, C. F. K. 2014. Arrhenius equation. Available from: http://chemwiki.ucdavis.edu/Physical_Chemistry/Kinetics/Modeling_Reaction_Kinetics/Temperature_Dependence_of_Reaction_Rates/The_Arrhenius_Law/Arrhenius_Equation (last accessed 12 March 2015).

Knight, G. C., Nicol, R. S. & McMeekin, T. A. 2004. Temperature step changes: a novel approach to control biofilms of *Streptococcus thermophilus* in a pilot plant-scale cheese-milk pasteurisation plant. *International Journal of Food Microbiology*, **93**, 305–18.

Picioreanu, C. & Van Loosdrecht, M. C. 2002. A mathematical model for initiation of microbiologically influenced corrosion by differential aeration. *Journal of the Electrochemical Society*, **149**, B211–23.

Picioreanu, C., Van Loosdrecht, M. C. M. & Heijnen, J. J. 1998. Mathematical modeling of biofilm structure with a hybrid differential-discrete cellular automaton approach. *Biotechnology and Bioengineering*, **58**, 101–16.

Picioreanu, C., Van Loosdrecht, M. C. M. & Heijnen, J. J. 2000. Modelling and predicting biofilm structure, In: Allison, D. G., Gilbert, P., Lappin-Scott, H. M. & Wilson M. (eds) Community Structure and Co-operation in Biofilms. Cambridge University Press, Cambridge, pp. 129–66.

Ratkowsky, D. A., Lowry, R. K., McMeekin, T. A., Stokes, A. N. & Chandler, R. E. 1983. Model for bacterial culture growth rate throughout the entire biokinetic temperature tange. *Journal of Bacteriology*, **154**, 1222–6.

Rittmann, B. E. & Sáez, P. B. 2004. Improved pseudoanalytical solution for steady-state biofilm kinetics. *Biotechnology and Bioengineering*, **32**, 379–85.

Robertson, G. L. 1998. Food Packaging: Principles and Practice. CRC Press, Boca Raton, FL.

Scott, S. A., Brooks, J. D., Rakonjac, J., Walker, K. M. R. & Flint, S. H. 2007. The formation of thermophilic spores during the manufacture of whole milk powder. *International Journal of Dairy Technology*, **60**, 109–17.

Stover, C. & Weisstein, E. W. 2013. Population growth. Available from: http://mathworld.wolfram.com/PopulationGrowth.html (last accessed 12 March 2015).

Vandermeer, J. 2010. How populations grow: the exponential and logistic equations. *Nature Education Knowledge Project*, **3**(10), 15. Available from: http://www.nature.com/scitable/knowledge/library/how-populations-grow-the-exponential-and-logistic-13240157 (last accessed 12 March 2015).

Verhulst, P.-F. 1845. Recherches mathématiques sur la loi d'accroissement de la population. *Nouveaux Mémoires de l'Académie Royale des Sciences et Belles-Lettres de Bruxelles*, **18**, 1–42.

Wanner, O., Eberl, H., Morgenroth, E., Noguera, D., Picioreanu, C., Rittman, B. & Van Loosdrecht, M. 2006. Mathematical Modeling of Biofilms: Scientific and Technical Report No. 18. International Water Association, London.

Weisstein, E. W. 2013. Logistic equation. Available from: http://mathworld.wolfram.com/LogisticEquation.html (last accessed 12 March 2015).

Zwietering, M. H., Jongenburger, I., Rombouts, F. M. & Van't Riet, K. 1990. Modeling of the bacterial growth curve. *Applied and Environmental Microbiology*, **56**, 1875–81.

Zwietering, M. H., De Koos, J. T., Hasenack, B. E., De Wit, J. C. & Van't Riet, K. 1991. Modeling of bacterial growth as a function of temperature. *Applied and Environmental Microbiology*, **57**, 1094–101.

12 Biofilm Control in Dairy Manufacturing Plants

Geoff Knight

Food Process Hygiene Solutions, Melbourne, Victoria, Australia

12.1 Introduction

Biofilm development is a concern on both the internal surfaces of processing equipment and on environmental surfaces located within dairy manufacturing plants. Biofilms that develop on internal surfaces, referred to as 'process biofilms', can lead to direct microbial contamination of product streams and result in microbial quality and safety issues for dairy products. Biofilms that develop on environmental surfaces, referred to as 'environmental biofilms', may harbour, and be potential sources of, pathogenic and spoilage bacteria. Although dairy products and product streams do not come into direct contact with environmental surfaces, environmental biofilms may be a source of indirect contamination of dairy products through, for example, the creation and spread of aerosols during cleaning processes.

12.2 Factors that influence growth and survival of bacteria in biofilms

The daily (or cyclic) production schedule employed in dairy manufacturing, comprising a period of production, followed by cleaning and sanitation, followed by a period of production and so on, leads to a cyclic pattern of biofilm development in dairy manufacturing plants (Figure 12.1). This pattern holds for biofilms that develop in processing equipment and on environmental surfaces. Microorganisms have the opportunity to grow during the production period and are inactivated and/or removed by cleaning and sanitation. Microorganisms may survive, grow or be inactivated, depending on the conditions, in the period before production resumes. A range of factors, of relevance for both groups of biofilms, influence how microorganisms grow and survive. Some of the most important are described in this section.

12.2.1 *Temperature*

The local temperature is one of the most influential factors affecting microbial growth. Typical temperature growth profiles for psychrophilic, psychrotrophic, mesophilic and thermophilic microorganisms are shown in Figure 12.2. Key values are the minimum (T_{min}), maximum

Biofilms in the Dairy Industry, First Edition.
Edited by Koon Hoong Teh, Steve Flint, John Brooks and Geoff Knight.

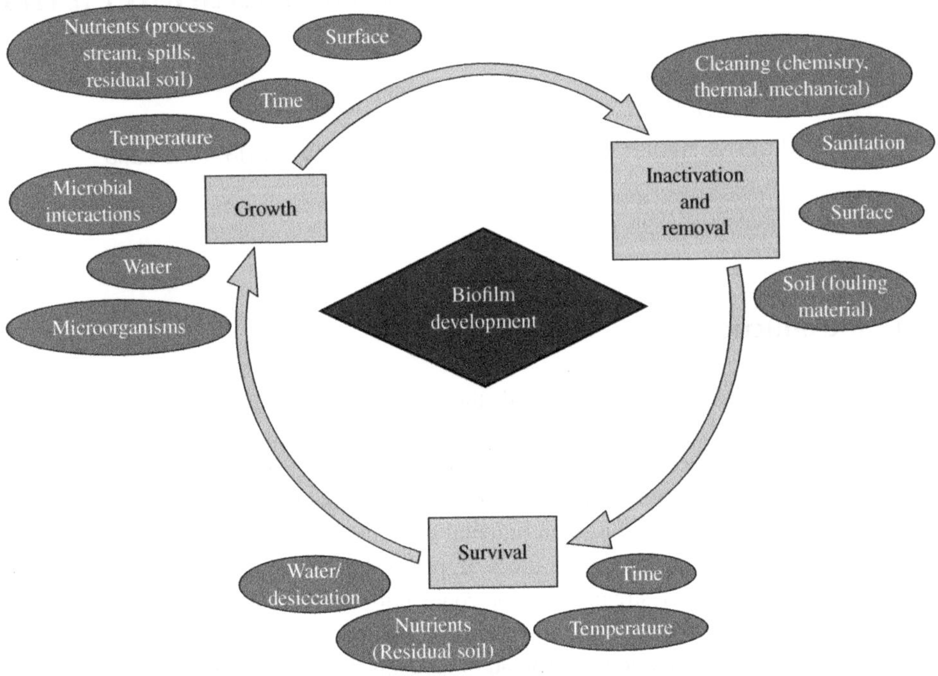

Figure 12.1 Pictorial representation of the cyclic pattern of biofilm development that occurs in dairy manufacturing plants. This includes periods of growth (during production), inactivation and removal (during cleaning and sanitation) and survival (in the time before production resumes). Major factors that influence the growth, inactivation, removal and survival of microorganisms are shown.

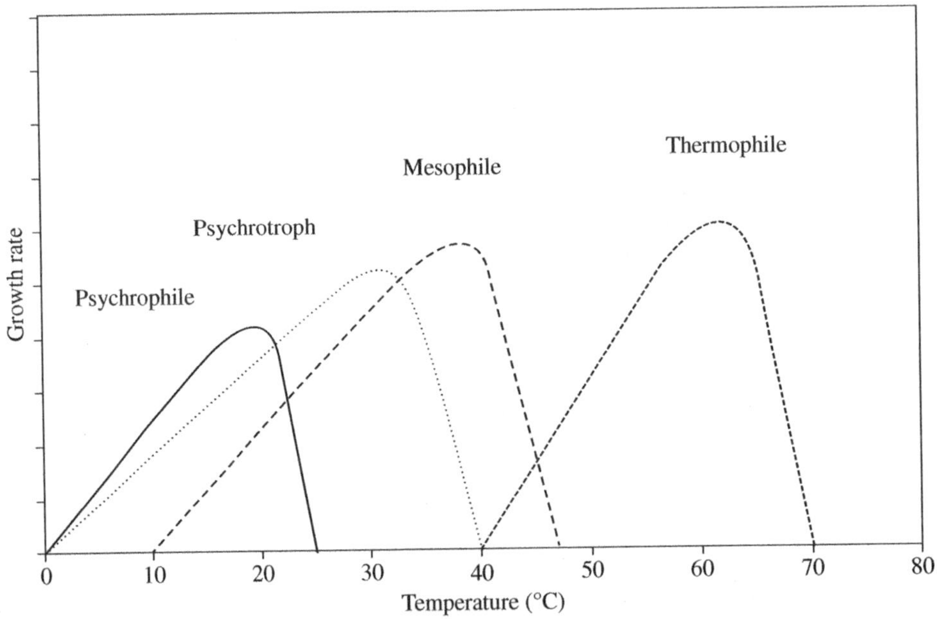

Figure 12.2 Typical temperature growth profiles for psychrophilic, psychrotrophic, mesophilic and thermophilic bacteria.

(T_{max}) and optimum (T_{opt}) growth temperatures. Important bacterial groups for the dairy industry include psychrotrophic bacteria, which can grow when the temperature is <5 °C, mesophilic bacteria, which have optimal growth temperatures between 30 and 40 °C and include many pathogenic bacteria, and thermophilic bacteria, which can grow when the temperature is between 45 and 70 °C.

The temperature throughout the dairy production chain varies from a typical storage temperature for raw milk and refrigerated dairy products of <5 °C to >140 °C for some ultra-high-temperature (UHT) processes. The temperature at each step or in each process along the dairy chain influences the types of bacteria that grow.

Raw milk typically arrives at a dairy manufacturing plant at <5 °C, although it can be higher depending on a country's regulations and whether the milk has been allowed to cool sufficiently on-farm prior to collection by the milk receival tankers. At some dairy manufacturing plants, raw milk is cooled as it is pumped from milk tankers into raw milk silos to ensure it is at <5 °C. In addition to the silos used for storage of raw milk, the transfer lines that transport raw milk to the processing equipment are also typically kept at <5 °C during production. Raw milk and dairy products are typically maintained or stored at refrigeration temperatures (<5 °C) to minimise microbial growth. This also limits the types of microorganism able to grow to psychrophilic and psychrotrophic microorganisms.

The processing of milk invariably includes a thermal treatment step. In Australia, New Zealand and many other countries, the minimum thermal treatment for milk is a pasteurisation step, in which raw milk is thermally treated at a minimum temperature of 72 °C for 15 seconds. In some manufacturing processes, such as for high-heat milk powders and UHT milk products, temperatures in excess of 100 °C are used.

When milk is heated in heat-exchange equipment, to achieve pasteurisation conditions or as part of another manufacturing process, it comes into contact with surfaces at increasing temperatures. For example, in a pasteurisation plant there are surfaces at <5, 10, 20, 30, 40, 50, 60, 70 and 72 °C (and all temperatures in between). Bacteria present in the raw milk, as they pass through the pasteurisation equipment, have the opportunity to attach to and grow on surfaces at temperatures that are favourable for their growth. For example, psychrophilic bacteria can attach and grow on surfaces at <5–25 °C, mesophilic bacteria at 20–45 °C and thermophilic bacteria at 45–70 °C. The growth of many of these bacteria on surfaces on the raw milk side of the pasteurisation equipment is not a major concern for dairy manufacturers, as cells that detach from surfaces and enter the product stream are inactivated when they are exposed to pasteurisation conditions (>72 °C for 15 seconds). Pasteurised milk is cooled back to <5 °C prior to packaging. The heat exchangers on the cooling side of the pasteurisation equipment provide additional surfaces at 72, 70, 60, 50 °C, and so on, which can also support bacterial growth, but only for bacteria in the milk stream that survive exposure to pasteurisation conditions. Similar opportunities for bacterial growth on surfaces arise in other types of heated dairy processing equipment.

The local temperature similarly affects bacterial growth on environmental surfaces in dairy manufacturing plants. The temperature can vary considerably between different locations. Many dairy manufacturing plant locations remain at ambient temperature, which can result in low temperatures in winter and warm temperatures in summer. Product storage and processing areas may also operate under cool or refrigeration conditions. In addition, there can be localised areas around heated equipment where surface temperatures are warm or hot and around refrigerated equipment where temperatures are low.

12.2.2 *Surface materials*

Interactions between bacteria and surfaces are discussed in detail in Chapter 2. Many of the opportunities for the control of biofilm development in dairy manufacturing plants revolve around influencing the ability of bacteria to attach to surfaces and the strength of this attachment.

12.2.3 *Nutrients*

Milk and milk products contain a range of components, including lactose, proteins, fats, minerals, vitamins and a variety of organic molecules, that are a source of nutrients for and can support the growth of a wide range of bacteria. Bacteria that attach to and form biofilms on surfaces in direct contact with milk or milk product streams will be able to access these milk components for growth. Fouling of processing equipment surfaces is an issue for dairy manufacturers, particularly with heated processing equipment. Small amounts of fouling material that remain associated with surfaces following cleaning can be a source of nutrients for bacteria that survive cleaning processes.

The availability of nutrients to microorganisms attached to environmental surfaces in production environments can be variable. Nutrients may be introduced into the factory environment by the transfer of soil from boots, clothing and equipment and through product spills. Small quantities of soil can remain associated with surfaces following cleaning, which may be sufficient to allow microbial growth to occur, particularly if surfaces have not been allowed to dry.

12.2.4 *Water*

Water is essential for microbial growth in foods and on surfaces, and its presence influences survival where growth does not occur. The availability of water in food is measured as water activity and in air this is measured as relative humidity (RH). Most bacteria grow optimally when the water activity is 0.98–0.99, although many can grow when it is as low as 0.92.

The water activity of milk and many other dairy products is sufficiently high to allow bacterial growth to occur. This means that bacteria can grow on processing equipment surfaces that are in contact with milk and many other dairy products. Some manufacturing processes are designed to reduce the water content of dairy products. For example, during the manufacture of skim milk concentrate by evaporation, the water content of the skim milk is reduced from approximately 91 to 55% w/w (corresponding to an increase in the total solids from approximately 9 to 45% w/w). This results in the water activity of skim milk decreasing to a level that is sufficient to prevent bacterial growth.

Water is common within dairy manufacturing environments and may be present either as films on surfaces or as pools that collect in locations with poor drainage. Water may originate from a number of sources. It is used to prepare cleaning solutions and sanitisers for application in dairy manufacturing plants and to rinse dairy products, cleaning solutions and sanitisers from manufacturing equipment and from environmental surfaces. Water may also be present in the air on days of high humidity and can collect on surfaces due to condensation. As a result, it is inevitable that water will be present on surfaces within manufacturing environments.

12.2.5 *Time*

For dairy manufacturers, the daily production cycle is divided between periods where product is manufactured, periods where cleaning and sanitation operations are performed and nonoperational periods. In the peak of the dairy season, production equipment is operated for as long as possible and cleaning and sanitation is performed over the shortest time frame that does not impact on product safety and quality.

The length of time for which dairy processing equipment operates continuously is limited by the accumulation of organic (primarily protein) and mineral fouling material and the accumulation of bacteria in biofilms on processing equipment surfaces (Bouman *et al.*, 1982; Refstrup, 2000; Fryer *et al.*, 2006; Burgess *et al.*, 2010). An increase in bacterial numbers in product exiting dairy processing equipment as a function of production time has been shown for thermoresistant streptococci (Bouman *et al.*, 1982; Knight *et al.*, 2004) and thermophilic spore-forming bacteria (Murphy *et al.*, 1999; Refstrup, 2000; Scott *et al.*, 2007). After a certain period of production, the levels of bacteria in the product increase to a point where they impact on product quality. The length of this period depends on the microorganism, the manufacturing process and the individual manufacturing plant. The accumulation of fouling material and the growth of bacteria in biofilms are only abated by performing cleaning and sanitation operations.

A similar concept applies to the manufacturing environment. Microbial counts for environmental surfaces can increase due to growth of microorganisms – under favourable conditions – during the period between cleaning and sanitation operations. The amount of growth that occurs on surfaces can be limited by reducing the time between cleaning and sanitation operations.

12.2.6 *Cleaning and sanitation*

Cleaning and sanitation of dairy processing equipment is performed to return surfaces to a state in which they are physically, chemically and microbiologically clean and will not have a detrimental impact on product quality (Dunsmore *et al.*, 1981; Graßhoff, 1997). During manufacturing, fouling material accumulates on the surfaces of dairy processing equipment, particularly on surfaces that are at elevated temperatures. Fouling material typically comprises organic (mostly protein) and mineral components (Jeurnink & Brinkman, 1994; Visser *et al.*, 1997), which are removed from surfaces using alkaline (1–3% w/w NaOH) and acidic (0.8–1.0% w/w nitric acid) cleaning solutions, respectively, at temperatures of 60–85 °C (Graßhoff, 1997; Jeurnink & Brinkman, 1994).

Cleaning is typically achieved by employing a cleaning-in-place (CIP) procedure (Graßhoff, 1997; Fryer *et al.*, 2006), which involves circulation of cleaning solutions through processing equipment under turbulent flow conditions and through large vessels, such as silos, with the aid of spray balls to ensure complete surface coverage. A typical cleaning procedure used in the dairy industry is described in Section 4.5.2; briefly, it includes a post-product water rinse, an alkali wash, a post-alkali water rinse, an acid wash, a post-acid water rinse, a sanitation step and a final water rinse. The acid wash may not be performed during every cleaning cycle in equipment in which milk fouling is not significant, such as raw milk silos and milk transfer lines.

The alkali and acid washes are generally considered effective at eliminating bacteria in biofilms from the surfaces of processing equipment, although they are not specifically designed to do this. For this reason, the sanitation step is often omitted from the cleaning regime. This has been questioned in recent years, and a number of studies have investigated the ability of cleaning treatments to inactivate bacteria and to remove bacteria and the biofilm matrix from surfaces (Flint *et al.*, 1999; Parkar *et al.*, 2004; Bremer *et al.*, 2006).

Cleaning of environmental surfaces in dairy manufacturing plants also focuses on the removal of organic (e.g. proteins and fats) and inorganic deposits. Cleaning products are selected primarily based on the types of deposit to be removed, but it is also important to be aware of the efficacy of the cleaning application for biofilms. Cleaning treatments applied to environmental surfaces are effective at reducing microbial numbers on surfaces, but are not typically as harsh on microorganisms as those applied during CIP of production equipment. This is primarily due to the lower thermal (temperature) and mechanical (turbulence) energy employed during cleaning of environmental surfaces. As a result, viable vegetative cells and spores may be present on environmental surfaces following cleaning.

There is an emphasis on controlling pathogenic bacteria on environmental surfaces in dairy manufacturing plants, in order to reduce the risk of product contamination from the environment. For this reason, it is common to include a separate sanitation step as part of the cleaning and sanitation regime. A range of sanitiser products are available for use within the dairy industry, with individual products selected based on biocidal activity against bacteria or microbial groups of concern, compatibility with surface materials within the dairy manufacturing plant and ease of rinsing from surfaces.

12.2.7 *Interactions between bacteria in biofilms*

Biofilms that develop on surfaces in natural aquatic systems typically include an array of prokaryotic and eukaryotic microorganisms. The microorganisms that colonise surfaces – both early and late colonisers – are those that can attach, grow and survive under the local environmental conditions (temperature, nutrient availability, pH, water activity etc.) and are diverse. The diversity and close proximity of the different microorganisms in biofilms means that there are likely to be interactions between them. The nature of these interactions can be beneficial (e.g. one microorganism produces matrix material, which assists another in attaching to the surface), competitive (e.g. microorganisms compete for the same nutrients), antagonistic (e.g. one microorganism releases a compound that is toxic to another) or neutral.

Environmental surfaces within dairy manufacturing plants are similarly colonised by a diverse array of microorganisms. Even under conditions designed to control or limit microbial growth (e.g. refrigerated production areas), a diverse range of microorganisms will be present. The microorganisms present will be those best able to grow and survive under the local (e.g. refrigerated) conditions. There will inevitably be interactions between the pathogenic bacteria of concern to dairy manufacturers (e.g. *Listeria monocytogenes*) and the other microorganisms present within the biofilms on environmental surfaces.

Biofilms that develop on the surfaces of pipes or processing equipment may be dominated by one or a few bacterial species, or they may contain a range of microorganisms. Raw milk contains a range of microorganisms, all of which can, under favourable conditions, attach to surfaces and form biofilms. Consequently, biofilms that develop on the surfaces of pipes and on processing equipment that comes into contact with raw milk will contain a range of bacteria originating in that raw milk. Local environmental conditions, such as the temperature, will influence which of these bacteria attach to the surfaces and form biofilms. For example, the microflora in biofilms that develop on surfaces where the local temperature is <5 °C (e.g. in silos) will differ to that which develops on surfaces at 40 °C (e.g. in heat exchangers).

An example of one or a few species being dominant is the growth of *Streptococcus thermophilus* in cheese-milk pasteurisation equipment. *S. thermophilus* survives pasteurisation and grows on the pasteurised milk side of heat exchange equipment at temperatures between 35 and 50 °C (Bouman *et al.*, 1982; Knight *et al.*, 2004). Pasteurisation reduces the total number of viable microorganisms present and eliminates many bacterial groups from milk. Whereas on the raw milk side of heat exchange equipment, *S. thermophilus* must compete with a range of other bacteria present in the raw milk, on the pasteurised milk side its ability to grow rapidly leads to it becoming the dominant microorganism (Bouman *et al.*, 1982; Knight *et al.*, 2004).

12.3 Controlling biofilm development in dairy processing equipment

Before designing approaches to control biofilms in production equipment, it is important to know which locations are susceptible to biofilm development, so that solutions can be targeted to them. This information can be gathered through line surveys and by direct examination of surfaces for evidence of biofilm formation. A significant amount of information has been obtained by which to identify the locations where *S. thermophilus* biofilms develop within pasteurisation equipment (Bouman *et al.*, 1982; Knight *et al.*, 2004) and biofilms of thermophilic spore-forming bacteria within milk powder production equipment (Murphy *et al.*, 1999; Refstrup, 2000; Scott *et al.*, 2007).

It is also useful to have a good understanding of the manufacturing process in order to know the limits of what can be changed within it. For example, in most dairy manufacturing processes, the nutrient levels and water activity of the product and the product stream cannot be altered without drastically altering the dairy products. Solutions may then be devised focusing on one or more factor, some of which are described in this section.

12.3.1 *Controlling biofilms with standard cleaning practices*

Cleaning, although primarily designed to remove fouling material, is one of the most important practices for controlling biofilms in processing equipment. The ability of cleaning solutions to remove fouling material is primarily dependent upon three factors: (i) the chemistry of the cleaning solution (to chemically transform fouling material); (ii) thermal energy (cleaning efficacy is enhanced at elevated temperatures); and (iii) mechanical energy (turbulent flow enhances removal of deposits) (Graßhoff, 1997; Fryer *et al.*, 2006). These factors also impact on the ability of cleaning solutions to eliminate (inactivate and/or remove) bacteria from equipment surfaces.

The alkali and acid wash steps are the most important for minimising levels of bacteria on surfaces (Dunsmore *et al.*, 1981; Carpentier & Cerf, 1993). There are two reasons for this. First, cleaning solutions eliminate a high proportion of bacteria from surfaces, both by removing them and by inactivating those that remain (Dunsmore *et al.*, 1981; Carpentier & Cerf, 1993). Second, the removal of soil (fouling material) from surfaces, which can interfere with the activity of sanitisers, allows the sanitiser to work more efficiently during the subsequent sanitation step (Dunsmore *et al.*, 1981; Zottola & Sasahara, 1994).

It is not clear how effective cleaning and sanitation procedures are at eliminating bacteria from the surfaces of processing equipment. Investigations in this area have been conducted using a number of different approaches, including immersion experiments, in which biofilms grown on test surfaces were immersed in cleaning solutions, and laboratory- and pilot-scale continuous-flow systems (Flint *et al.*, 1999; Parkar *et al.*, 2004; Bremer *et al.*, 2006). Each approach differs in how closely cleaning conditions can be replicated. For example, with immersion experiments, it can be relatively easy to replicate cleaning temperatures and cleaning chemical concentrations, but it is more difficult to replicate the mechanical energy applied at surfaces (i.e. the flow conditions). Such limitations should be considered when interpreting results.

Some very important work in this field was carried out by Dunsmore and colleagues (Dunsmore, 1981; Dunsmore & Thomson, 1981; Dunsmore *et al.*, 1981), who utilised a system that simulated fouling and cleaning of on-farm milking equipment to demonstrate the impact of cleaning steps (pre-rinse, alkali wash, post-rinse, sanitiser application) on the accumulation of bacteria and dairy soil on surfaces over multiple fouling and cleaning cycles. The cleaning treatment employed (0.3% w/v NaOH at 50 °C) was very mild and allowed dairy foulant and bacteria to accumulate on surfaces. Nevertheless, this work demonstrated a couple of important trends. First, cleaning solutions and sanitisers became less effective at eliminating bacteria as foulant material accumulated on surfaces, which highlights the importance of the removal of this material during cleaning. Second, the efficacy of the sanitation step was dependent on when the sanitiser was applied. For example, the sanitiser was more efficacious when left in the equipment for the entire intercycle period (the period between the end of cleaning and the start of the next soiling step) than when applied immediately after cleaning (second most effective) or immediately prior to fouling (least effective).

Preventing bacterial growth during the intercycle period is very important. At the peak of the dairy season, the length of the intercycle period can be minimal, with production resuming as soon as cleaning and sanitation is completed. However, for processing equipment that is operated infrequently, significant levels of bacterial growth may occur during the intercycle period, which can have major consequences for product safety and quality. One solution, as suggested by the study of Dunsmore & Thomson (1981), is to leave a sanitiser in contact with equipment for the entire intercycle period. Careful consideration must be given to the selection of the sanitiser, as it may need to maintain biocidal activity for the entire period.

Studies investigating the ability of cleaning treatments to remove bacteria from surfaces have made use of immersion experiments (Flint *et al.*, 1999, Parkar *et al.*, 2004). Flint *et al.* (1999) found that alkali (2% NaOH at 75 °C for 30 minutes) and acid (1.8% HNO_3 at 75 °C for 30 minutes) treatments did not remove cells of *S. thermophilus* (attached or grown in biofilms) from surfaces. Parkar *et al.* (2004) applied these same treatments to biofilms of *Anoxybacillus flavithermus*, although they applied them sequentially, achieving complete inactivation and removal of vegetative cells from surfaces. Parkar *et al.* (2004) also evaluated alkali and acid treatments using reduced temperatures and cleaning chemical concentrations, and in these cases achieved complete inactivation but not complete removal of cells from surfaces. The results of these two studies indicate that the cleaning treatments typically employed in the dairy industry are able to inactivate bacteria in biofilms, but it is not clear whether these treatments remove cells from surfaces in the absence of mechanical energy (turbulent flow), which is typically experienced during cleaning of dairy equipment.

Both of these investigations (Flint *et al.*, 1999; Parkar *et al.*, 2004) also utilised a modified Robbins device (MRD), connected in-line with pilot-scale pasteurisation equipment, to demonstrate the effects of cleaning on bacterial biofilms. Flint *et al.* (1999) inoculated *S. thermophilus* cells on to test surfaces, transferred these to the MRD, allowed biofilms to develop on test surfaces while processing milk for 8 hours, and subjected the biofilms to cleaning (1.8% NaOH at 75 °C for 30 minutes followed by 1.0% HNO_3 at 75 °C for 30 minutes). In this case, low numbers of viable cells were detected on the surfaces by an impedance method and cells were detected on the surfaces by fluorescence microscopy. The procedure used by Parkar *et al.* (2004) differed in that biofilms were grown on test surfaces in the laboratory, transferred to the MRD and immediately subjected to cleaning. In this case, viable cells were not detected on surfaces following cleaning using the impedance method and cells were not observed by fluorescence microscopy. It is possible that *S. thermophilus* biofilms are more difficult to eliminate from surfaces or that allowing *S. thermophilus* biofilms to develop in the MRD, where conditions were closer to those experienced in production equipment, resulted in development of more robust biofilms.

Other studies have looked at the ability of cleaning solutions to remove bacteria from surfaces using laboratory-scale continuous-flow systems (Dufour *et al.*, 2004; Bremer *et al.*, 2006). Bremer *et al.* (2006) subjected mixed bacterial biofilms, grown under continuous-flow conditions in reconstituted skim milk powder at approximately 55 °C for 18 hours, to a standard cleaning procedure (1.0% NaOH at 65 °C for 10 minutes followed by 1.0% HNO_3 at 65 °C for 10 minutes) and to a range of caustic and nitric acid-based cleaning products. This system, which used treatment times much shorter than those typically employed in the dairy industry, was not designed to eliminate all bacteria from surfaces but rather to identify the most effective combination of cleaning products and cleaning additives with which to reduce biofilms. The types of bacteria in biofilms were undefined, and it is not clear whether bacterial spores were present in the biofilms, which could have affected the results. Nevertheless, while differences were observed between cleaning treatments, viable bacteria were detected on test surfaces in all cases. This system could be adapted to investigate the effects of cleaning treatments on target micro-organisms in biofilms, as well as approaches to enhancing cleaning treatments in order to achieve complete elimination of bacteria from surfaces.

Effect of cleaning solutions on bacterial endospores

Biofilms that contain spore-forming bacteria are particularly difficult to eliminate from the surfaces of dairy processing equipment due to the resistance of bacterial spores to inactivation by cleaning solutions. Several studies have investigated the ability of hot alkali solutions to inactivate bacterial spores (Stadhouders, 1964; Te Giffel *et al.*, 1997; Knight & Weeks, 2008). Stadhouders (1964) exposed spores of *Bacillus subtilis, Bacillus circulans* and *Bacillus cereus* to 0.5, 1.0 and 2.0% NaOH, at temperatures of 70, 80 and 90 °C. The levels of inactivation achieved for *B. subtilis* and *B. circulans* spores exposed to 1.0% NaOH at 70 °C for 30 minutes were 2.5 and 3.4 $\log_{10}$ CFU/ml, respectively. Spores of the thermophilic bacterium *Geobacillus stearothermophilus* appeared more resistant to hot alkali solutions, with an inactivation level of 2.2 $\log_{10}$ CFU/ml achieved following exposure to 1.0% NaOH at 70 °C for 60 minutes (Knight & Weeks, 2008). Both of these studies demonstrated that spore inactivation increased with NaOH concentration and treatment temperature. The implication of these results is that the alkaline cleaning step (combination of NaOH concentration, temperature and time) should be designed around elimination of the potential spore loads present in dairy processing equipment.

The sporicidal activities of acidic cleaning solutions have not been investigated to the same extent as those of alkaline solutions. Strong inorganic acids (e.g. hydrochloric, nitric, sulphuric and phosphoric acids) at high concentrations are known to be sporicidal but there are few detailed studies. Setlow *et al.* (2002) investigated the sporicidal activity of 0.5–1.0 M HCl towards spores of *B. subtilis*. They demonstrated that counts for *B. subtilis* spores exposed to 0.5 M HCl at 24 °C were reduced by 90 and >95% after 60 and 120 minutes, respectively. Spores were inactivated more rapidly following exposure to 1.0 M HCl, with counts reduced by >99% within 40 minutes. Setlow *et al.* (2002) also reported that counts for *B. subtilis* spores exposed to 3.0 M phosphoric acid were reduced by 50% within 90 minutes, demonstrating that strong inorganic acids can have substantially different sporicidal activities.

There is also evidence that low concentrations of inorganic acids, when combined with mild heat, can be sporicidal. Acid titration of spores, a procedure in which spores are exposed to low concentrations of HCl in the presence of mild heat (e.g. 0.033 M HCl at 60 °C), results in the exchange of spore cations with hydrogen ions (Bender & Marquis, 1985). This procedure results in the generation of H-form spores, which have a significantly lower resistance to inactivation by moist heat than native spores (Bender & Marquis, 1985; Palop *et al.*, 1999). Bender and Marquis (1985) reported that, depending on the bacterial species, acid titration procedures could lead to spore inactivation, although they did not indicate the levels of inactivation that occurred. Results obtained by the author (G. Knight, unpublished) with four separate strains of *Geobacillus stearothermophilus* demonstrated that spore counts were not reduced following exposure to a 1.0% v/v nitric acid-based cleaning solution at 20 °C for 60 minutes, but were reduced by >4.0 $\log_{10}$ CFU/ml following exposure to this cleaning solution at 70 °C for 10 minutes. While the information available is limited, these studies indicate that inorganic acids can be sporicidal, particularly at high concentrations and in combination with high temperatures, and that nitric acid-based cleaning products are sporicidal when employed under typical application conditions.

Locations prone to biofilm formation

A range of locations within dairy processing equipment are considered risks to process hygiene because they are difficult to clean, including gaskets, pump seals, dead legs, end caps, contact points and other areas where the flow rate may be low. Many of these risks have been addressed through improvements in equipment design and manufacturing practices, which are covered by the standards and guidelines produced by 3-A Sanitary Standards, Incorporated (3-A SSI) in the United States and European Hygienic Engineering and Design Group (EHEDG) in Europe. However, issues still remain, particularly with older manufacturing plants.

A number of studies have identified the locations where gaskets are inserted into equipment as susceptible to biofilm formation (Czechowski, 1990; Austin & Bergeron, 1995; Mettler & Carpentier, 1997). These studies demonstrated that biofilms were more common and extensive on the surfaces of stainless steel and gasket materials where they contacted with each other (the contact points) than on surfaces directly exposed to the bulk flow. Gasket materials, including those made of Buna-n, EPDM and polytetrafluoroethylene (PTFE), are subject to wear, as a result of exposure to cleaning solutions and the expansion/contraction that occurs with extreme changes in temperature. This wearing includes pitting, cracking and formation of crevices, all of which provide bacteria with locations where they can attach and grow and where they receive protection from the action of cleaning solutions (Czechowski, 1990; Austin & Bergeron, 1995; Mettler & Carpentier, 1997). Growth of bacteria in contact points, and in the cracks and crevices of gaskets, is supported by migration of nutrients from the bulk liquid during the production period. In addition, the relatively short times allocated to cleaning, compared with the length of a production run, are insufficient to enable cleaning chemicals to penetrate to the deeper locations within contact points (Austin & Bergeron, 1995). The most effective approaches to the control of biofilms associated with gaskets and contact points are to ensure CIP systems are working effectively and to have a programme that ensures gaskets are changed regularly (Czechowski, 1990; Austin & Bergeron, 1995).

A major impediment to the elimination of thermophilic spore-forming bacteria from processing equipment is that locations where they grow are also susceptible to dairy fouling. Due to the complexity of the design of dairy evaporators, there are locations where the flow rate is low and where fouling material accumulates during production. Scott *et al.* (2007) removed fouling material containing high levels of thermophilic spore-forming bacteria from three locations within a dairy evaporator, after the equipment had been cleaned using a typical CIP regime. This foulant material is likely to act as a source for these bacteria in subsequent production runs. Hinton *et al.* (2002) demonstrated, using a laboratory-scale flow system, that fouling material enhanced accumulation of *G. stearothermophilus* on stainless steel surfaces. The fouling material also protected cells and spores from inactivation when test surfaces were immersed in a 2% NaOH solution at 65 °C for 15 minutes. Te Giffel *et al.* (1997) also demonstrated that *B. cereus* spores were protected from inactivation by a hot alkaline cleaning solution when spores were attached to a stainless steel surface in the presence of dairy deposits.

Application of enzyme-based cleaning and sanitation products

After finding that a typical cleaning and sanitation regime was unable to remove *S. thermophilus* cells from stainless steel surfaces, Flint *et al.* (1999) suggested that a useful approach to the control of biofilms might be to focus on methods of detaching bacteria from surfaces. They trialled a commercial proteolytic-based cleaning product against *S. thermophilus* biofilms, using an MRD connected in-line with pilot-scale pasteurisation equipment, and demonstrated that lower numbers of cells remained on test surfaces following application of the proteolytic-based cleaning product compared with a standard cleaning regime. Parkar *et al.* (2004) trialled the same proteolytic-based cleaning product against biofilms of *A. flavithermus,* using the same equipment, and found the cleaning treatment removed all cells (total and viable) from test surfaces. Proteolytic-based cleaning products are commercially available for the cleaning of processing equipment and other equipment that comes into contact with raw milk. They are not currently employed for the cleaning of heated processing equipment, but they are employed for cleaning with some membrane processes, where surfaces can be sensitive to standard cleaning chemicals. The additional cost of enzyme cleaners compared with acid and caustic cleaners is the main reason why they are not used more widely in the dairy industry. Another concern is the effect that any residual enzyme might have on product quality.

The complexity and diversity of the extracellular polymeric substances (EPS) that make up the biofilm matrix pose a problem when selecting which enzymes to use to break down the biofilm matrix and aid the dispersal of bacteria. Biofilms may contain a number of different types of polysaccharide, which will vary depending on the bacterial species and genera present. Johansen *et al.* (1997) used a commercial product consisting of a mixture of polysaccharide hydrolysing enzymes (polysaccharidases) to disperse biofilms and had limited success in removing cells of *Staphylococcus aureus*, *Staphylococcus epidermidis*, *Pseudomonas aeruginosa* and *Pseudomonas fluorescens* from stainless steel surfaces. The limited success achieved may relate to differences between the polysaccharides produced by the different bacterial species and the extent to which other EPS components, such as proteins, DNA and lipids, contribute to maintenance of the biofilm structure.

Lequette *et al.* (2010) investigated biofilm removal of a number of bacterial species using a variety of proteases and polysaccharidases. Both groups of enzymes were able to remove biofilm biomass from surfaces, but proteases proved more effective against a wider range of bacterial species. Lequette *et al.* (2010) were also able to enhance the effectiveness of enzymes at removing biofilms through the addition of cleaning additives, such as surfactants and dispersing and chelating agents (which are also currently used with alkali-based cleaning products).

To be effective at biofilm removal, enzyme-based cleaning products need to demonstrate activity towards biofilms containing multiple EPS components and should be formulated with appropriate cleaning additives. However, combining different enzyme groups may prove difficult due to compatibility issues. For example, proteases will probably demonstrate activity towards polysaccharidases. It should also be noted that such products do not need to perform a traditional cleaning function (i.e. removal of protein and mineral deposits), but may be used as a supplement treatment in an existing cleaning and sanitation regime.

Conclusions on the efficacy of cleaning and sanitation

It is clear that alkali and acid cleaning solutions will inactivate vegetative cells and spores of bacteria when applied under conditions that are typically employed for the cleaning of dairy processing equipment. A question still remains over whether residing within a biofilm provides cells and spores with protection against cleaning solutions. Results obtained by Flint *et al.* (1999) and Parkar *et al.* (2004) are conflicting, with the former demonstrating that bacteria in biofilms can survive cleaning processes and the latter demonstrating the opposite. This area needs further study to clarify this point.

A second important point is that accumulation of fouling material on processing equipment surfaces benefits bacteria by enhancing attachment to surfaces and protecting the bacteria from inactivation and removal. This is particularly apparent for thermophilic spore-forming bacteria, which have been found in foulant material removed from processing equipment following cleaning (Scott *et al.*, 2007). In addition to improving cleaning regimes, a good approach to reducing issues with thermophilic spore-forming bacteria is to change the design of processing equipment to eliminate locations where fouling material accumulates during production. This can be achieved, for example, by changing the design of distributor plates used in evaporators or the locations of support struts, or by adding spray balls to assist in cleaning of particular locations.

Another point can be made about the use of gaskets in processing equipment, as these appear to be prone to colonisation by bacteria and biofilm formation. Again, there may be benefits in changing the design of processing equipment to avoid the presence of gaskets in areas where biofilm formation is known to occur. For thermophilic spore-forming bacteria, this can be anywhere that the temperature is between 40 and 70 °C (particularly between 50 and 65 °C, where growth rates are highest). In some cases, such as for plate heat exchangers, this will not be practical and other approaches may be necessary.

Finally, it can be difficult to assess the effectiveness of cleaning and sanitation for the control of biofilms in processing equipment. In the case of thermophilic spore-forming bacteria, typing studies have shown that the same strains consistently contaminate product from milk powder manufacturing plants (Ronimus *et al.*, 2003; Scott *et al.*, 2007). The implications here are that the same strains are introduced into processing equipment in every production run or, as the foulant material contaminated with thermophiles found by Scott *et al.* (2007) suggests, that residues that remain in the processing equipment following cleaning are sources of thermophilic bacteria. One approach that may be used to determine the efficacy of cleaning procedures is to monitor counts of thermophilic bacteria in a product over a series of production runs. An example of this is shown in Figure 12.3, where total and spore counts for thermophilic bacteria in milk powder are determined for a set of six sequential production runs (G. Knight, unpublished). For each production run, there is a trend of increasing total and spore counts with time of production. The total and spore counts generally decrease between the end of one run and the start of the next, due to the effects of cleaning. However, in some cases, total and spore counts are not lower at the beginning of a production run, and counts either continue to increase or decrease for a few hours of production before increasing again. Such a pattern indicates that the processing equipment has not been properly cleaned at the end of the previous production run and that residues containing thermophilic bacteria are present in the equipment at the start of production.

12.3.2 *Changing equipment design*

The initial designs for dairy processing equipment are often based solely on engineering principles, with an eye to maximising processing and energy efficiency. However, equipment designs have evolved as engineers and microbiologists have recognised and attempted to

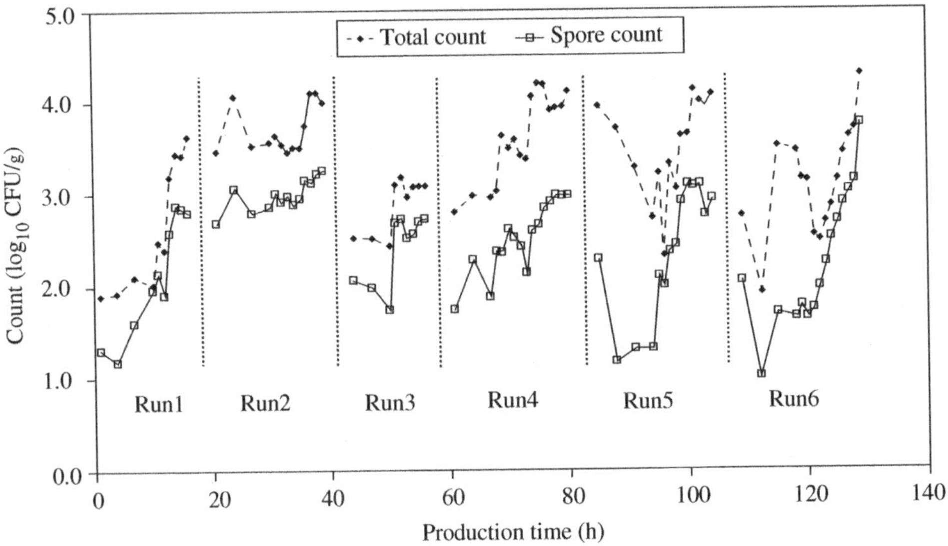

Figure 12.3 Total and spore counts for thermophilic bacteria in milk powder for six sequential milk powder production runs. Milk powder samples were obtained every 3 hours at the beginning of each production run and then every hour. Vertical lines indicate a cleaning and sanitation operation was performed.

solve process hygiene issues. As mentioned earlier, many process hygiene issues have been addressed by the standards and guidelines produced by 3-A SSI in the United States and EHEDG in Europe. However, these standards and guidelines still do not address many process hygiene issues, and some ingenuity is required to resolve these. Recent developments in our understanding of process hygiene have led to innovative designs for processes and processing equipment that have the potential to eliminate or substantially reduce some specific process hygiene issues.

The dairy industry has experienced problems with thermophilic spore-forming bacteria in milk powders dating back to at least the 1960s (Galesloot & Stadhouders, 1968). The role that biofilms produced by thermophilic spore-forming bacteria play in this issue is now recognised and the locations where biofilms develop in milk powder production equipment have been identified (Murphy *et al.*, 1999; Refstrup, 2000; Scott *et al.*, 2007). One approach to controlling this issue is to redesign equipment and processes to reduce or eliminate the surface area on which biofilms can develop. Murphy *et al.* (1999) suggested bypassing the second and third preheater sections of an evaporator, which heated milk from 45 to 65 °C, and achieving this heating using a direct steam injection (DSI) unit. When the evaporator was operated in this configuration, growth of thermophilic bacteria was not detected in a 20-hour production trial. When making such a change, energy efficiency has to be considered. In the original configuration, the heating achieved by the second and third preheaters made use of excess heat from the evaporation process. Using a DSI unit to achieve this heating requires additional heat input and the costs for this, which can be significant, have to be weighed up against the benefits of obtaining milk powder with lower thermophile levels. A slightly different design was proposed by Refstrup (2000), which involved again bypassing the preheaters, but heated the milk to 67–70 °C using a direct-contact preheating (DCP) system, which uses vapour collected from the first evaporator effect as a source of heat. Operating the evaporator with this configuration is similarly able to prevent the growth of thermophilic bacteria.

Another example of a change to a process is that suggested by Knight *et al.* (2004) to control the growth of thermoresistant streptococci in a cheese-milk pasteurisation plant. In this case, the locations in the regenerative heat exchange section where growth of thermoresistant streptococci occurs (35–50 °C) were periodically (every 60 minutes) subjected to a higher temperature (55 °C) for approximately 10 minutes. This change in the operation of the pasteuriser resulted in an increase in the production time at which growth of thermoresistant streptococci was detected from 8–10 hours to >20 hours. As per the previous examples, losses in energy efficiency during production must be weighed up against gains from longer production times and lower cleaning costs.

12.4 Controlling biofilm development on environmental surfaces

A major improvement in the hygiene of dairy manufacturing has been the practice of physically separating locations where raw materials are stored and handled from locations where processing occurs and where products are stored. Particularly important has been the introduction of critical hygiene areas, which are locations within a manufacturing plant that enclose processing steps considered vulnerable to contamination (e.g. the filling step for packaging of a product). Access to these areas is restricted to production personnel and dedicated clothing and footwear is provided for personnel to wear. The separation of production areas has enabled manufacturers to improve control over the hygiene of the manufacturing environment by, for example, limiting the introduction and movement of microorganisms into and around the manufacturing plant. A further improvement has been the widespread implementation of environmental monitoring programmes in dairy manufacturing plants. A high level of importance is now placed on these programmes to demonstrate control over manufacturing plant hygiene, and they are used to identify potential hazards before they occur and to identify locations susceptible to colonisation by pathogenic bacteria, which may require targeted sanitation treatments or a long-term solution.

Both of these improvements in manufacturing practices influence the ability of microorganisms, including pathogenic and spoilage microorganisms, to colonise surfaces in dairy manufacturing environments. There are a number of additional factors that dairy manufacturers can consider to reduce the impact that biofilms in the manufacturing environment have on product safety and quality.

12.4.1 *Standard cleaning and sanitation practices*

Environmental cleaning products are designed to remove organic and inorganic deposits that accumulate on environmental surfaces within the production environment. The cleaning products employed are typically chosen based on the types of deposit to be removed. The most common environmental cleaning products used in the dairy industry are chlorinated and nonchlorinated alkaline cleaning products, which are designed to remove organic deposits. Acid-based cleaning solutions are used to remove inorganic deposits, while neutral pH detergent-based products may be used where surface materials are sensitive to acidic or alkaline solutions.

It is important to recognise that cleaning products do affect biofilms. This includes the removal of bacteria from surfaces as part of the normal cleaning process, as well as the inactivation of bacteria that remain on surfaces. Many cleaning products include components that demonstrate biocidal activity (e.g. chlorine or anionic surfactants), which allows the products to be used as combined cleaning/sanitising agents.

Cleaning solutions and sanitisers can influence the microorganisms that survive on surfaces because they demonstrate different biocidal activities towards different microbial groups. For example, sanitisers based on quaternary ammonium compounds (QUATS) are highly effective against Gram-positive bacteria, including *Listeria monocytogenes* and *Staphylococcus aureus*, but are less effective against Gram-negative bacteria such as *Pseudomonas* spp. (Van Klingeren *et al.*, 1998; McDonnell & Russell, 1999; Morente *et al.*, 2013). There are several examples in the literature of a sanitiser being shown to have a greater impact on individual bacterial groups within a mixed bacterial biofilm. For example, Kinniment *et al.* (1996) generated mixed bacterial biofilms containing nine different bacterial species in a constant-depth film fermenter (CDFF) and exposed these to the sanitiser chlorhexidine (0.125%). The overall biofilm population was reduced by approximately 2.8 $\log_{10}$CFU/plug following exposure to the sanitiser, but some species were affected to a greater or lesser extent. *Veillonella dispar* went from representing <0.01% of the population to representing 8.7% of the population, whereas *Porphyromonas gingivalis* went from representing 19.1% of the population to representing 3.0% of the population. Another example is the study of Norwood and Gilmour (2000), in which mixed bacterial biofilms comprising *Pseudomonas fragi*, *Staphylococcus xylosus* and *L. monocytogenes* were generated in a CDFF. Treatment of the biofilms with sodium hypochlorite (1000 ppm free chlorine) caused significantly greater reductions in counts for *P. fragi* and *S. xylosus* compared with *L. monocytogenes* (reductions of 2.84, 2.56 and 1.75 $\log_{10}$CFU/plug, respectively). Some additional examples include the studies of Moore *et al.* (2008) and Knight & Craven (2010).

An important implication of sanitisers exhibiting different biocidal activities towards different microbial groups is that the cleaning and sanitation regime will impose a selective pressure and potentially lead to the development of a resident microflora that is adapted to the regime. This would be a particular concern if the resident microflora included a pathogenic microorganism. One approach to preventing the development of an adapted resident microflora is to alternate between different types of sanitiser (Langsrud & Sundheim, 1997).

There have been some recent developments in the testing methodology used to determine the efficacy of sanitisers. Historically, sanitisers have been evaluated against microbial cells in suspensions (Gibson *et al.*, 1995; Van Klingeren *et al.*, 1998). It is now recognised that cells within food processing environments are typically associated with surfaces and that results obtained using suspension tests are not a good indicator of sanitiser efficacy against cells on surfaces. For this reason, standardised testing procedures have been developed to evaluate sanitisers against cells dried on to test surfaces (Holah *et al.*, 1998; Van Klingeren *et al.*, 1998). More recently, studies have attempted to replicate 'in-use' conditions using model systems and have subjected biofilms developed on test surfaces to multiple cycles of growth, cleaning treatments and recovery (e.g. Verran *et al.*, 2001; Pan *et al.*, 2006; Peneau *et al.*, 2007; Knight & Craven, 2010; Marouani-Gadri *et al.*, 2010). Such model systems enable the influence of cleaning products and sanitisers to be evaluated against microbial biofilms under practical conditions and over a longer term. And there is scope to adapt these model systems to include a model resident microflora and include pathogenic microorganisms that are of particular concern. So, while suspension and basic surface tests are still valuable tools for evaluating sanitiser efficacy, model systems that replicate 'in-use' conditions are likely to find greater use in evaluating the effectiveness of environmental cleaning and sanitation procedures.

12.4.2 *Moisture*

Water has a significant influence on the growth and survival of microorganisms on surfaces in the processing environment. Water may be introduced into the processing environment from a variety of sources, including condensation, product spills (and associated clean-up activities) and cleaning and sanitation operations. Manufacturing plants should be designed to aid the removal of water from surfaces through the installation of correctly sloped floors, effective drainage systems and air handling systems that encourage evaporation. Manufacturing plant personnel should also limit the introduction of water and allow surfaces to dry through evaporation.

The intense use of water can lead to excess moisture in the atmosphere in dairy manufacturing plant environments. Moisture may also be present in the atmosphere simply due to humid weather conditions. Survival of microorganisms on surfaces tends to be the greatest when the humidity level is close to 100% RH and to decrease as the RH decreases. However, depending on the microbial group and other environmental factors, such as the temperature, survival may be significant at humidity levels as low as 75% RH. For example, Helke and Wong (1994) investigated the survival of *L. monocytogenes* spotted on to stainless steel and Buna-n rubber surfaces during storage at temperatures of 6 and 25 °C and RH levels of 32.5 and 75.5%. Survival of *L. monocytogenes* was greatest during storage at 6 °C and 75.5% RH, with viable counts decreasing by approximately $1.0 \log_{10}$ CFU/cm^2 after 10 days. In comparison, viable counts for *L. monocytogenes* stored at 6 °C and 32.5% RH decreased by 3–4 $\log_{10}$ CFU/cm^2 after 10 days. Temperature is also an important factor affecting survival. Counts for *L. monocytogenes* decreased more rapidly when the storage temperature was 25 °C (Helke & Wong, 1994).

It is important to recognise that different microbial groups can have slightly different responses to RH levels. Møretrø *et al.* (2010) investigated the survival of Shiga toxin-producing *E. coli* (STEC) inoculated on to stainless steel surfaces during storage at 20 °C at RH levels of 35, 44, 70, 85 and 98%. Survival was greatest during storage at 98% RH and poorest during storage at 70 and 85% RH. In fact, STEC strains survived better during storage at low RH levels (35 and 44%) than at 70 and 85% (Møretrø *et al.*, 2010). Similar observations were made by Kim *et al.* (2008), who investigated survival of *Enterobacter sakazakii* (now known as *Cronobacter* spp.) dried on to stainless steel surfaces during storage at a range of RH levels. The survival of *E. sakazakii* was greatest at a level of approximately 100% RH and poorest at 85% RH. Like the STEC strains, *E. sakazakii* survived better at low RH levels (23 and 43%) than at 85% RH (Kim *et al.*, 2008).

The presence of soiling material (organic and inorganic) will also influence the survival of microorganisms on surfaces. These same studies investigating the influence of RH on microorganism survival also investigated the influence of the presence of soiling material. In each case, survival of cells was greatest, under all RH levels investigated, when cells were suspended with organic material, such as microbial growth medium (brain heart infusion broth), individual components of a growth medium (glucose or peptone), infant formula and milk (Helke & Wong, 1994; Kim *et al.*, 2008; Møretrø *et al.*, 2010). With some combinations of RH levels, temperature and nutrients, the microorganisms grew on the test surfaces during the storage period.

Investigations of the influence of RH levels on the survival of microorganisms suggest that there is a humidity range within which factories should operate so as to reduce the survival of microorganisms on surfaces. Given that there is some variability between the responses of different microbial groups, the safest option is to operate at the lowest humidity level that can be achieved. This will limit the amount of growth that occurs, even if water is periodically introduced. Whatever humidity level is selected, there will always be one or more microbial groups favoured by the conditions.

12.4.3 *Interactions with other microorganisms*

The microbial groups considered most important on environmental surfaces in dairy manufacturing plants are pathogenic bacteria, such as *L. monocytogenes* and *Salmonella* spp. Pathogenic bacteria are unlikely to be present alone as pure cultures on surfaces within food processing environments, but rather will be present together with the resident microflora (Jeong & Frank, 1994a; Bremer *et al.*, 2001). For this reason, pathogenic bacteria are likely to interact with any other microorganisms present, which may have a significant influence on their ability to grow, survive and persist on surfaces within the processing environment.

Due to its importance in food safety and its ability to survive and persist within food processing environments, *L. monocytogenes* has been the subject of many studies on biofilm formation in food processing plant environments (e.g. Sasahara & Zottola, 1993; Jeong & Frank, 1994a,b; Bremer *et al.*, 2001; Carpentier & Chassaing, 2004). *L. monocytogenes* is not very good at forming biofilms by itself and has been shown to require other bacteria to efficiently colonise surfaces (Sasahara & Zottola, 1993; Bremer *et al.*, 2001; Kalmokoff *et al.*, 2001). A number of studies have demonstrated that strains of other bacterial species reduce, have no effect on or enhance biofilm formation by *L. monocytogenes* (Jeong & Frank, 1994a,b; Carpentier & Chassaing, 2004). Bremer *et al.* (2001) also showed that survival of *L. monocytogenes* on stainless steel surfaces during storage at 75% RH, at 4 or 15 °C, was enhanced when grown in a mixed bacterial biofilm with strains of *Flavobacterium* spp. A recent investigation employed a confocal laser scanning microscope to show the spatial organisation in mixed-bacteria biofilms comprising strains of *L. monocytogenes*, labelled with green fluorescent protein, and *Lactococcus lactis* (Habimama *et al.*, 2011). In this work, *L. monocytogenes* was shown to grow exclusively at the base of the biofilm, in contact with the substratum, while the *L. lactis* strain formed a thick confluent layer over the top. Similar observations were made in a study using an STEC strain grown in biofilms with an environmental strain of *Acinetobacter calcoaceticus* (Habimama *et al.*, 2010). In this case, colonisation of test surfaces by the STEC strain was significantly enhanced when it was grown together with the *A. calcoaceticus* strain.

It is clear that pathogenic bacteria interact with other microorganisms and that this affects their ability to grow and survive on surfaces in biofilms. However, it is not clear whether these interactions influence the ability of pathogenic bacteria to survive cleaning and sanitation processes. Kostaki *et al.* (2012) generated mixed bacterial biofilms with *L. monocytogenes* and *S. enterica* strains, but found the strains employed did not influence each other's growth or susceptibility to sanitisers. In contrast, Van Der Veen and Abee (2011) found that strains of *L. monocytogenes* and *L. plantarum* both survived sanitiser treatments better when grown together in mixed bacterial biofilms. Further work is required in this area to elucidate the full extent of the influence of the resident microflora. And, rather than co-culturing *L. monocytogenes* with just one other species, it would be more appropriate to include strains belonging to a number of species, preferably isolated from a relevant food processing environment, to replicate the typical resident microflora of a food processing plant.

The interactions between pathogens and resident microflora suggest another approach that might be exploited to reduce the impact of pathogens in the environment, namely the use of competitive exclusion cultures. Zhao *et al.* (2004) isolated yeast and bacterial strains from biofilms present in the drains of food processing facilities with a history of being free of *L. monocytogenes* and were able to identify a number of strains that possessed antilisterial properties. Two isolates demonstrated a strong ability to inhibit biofilm formation by a five-strain cocktail of *L. monocytogenes* on stainless steel surfaces. These isolates were trialled as competitive exclusion cultures and were shown to significantly reduce the levels of *Listeria* spp. detected in the drains of a poultry processing plant (Zhao *et al.*, 2006). This approach is equally applicable to dairy manufacturing plants, although it would be necessary to obtain strains that were capable of surviving and demonstrating antilisterial properties within dairy manufacturing plant environments.

12.5 Conclusion

Biofilm formation is a complex process that is influenced by many factors. Controlling one factor may not be sufficient to prevent the occurrence of a biofilm-related issue and it is often necessary to focus on several factors at once. There are many potential solutions available to dairy food manufacturers, each of which has associated costs. These costs must be weighed up against the control that dairy food manufacturers want to have over biofilms in manufacturing processes and environments and against any associated product quality and safety issues. Of the current control measures employed by the dairy industry, cleaning and sanitation is the most effective, and this is unlikely to change in the near future.

The dairy industry continues to design and develop new dairy manufacturing processes, which come with their own biofilm-related issues. Most of these issues come down to a time–temperature relationship. If you want to operate a process at a certain temperature, there will be a maximum time associated with that operation before a biofilm-related issue arises. Many such issues can be predicted if enough is known about the process.

References

Austin, J. W. & Bergeron, G. 1995. Development of bacterial biofilms in dairy processing lines. *Journal of Dairy Research*, **62**, 509–19.

Bender, G. R. & Marquis, R. E. 1985. Spore heat resistance and specific mineralisation. *Applied and Environmental Microbiology*, **50**, 1414–21.

Bouman, S., Lund, D. B., Driessen, F. M. & Schmidt, D. G. 1982. Growth of thermoresistant streptococci and deposition of milk constituents on plates of heat exchangers during long operating times. *Journal of Food Protection*, **45**, 806–12.

Bremer, P. J., Monk, I. & Osborne, C. M. 2001. Survival of *Listeria monocytogenes* attached to stainless steel surfaces in the presence or absence of *Flavobacterium* spp. *Journal of Food Protection*, **64**, 1369–76.

Bremer, P. J., Fillery, S. & McQuillan, A. J. 2006. Laboratory scale clean-in-place (CIP) studies on the effectiveness of different caustic and acid wash steps on the removal of dairy biofilms. *International Journal of Food Microbiology*, **106**, 254–62.

Burgess, S. A., Lindsay, D. & Flint, S. H. 2010. Thermophilic bacilli and their importance in dairy processing. *International Journal of Food Microbiology*, **144**, 215–25.

Carpentier, B. & Cerf, O. 1993. Biofilms and their consequences, with particular reference to hygiene in the food industry. *Journal of Applied Bacteriology*, **75**, 499–511.

Carpentier, B. & Chassaing, D. 2004. Interactions in biofilms between *Listeria monocytogenes* and resident microorganisms from food industry premises. *International Journal of Food Microbiology*, **97**, 111–22.

Czechowski, M. H. 1990. Bacterial attachment to Buna-n gaskets in milk processing equipment. *Australian Journal of Dairy Technology*, **45**, 113–14.

Dufour, M., Simmonds, R. S. & Bremer, P. J. 2004. Development of a laboratory scale clean-in-place system to test the effectiveness of 'Natural' antimicrobials against dairy biofilms. *Journal of Food Protection*, **67**, 1438–43.

Dunsmore, D. G. 1981. Bacteriological control of food equipment surfaces by cleaning systems. I. Detergent effects. *Journal of Food Protection*, **44**, 15–20.

Dunsmore, D. G. & Thomson, M. A. 1981. Bacteriological control of food equipment surfaces by cleaning systems. II. Sanitizer effects. *Journal of Food Protection*, **44**, 21–7.

Dunsmore, D. G., Twomey, A., Whittlestone, W. G. & Morgan, H. W. 1981. Design and performance of systems for cleaning product-contact surfaces of food equipment: a review. *Journal of Food Protection*, **44**, 220–40.

Flint, S. H., Van Den Elzen, H., Brooks, J. D. & Bremer, P. J. 1999. Removal and inactivation of thermo-resistant streptococci colonising stainless steel. *International Dairy Journal*, **9**, 429–36.

Fryer, P. J., Christian, G. K. & Liu, W. 2006. How hygiene happens: physics and chemistry of cleaning. *International Journal of Dairy Technology*, **59**, 76–84.

Galesloot, T. E. & Stadhouders, J. 1968. The microbiology of spray-dried milk products with special reference to *Staphylococcus aureus* and Salmonellae. *Netherlands Milk and Dairy Journal*, **22**, 158–72.

Gibson, H., Elton, R., Peters, W. & Holah, J. T. 1995. Surface and suspension testing: conflict or complementary. *International Biodeterioration and Biodegradation*, **36**, 375–84.

Graßhoff, A. 1997. Cleaning of heat treatment equipment. *Bulletin of the International Dairy Federation*, **328**, 32–44.

Habimama, O., Heir, E., Langsrud, S., Åsli, A. W. & Møretrø, T. 2010. Enhanced surface colonization by *Escherichia coli* O157:H7 in biofilms formed by an *Acinetobacter calcoaceticus* isolate from meat-processing environments. *Applied and Environmental Microbiology*, **76**, 4557–9.

Habimama, O., Guillier, L., Kulakauskas, S. & Briandet, R. 2011. Spatial competition with *Lactococcus lactis* in mixed-species continuous-flow biofilms inhibits *Listeria monocytogenes* growth. *Biofouling*, **27**, 1065–72.

Helke, D. M. & Wong, A. C. L. 1994. Survival and growth characteristics of *Listeria monocytogenes* and *Salmonella typhimurium* on stainless steel and Buna-n rubber. *Journal of Food Protection*, **57**, 963–8.

Hinton, A. R., Trinh, K. T., Brooks, J. D. & Manderson, G. J. 2002. Thermophile survival in milk fouling and on stainless steel during cleaning. *Transactions of the Institution of Chemical Engineers*, **80**(C), 299–304.

Holah, J. T., Lavaud, A., Peters, W. & Dye, K. A. 1998. Future techniques for disinfectant efficacy testing. *International Biodeterioration and Biodegradation*, **41**, 273–9.

Jeong, D. K. & Frank, J. F. 1994a. Growth of *Listeria monocytogenes* at 10 °C in biofilms with microorganisms isolated from meat and dairy processing environments. *Journal of Food Protection*, **57**, 576–86.

Jeong, D. K. & Frank, J. F. 1994b. Growth of *Listeria monocytogenes* at 21 °C in biofilms with microorganisms isolated from meat and dairy processing environments. *Lebensmittel-Wissenschaft and Technologie*, **27**, 415–24.

Jeurnink, T. J. M. & Brinkman, D. W. 1994. The cleaning of heat exchangers and evaporators after processing milk or whey. *International Dairy Journal*, **4**, 347–68.

Johansen, C., Falholt, P. & Gram, L. 1997. Enzymatic removal and disinfection of bacterial biofilms. *Applied and Environmental Microbiology*, **63**, 3724–8.

Kalmokoff, M. L., Austin, J. W., Wan, X.-D., Sanders, G., Banerjee, S. & Farber, J. M. 2001. Adsorption, attachment and biofilm formation among isolates of *Listeria monocytogenes* using model conditions. *Journal of Applied Microbiology*, **91**, 725–34.

Kim, H., Bang, J., Beuchat, L. R. & Ryu, J.-H. 2008. Fate of *Enterobacter sakazakii* attached to or in biofilms on stainless steel upon exposure to various temperatures or relative humidities. *Journal of Food Protection*, **71**, 940–5.

Kinniment, S. L., Wimpenny, J. W. T, Adams, D. & Marsh, P. D. 1996. The effect of chlorhexidine on defined, mixed culture oral biofilms grown in a novel model system. *Journal of Applied Microbiology*, **81**, 120–5.

Van Klingeren, B., Koller, W., Bloomfield, S. F., Böhm, R., Cremieux, A., Holah, J., Reybrouck, G. & Rödger, H.-J. 1998. Assessment of the efficacy of disinfectants on surfaces. *International Biodeterioration and Biodegradation*, **41**, 289–96.

Knight, G. C. & Craven, H. M. 2010. A model system for evaluating surface disinfection in dairy factory environments. *International Journal of Food Microbiology*, **137**, 161–7.

Knight, G. C. & Weeks, M. G. 2008. Conditions for inactivation of thermophilic spores in NaOH solutions for reuse applications. *Australian Journal of Dairy Technology*, **63**, 82–6.

Knight, G. C., Nicol, R. S. & McMeekin, T. A. 2004. Temperature step changes: a novel approach to control biofilms of *Streptococci thermophilus* in a pilot plant-scale cheese-milk pasteurisation plant. *International Journal of Food Microbiology*, **93**, 305–18.

Kostaki, M., Chorianopoulos, N., Braxou, E., Nychas, G.-J. & Giaouris, E. 2012. Differential biofilm formation and chemical disinfection resistance of sessile cells of *Listeria monocytogenes* strains under monospecies and dual-species (with *Salmonella enterica*) conditions. *Applied and Environmental Microbiology*, **78**, 2586–95.

Langsrud, S. & Sundheim, G. 1997. Factors contributing to the survival of poultry associated *Pseudomonas* spp. exposed to a quaternary ammonium compound. *Journal of Applied Microbiology*, **82**, 705–12.

Lequette, Y., Boels, G., Clarisse, M. & Faille, C. 2010. Using enzymes to remove biofilms of bacterial isolates sampled in the food industry. *Biofouling*, **26**, 421–31.

Marouani-Gadri, N., Firmesse, O., Chassaing, D., Sandris-Nielsen, D., Arneborg, N. & Carpentier, B. 2010. Potential of *Escherichia coli* O157:H7 to persist and form viable but non-culturable cells on a food-contact surface subjected to cycles of soiling and chemical treatment. *International Journal of Food Microbiology*, **144**, 96–103.

McDonnell, G. & Russell, A. D. 1999. Antiseptics and disinfectants: activity, action and resistance. *Clinical Microbiology Reviews*, **12**, 147–79.

Mettler, E. & Carpentier, B. 1997. Location, enumeration and identification of the microbial contamination after cleaning of EPDM gaskets introduced into a milk pasteurisation line. *Lait*, **77**, 489–503.

Moore, L. E., Ledder, R. G., Gilbert, P. & McBain, A. J. 2008. In vitro study of the effect of cationic biocides on bacterial population dynamics and susceptibility. *Applied and Environmental Microbiology*, **74**, 4825–34.

Møretrø, T., Heir, E., Mo, K. R., Habimana, O., Abdelgani, A. & Langsrud, S. 2010. Factors affecting survival of Shigatoxin-producing *Escherichia coli* on abiotic surfaces. *International Journal of Food Microbiology*, **138**, 71–7.

Morente, E. O., Fernández-Fuentes, M. A., Burgos, M. J. G., Abriouel, H., Pulido, R. P. & Gálvez, A. 2013. Biocide tolerance in bacteria. *International Journal of Food Microbiology*, **162**, 13–25.

Murphy, P. M., Lynch, D. & Kelly, P. M. 1999. Growth of thermophilic spore forming bacilli in milk during the manufacture of low heat powders. *International Journal of Dairy Technology*, **52**, 45–50.

Norwood, D. E. & Gilmour, A. 2000. The growth and resistance to sodium hypochlorite of *Listeria monocytogenes* in a steady-state multispecies biofilm. *Journal of Applied Microbiology*, **88**, 512–20.

Palop, A., Sala, F. J. & Condón, S. 1999. Heat resistance of native and demineralised spores of *Bacillus subtilis* sporulated at different temperatures. *Applied and Environmental Microbiology*, **65**, 1316–19.

Pan, Y., Breidt, F., Jr. & Kathariou, S. 2006. Resistance of *Listeria monocytogenes* biofilms to sanitizing agents in a simulated food processing environment. *Applied and Environmental Microbiology*, **72**, 7711–17.

Parkar, S. G., Flint, S. H. & Brooks, J. D. 2004. Evaluation of the effect of cleaning regimes on biofilms of thermophilic bacilli on stainless steel. *Journal of Applied Microbiology*, **96**, 110–16.

Peneau, S., Chassaing, D. & Carpentier, B. 2007. First evidence of division and accumulation of viable but nonculturable *Pseudomonas fluorescens* cells on surfaces subjected to conditions encountered at meat processing premises. *Applied and Environmental Microbiology*, **73**, 2839–46.

Refstrup, E. 2000. Evaporation and drying technology developments. *International Journal of Dairy Technology*, **53**, 163–7.

Ronimus, R., Parker, L. E., Turner, N., Poudel, S., Rückert, A. & Morgan, H. W. 2003. A RAPD-based comparison of thermophilic bacilli from milk powders. *International Journal of Food Microbiology*, **85**, 45–61.

Sasahara, K. C. & Zottola, E. A. 1993. Biofilm formation by *Listeria monocytogenes* utilizes a primary colonizing microorganism in flowing systems. *Journal of Food Protection*, **56**, 1022–8.

Setlow, B., Loshon, C. A., Genest, P. C., Cowan, A. E., Setlow, C. & Setlow, P. 2002. Mechanisms of killing spores of *Bacillus subtilis* by acid, alkali and ethanol. *Journal of Applied Microbiology*, **92**, 362–75.

Scott, S. A., Brooks, J. D., Rakonjac, J., Walker, K. M. R. & Flint, S. H. 2007. The formation of thermophilic spores during the manufacture of whole milk powder. *International Journal of Dairy Technology*, **60**, 109–17.

Stadhouders, J. 1964. The destruction of spores of aerobic sporing bacilli in hot caustic solutions. *Netherlands Milk and Dairy Journal*, **18**, 182–203.

Te Giffel, M. C., Beumer, R. R., Langeveld, L. P. M. & Rombouts, F. M. 1997. The role of heat exchangers in the contamination of milk with *Bacillus cereus* in dairy processing plants. *International Journal of Dairy Technology*, **50**, 43–7.

Van Der Veen, S. & Abee, T. 2011. Mixed species biofilms of *Listeria monocytogenes* and *Lactobacillus plantarum* show enhanced resistance to benzalkonium chloride and peracetic acid. *International Journal of Food Microbiology*, **144**, 421–31.

Verran, J., Boyd, R. D., Hall, K. & West, R. H. 2001. Microbiological and chemical analyses of stainless steel and ceramics subjected to repeated soiling and cleaning treatments. *Journal of Food Protection*, **64**, 1377–87.

Visser, H., Jeurnink, T. J. M., Schraml, J. E., Fryer, P. & Delplace, F. 1997. Fouling of heat treatment equipment. *Bulletin of the International Dairy Federation*, **328**, 7–31.

Zhao, T., Doyle, M. P. & Zhao, P. 2004. Control of *Listeria monocytogenes* in a biofilm by competitive-exclusion microorganisms. *Applied and Environmental Microbiology*, **70**, 3996–4003.

Zhao, T., Podtburg, T. C., Zhao, P., Schmidt, B. E., Baker, D. A., Cords, B. & Doyle, M. P. 2006. Control of *Listeria* spp. by competitive-exclusion bacteria in floor drains of a poultry processing plant. *Applied and Environmental Microbiology*, **72**, 3314–20.

Zottola, E. A. & Sasahara, K. C. 1994. Microbial biofilms in the food processing industry – should they be a concern? *International Journal of Food Microbiology*, **23**, 125–48.

Index

Biofilms in the Dairy Industry, First Edition.
Edited by Koon Hoong Teh, Steve Flint, John Brooks and Geoff Knight.

Printed and bound by CPI Group (UK) Ltd, Croydon, CR0 4YY
20/09/2026
14965795-0001